SEMICONDUCTOR CIRCUIT APPROXIMATIONS

An Introduction to Transistors and Integrated Circuits
Fourth Edition

ALBERT PAUL MALVINO, Ph.D.

McGRAW-HILL BOOK COMPANY

New York • Atlanta • Dallas • St. Louis • San Francisco • Auckland • Bogotá • Guatemala • Hamburg • Johannesburg • Lisbon • London • Madrid • Mexico • Montreal • New Delhi • Panama • Paris • San Juan • São Paulo • Singapore • Sydney • Tokyo • Toronto

Sponsoring Editor: Paul Berk
Editing Supervisor: Mitsy Kovacs
Design and Art Supervisor; Text and Cover Designer: Frances Conte Saracco
Production Supervisor: Laurence Charnow

It is the mark of an instructed mind
to rest satisfied
with that degree of precision
which the nature of the subject admits,
and not to seek exactness
where only an approximation
of the truth is possible.

ARISTOTLE

TO MY WILD IRISH ROSE

Other Books by Albert Paul Malvino   Electronic Principles • Digital Computer Electronics • Resistive and Reactive Circuits • Electronic Instrumentation Fundamentals • Experiments for Electronic Principles • Experiments for Transistor Circuit Approximations • Digital Principles and Applications (with D. Leach)

Library of Congress Cataloging in Publication Data

Malvino, Albert Paul.
   Semiconductor circuit approximations.

   Fourth ed. of: Transistor circuit approximations.
3rd ed. © 1980.
   Includes index.
   1. Transistor circuits.   I. Malvino, Albert Paul.
Transistor circuit approximations.   II. Title.
TK7871.9.M316   1985   621.3815'3042   84-25036
ISBN 0-07-039898-4

Semiconductor Circuit Approximations: An Introduction to Transistors and Integrated Circuits.
Fourth Edition

   3 4 5 6 7 8 9 0 DON DON 8 9 1 0 9 8 7 6

ISBN 0-07-039898-4

Contents

reface

The philosopher Rene Descartes divided every problem into as many parts as possible. Then, he analyzed the simplest part before proceeding to the next more difficult item. Going from the simple to the complex is a key idea in Descartes' scientific method.

Semiconductor Circuit Approximations: An Introduction to Transistors and Integrated Circuits (formerly called *Transistor Circuit Approximations*) is an example of the Cartesian approach. It starts with ideal approximations (the simplest) and proceeds to second and third approximations (the complex). With the ideal-transistor approximation described in this fourth edition, you can sail through problems and arrive at an ideal solution. Then, when necessary, higher approximations can improve accuracy. This idealize-and-improve approach is the most efficient way to solve the problems encountered in troubleshooting and designing.

Semiconductor Circuit Approximations is intended for a course in electronic devices and circuits, particularly in connection with the technician training programs offered by two-year colleges and technical institutes. A prerequisite course in dc and ac circuits is assumed. As with the previous edition, you can use either electron flow or conventional flow with this book.

Each chapter contains end-of-chapter study aids including a summary and a glossary, as well as review questions and problems. A correlated laboratory manual, *Experiments for Semiconductor Circuits Approximations,* is also available.

Albert Paul Malvino

Semiconductor Physics

Bohr: Scoffed at and ridiculed by some colleagues for suggesting electrons orbit the nucleus, he later received the Nobel prize for his imaginative model of the atom. Although no longer satisfactory to the atomic physicist, the Bohr model is still a useful approximation for the atom.

To understand how diodes, transistors, and integrated circuits work, you first have to study *semiconductors,* materials that are neither conductors nor insulators. Conductors contain many free electrons, while insulators contain almost none. Semiconductors contain some free electrons, but what makes them unusual is the presence of *holes.* In this chapter you will learn more about semiconductors, holes, and other related topics.

1-1 ELECTRON ORBITS

About 600 B.C., the Greeks discovered that rubbed amber had one type of electric charge and rubbed glass another. In 1750, Franklin arbitrarily called the first type a *positive* charge and the second type a *negative* charge. In 1897, Thomson discovered the electron and proved that it had a negative charge. This was followed by the discovery of the proton (positive charge) and the neutron (neutral charge). As a result, we now know that matter is made up of atoms with a central nucleus and orbiting electrons. Because the nucleus contains protons and neutrons, it has a positive charge. When an atom is electrically neutral, the number of orbiting electrons equals the number of nuclear protons.

Boron Atom

Internal atomic forces confine the orbiting electrons to three-dimensional regions called *shells.* For instance, Fig. 1-1*a* shows a boron atom with two electrons in the first shell and three electrons in the second shell. These electrons are like satellites in stable orbits because the inward attraction of the positive nucleus is offset by the outward push of centrifugal force. Notice that two electrons are in small orbits, while three others are in larger orbits.

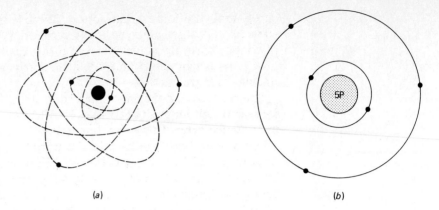

Fig. 1-1.
Boron atom. (a) Three-dimensional. (b) Two-dimensional.

(a)   (b)

Three-dimensional drawings like Fig. 1-1*a* are inconvenient for complicated atoms like copper and silicon. For this reason, we will be using two-dimensional drawings like Fig. 1-1*b*. This simplified figure of the boron atom gives the same information and is much easier to draw. Notice the five protons in the nucleus, the two electrons in small orbits, and the three in larger orbits. When drawn in two dimensions, a shell appears as a single orbital ring. From now on, we will refer to the first shell as the first orbit, the second shell as the second orbit, and so on.

Conductors   Silver (abbreviated Ag) has the highest electrical conductivity of all metals. Copper (Cu) has the second highest conductivity. Gold (Au) ranks third. The reason for copper's high conductivity is clear when we look at its atomic structure, shown in Fig. 1-2*a*. The nucleus contains 29 protons. When a copper atom is electrically neutral, 2 electrons are in the first orbit, 8 in the second, 18 in the third, and 1 in the fourth. The positive nucleus attracts the nearest electrons with the greatest force. This attraction decreases for larger orbits. In fact, the single electron in the outer orbit is so far from the nucleus that it

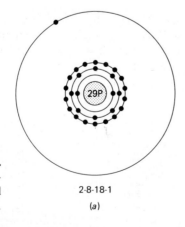

2-8-18-1

(a)

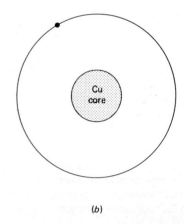

(b)

Fig. 1-2.
Copper atom. (a) Two-dimensional. (b) Simplified drawing.

barely feels the force of attraction. Because the nuclear attraction is so weak, the outer electron is often called a *free electron*. In a piece of copper wire, free electrons can move easily from one atom to the next. This is why the slightest voltage across a copper conductor can produce a large current.

The nucleus and the inner electrons are of little interest to us. Our focus throughout most of this book will be the outer orbit, also called the *valence* orbit. This orbit determines how an atom combines with other atoms, how conductive a material is, etc. To emphasize the importance of the valence orbit, we will draw a copper atom as shown in Fig. 1-2b. In this simplified drawing, the core represents the nucleus and inner electrons. The Cu core has a net charge of $+1$ because it contains 29 protons and 28 inner electrons. Our final picture of a copper atom is a small core with a charge of $+1$ and a valence electron with a charge of -1. Since the valence electron is in a large orbit around a net charge of only $+1$, the inward pull is very small.

The best conductors (silver, copper, and gold) have a core diagram like Fig. 1-2b. The key idea is the single outer electron in a large orbit around the nucleus. Because of the weak nuclear attraction, this isolated valence electron is free to move easily from one atom to the next.

Semiconductors

Germanium (Ge) and silicon (Si) are examples of semiconductors, materials that are neither conductors nor insulators. Figure 1-3a shows a germanium atom. In the center is a nucleus with 32 protons. The revolving electrons distribute themselves in different orbits, following the pattern of

$$2, 8, 18, \ldots, 2n^2$$

where n is the orbit number. The foregoing numbers represent the maximum possible number of electrons in the nth shell. In other words, there are 2 electrons in the first orbit, 8 in the second orbit, and 18 in the third. The last 4 electrons are in the outer, or valence, orbit.

In a similar way, we find that an isolated atom of silicon has 14 protons in the nucleus and 14 electrons in orbit. As shown in Fig. 1-3b, the first orbit contains 2 electrons and the second orbit contains 8 electrons. The 4 remaining electrons are in the outer orbit.

Fig. 1-3.
Semiconductor atoms.
(a) Germanium. (b) Silicon.
(c) Germanium. (d) Silicon.

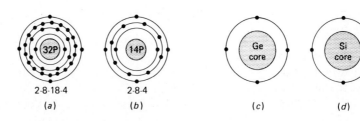

2-8-18-4   2-8-4

(a)   (b)   (c)   (d)

As before, the nucleus and inner electrons are not important for our discussion. This is why we prefer the simplified diagrams of Fig. 1-3c and d to represent germanium and silicon atoms. Notice that both germanium and silicon have four valence electrons. This is how you can distinguish a semiconductor from a conductor.

Whenever the valence orbit contains eight electrons, the substance acts like an insulator. Therefore, the number of electrons in the valence orbit is key to electrical conductivity. Conductors have one valence electron, semiconductors have four valence electrons, and insulators have eight valence electrons.

A final point. Although germanium is still used in the production of semiconductor devices, silicon dominates the industry because of certain advantages that will be discussed in later chapters. For this reason, we will emphasize silicon throughout this book.

1-2 FORBIDDEN ORBITS

When a satellite orbits the earth, it travels with the right velocity to balance the inward pull of gravity and the outward push of centrifugal force. A satellite can travel in an orbit of any size, provided its velocity has the correct value to balance gravitational and centrifugal forces.

Electrons are like satellites because they travel in stable orbits around the nucleus, and yet they are different because some orbit sizes are *forbidden*. For instance, the smallest stable orbit in a hydrogen atom has a radius of

$$r_1 = 0.53(10^{-10}) \text{ m}$$

The next stable orbit has a radius of

$$r_2 = 2.12(10^{-10}) \text{ m}$$

All values between r_1 and r_2 are forbidden. No matter what the velocity of the electron, it cannot find a stable orbit between r_1 and r_2. (The existence of forbidden orbits is one of the discoveries of *quantum physics,* the study of atomic structure.)

The main idea we need for our work is this: There are empty spaces between permitted orbits. An electron cannot find a stable orbit in these forbidden regions. It is possible for an electron to move from one permitted orbit to the next higher permitted orbit, or vice versa. In this case, the electron temporarily moves through a forbidden region, but it is impossible for an electron to remain indefinitely in a forbidden region.

Figure 1-4 summarizes the idea of forbidden regions in a silicon atom. The two electrons in the first orbit travel with just the right speed to maintain a stable orbit. The eight electrons in the second orbit travel more slowly, but the speed is still correct for this larger orbit. Likewise, the four valence electrons travel the most slowly of

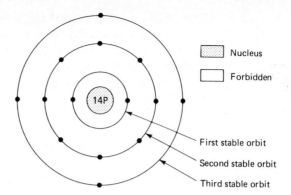

Fig. 1-4.
Stable orbits and forbidden regions.

Nucleus

Forbidden

First stable orbit

Second stable orbit

Third stable orbit

all because the nuclear attraction is weakest for this outer orbit. The empty areas between the stable orbits are forbidden regions. Electrons cannot find stable orbits in these forbidden areas.

1-3
ENERGY LEVELS

It takes energy to lift a stone off the earth. The higher you lift the stone, the more energy you expend. Where does the energy go? The stone absorbs it in the form of potential energy with respect to the earth. If you release the stone, it will fall and give back its acquired energy when it strikes the earth.

A similar idea applies to the electrons in an atom. Suppose we magnify a silicon atom until all we see is part of each orbit as shown in Fig. 1-5a. To move an electron from the first to the second orbit requires energy to overcome the attraction of the nucleus. When an electron moves into a larger orbit, it acquires potential energy with respect to the nucleus. If the electron is allowed to fall back to its original orbit, it will give up this energy.

As a convenience in drawing, everyone visualizes the curved orbits of Fig. 1-5a like the horizontal lines of Fig. 1-5b. The first orbit is called the *first energy level,* the second orbit is the *second energy level,* and so on. The larger the orbit, the higher the energy level. If external radiation bombards an atom, it can add energy to an electron

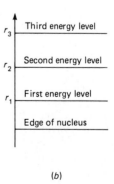

Fig. 1-5.
(a) Electron orbits.
(b) Equivalent energy levels.

r_3 Third orbit

r_2 Second orbit

r_1 First orbit

Nucleus

(a)

r_3 Third energy level

r_2 Second energy level

r_1 First energy level

Edge of nucleus

(b)

SEMICONDUCTOR PHYSICS

and lift it to a higher energy level. The atom is then said to be in a state of *excitation*. This state has a limited lifetime (typically nanoseconds), because the energized electron soon falls back to its original energy level. As it falls, the electron gives back the acquired energy in the form of heat, light, or other radiation.

1-4
CRYSTALS

Up to now, we have been discussing isolated atoms. For instance, we know a silicon atom has two electrons in the first orbit, eight in the second, and four in the third. This is true for a single isolated silicon atom. When more than one silicon atom is involved, something unusual happens.

Valence Orbit Needs Eight Electrons

When silicon atoms combine to form a solid, they automatically arrange themselves into an orderly pattern called a *crystal*. Each silicon atom combines in such a way as to have eight electrons in its valence orbit. Whenever an atom has eight electrons in its valence orbit, it becomes chemically stable. (There is no simple explanation for why the number eight is so special, except to say it is a law of nature. Quantum physics attempts to describe this phenomenon with mathematical equations, but ultimately it is an experimental fact, similar to the law of gravity.)

How does a silicon atom acquire eight electrons in its valence orbit? By positioning itself between four other silicon atoms as shown in Fig. 1-6a. Each neighboring atom shares an electron with the central atom. In this way, the central atom has picked up four additional electrons, making a total of eight electrons in its valence orbit. Actually, the electrons no longer belong to a single atom; they are shared by adjacent atoms.

Covalent Bonds

In Fig. 1-6a, each core represents a charge of +4 (14 protons and 10 inner electrons). Look at the central core and the one on the right. These two cores attract the pair of electrons between them with equal and opposite forces. This pulling in opposite directions is what holds the silicon atoms together in a solid. The idea is similar to tug-of-war

Fig. 1-6.
(a) 8-valence electrons.
(b) Covalent bonds.
(c) Bonding diagram of a crystal.

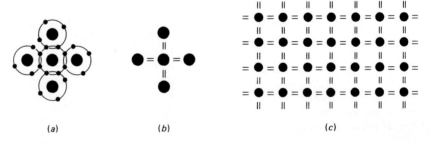

(a)   (b)   (c)

teams pulling on a rope. As long as both teams pull with equal and opposite forces, they remain bonded together.

Figure 1-6b symbolizes the mutual sharing and pulling on electrons. Each line represents a shared electron. Each shared electron establishes a bond between the central atom and its neighbor. For this reason, we call each line a *covalent bond*. In other words, each silicon atom in a crystal forms eight covalent bonds with its neighbors. It is these covalent bonds that hold the crystal together, that give it solidity.

Figure 1-6c shows the bonding diagram of a silicon crystal. Each atom has eight electrons traveling in its valence orbit. As previously explained, eight electrons in the valence orbit produce a chemical stability that allows the silicon crystal to exist in nature. The valence orbit can hold no more than eight electrons. Because of this, it is said to be *filled* or *saturated* when it contains eight electrons.

Incidentally, an isolated germanium atom also has four valence electrons. When germanium atoms combine to form a solid, they too produce a crystalline structure with a bonding diagram like Fig. 1-6c.

Free Electrons and Holes

The *ambient* temperature is the temperature of the environment. When the ambient temperature is above absolute zero ($-273°C$ or $-460°F$), the incoming heat energy causes the atoms of a silicon crystal to vibrate. The higher the ambient temperature, the stronger the mechanical vibrations of these atoms. If you pick up a warm crystal, the warmth you feel is caused by vibrating atoms.

The chaotic vibrations of the silicon atoms can occasionally dislodge an electron from the valence orbit as shown in Fig. 1-7. When this happens, the released electron gains enough energy to go into the next higher permitted orbit. In this larger orbit, the electron becomes a free electron because the nuclear attraction is almost negligible. As a result, the free electron can move easily throughout the crystal. Furthermore, the departure of the electron leaves a vacancy in the valence orbit that is called a *hole*. This hole behaves like a

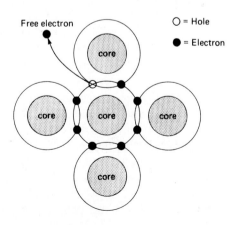

Fig. 1-7.
Thermal energy produces a free electron and a hole.

positive charge in the sense that it will attract and capture any electron in the immediate vicinity.

Recombination and Lifetime

In a crystal of pure silicon, equal numbers of free electrons and holes are created by *thermal* (heat) energy. The free electrons move randomly throughout the crystal. Occasionally, a free electron will approach a hole, feel the attraction, and fall into the hole. This merging of a free electron and a hole is called *recombination*. The average amount of time between the creation and disappearance of a free electron is called the *lifetime*. It varies from a few nanoseconds to several microseconds, depending on how perfect the crystalline structure is, and other factors.

At any instant, the following occurs within a silicon crystal:

1. Some free electrons and holes are being generated by thermal energy.
2. Other free electrons and holes are recombining.
3. Some free electrons and holes exist in an in-between state; they were previously generated and have not yet recombined.

1-5 ENERGY BANDS

When silicon atoms combine to form a solid, the orbit of an electron is influenced by neighboring atoms as well as the original atom. Naturally, the nearest charges have the greatest effect, but even faraway charges will have some effect on the orbit of an electron. In other words, the energy level of each electron is controlled to some extent by every charge in the crystal. Since each electron has a different position inside the crystal, no two electrons have exactly the same charge environment. This is equivalent to saying that no two electrons in a crystal have exactly the same energy level. (In quantum physics, this is known as the Pauli exclusion principle.)

Absolute Zero

Figure 1-8a shows how to visualize the energy levels of a silicon crystal at absolute zero temperature. All electrons traveling in first orbits have slightly different energy levels because no two see exactly the same charge environment. Since there are billions of first-orbit electrons, the slightly different energy levels form a cluster or band. Similarly, the billions of second-orbit electrons, all with slightly different energy levels, form the second energy band shown. And all third-orbit electrons form the valence band. Notice the new band of energy levels labeled *conduction band*. It represents the next group of permitted orbits above the valence band. This is the energy band of free electrons.

The first three energy bands are shown as dark areas. This will be our way of indicating filled or saturated bands; that is, all available orbits are occupied by electrons. If we leave part of a band unshaded, it means some orbits are empty, equivalent to vacant energy levels.

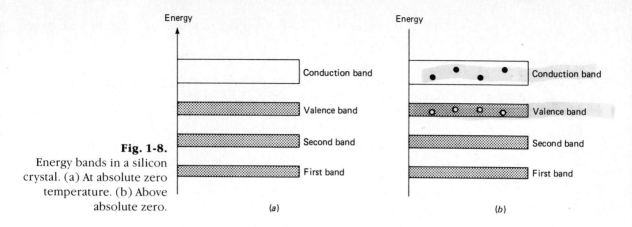

Fig. 1-8.
Energy bands in a silicon crystal. (a) At absolute zero temperature. (b) Above absolute zero.

Energy

Conduction band
Valence band
Second band
First band

(a)

Energy

Conduction band
Valence band
Second band
First band

(b)

This is why the conduction band is unshaded. At absolute zero temperature, there are no free electrons in a silicon crystal. The empty spaces between the energy bands are called *forbidden gaps* because they represent unstable orbits or energy levels.

Above Absolute Zero

When the ambient temperature is greater than absolute zero, the incoming thermal energy breaks some covalent bonds, as previously described. These valence electrons escape into the conduction band (see Fig. 1-8b). In this way, we get a limited number of conduction-band electrons. These free electrons travel in large orbits; so they are only loosely held by the silicon atoms. This means they can move easily from one atom to the next. In Fig. 1-8b, each time an electron is bumped up to the conduction band, a hole is created in the valence band. Therefore, the valence band is no longer filled or saturated; each hole represents an available orbit of rotation. The higher the ambient temperature, the greater the number of electrons kicked up to the conduction band.

**1-6
INTRINSIC
CONDUCTION**

An *intrinsic* semiconductor is a pure semiconductor. For instance, a silicon crystal is an intrinsic semiconductor if every atom in the crystal is a silicon atom. In this section we will examine how an intrinsic silicon crystal conducts.

Absolute Zero

Figure 1-9a shows an intrinsic silicon crystal with metal end surfaces. An external voltage source sets up an electric field between the ends of the crystal. Is there any current? This depends on the temperature. If the ambient temperature is at absolute zero, each silicon atom has eight electrons in its valence orbit. Because of the covalent bonds, these valence electrons are tightly held and cannot leave their atoms. Therefore, despite the applied voltage, the silicon crystal acts like an insulator because there are no free electrons to produce a current.

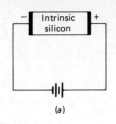

(a)

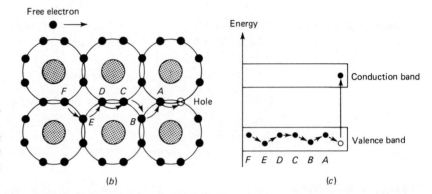

(b)

(c)

Fig. 1-9.
Current in a silicon crystal.
(a) Circuit. (b) Free
electron moves right, hole
moves left. (c) Equivalent
energy diagram.

Above Absolute Zero As the temperature increases above absolute zero, the atoms begin to vibrate. Every so often, a valence electron gains enough energy to escape into the conduction band. Because a few free electrons exist, a small current will appear. The higher the temperature, the larger the current. At room temperature (approximately 25°C or 77°F) the current is quite small compared with the current through a conductor. This is why a silicon crystal is known as a *semiconductor*.

Electron Flow Figure 1-9b shows part of the silicon crystal in Fig. 1-9a. Assume that thermal energy has produced a free electron and a hole. The free electron is in a very large orbit. Because of the external voltage source, the free electron tends to move to the right. It does this by moving from one large orbit to the next. In this way, the free electron participates in the overall flow of electrons through the silicon crystal. This is why we can visualize a steady stream of free electrons flowing from the negative source terminal to the positive source terminal in Fig. 1-9a.

Hole Flow A conductor has only free electrons; it does not have any holes. Because of this, the only charges flowing in a conductor are the free electrons. A semiconductor has both free electrons and holes. Both of these can and do flow when an external voltage is applied. This is what makes a semiconductor distinctly different from a conductor.

Hole flow is new to us; so let us take a close look at how it exists in a semiconductor. Notice the hole at the extreme right of Fig. 1-9b. This hole attracts the valence electron at A. This attraction plus the effect of the externally applied voltage causes the valence electron at *A* to move into the hole. When this happens, the original hole disappears and a new one appears at position *A*. The new hole at *A* can now attract and capture the valence electron at *B*. When the valence electron moves from *B* to *A,* the hole moves from *A* to *B*. This motion can continue with valence electrons moving along the path shown by the arrows. The hole, on the other hand, moves in the opposite direction, from *A* to *B* to *C,* and so forth.

Energy Levels

Above absolute zero, thermal energy bumps an electron from the valence band into the conduction band. This produces a free electron in the conduction band and a hole in the valence band as shown in Fig. 1-9c. Because of the external voltage, the free electron in the conduction band moves to the right. Similarly, the hole in the valence band moves to the left. Since thermal energy produces many electron-hole pairs, we can visualize a steady flow of free electrons through the conduction band (to the right) and a steady flow of holes through the valence band (to the left).

Holes Act Like Positive Charges

Strictly speaking, a hole is not a positive charge because it is only a vacancy in the valence band, a place where a valence electron once orbited its atom. But certain experiments indicate that a hole acts and moves exactly the same as a positive charge. In particular, a phenomenon called the *Hall effect* shows that it is more accurate to treat valence current as a flow of positive charges in the direction of the holes rather than negative charges in the direction of the valence electrons.

   As an example, Fig. 1-10 shows two ways to visualize intrinsic conduction. We can think of free electrons and valence electrons moving to the right (Fig. 1-10*a*), or we can think of free electrons moving to the right and holes moving to the left (Fig. 1-10*b*). From now on, we will use the latter convention because it is preferred by scientists and engineers.

Points to Remember

A semiconductor differs from a conductor because it has two atomic paths that charges can flow through. First, it has the ordinary path followed by free electrons in the conduction band. Second, it has the

Fig. 1-10.
Two atomic paths for current. (a) Free electron and valence electrons move right. (b) Free electrons move right and holes move left.

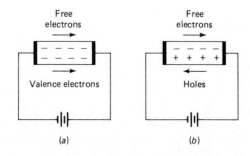

extraordinary path followed by holes in the valence band. The presence of these holes in semiconductors allows manufacturers to produce diodes, transistors, integrated circuits, and all kinds of *solid-state* (semiconductor) devices.

1-7
EXTRINSIC SEMI-CONDUCTORS

At room temperature the number of free electrons and holes in an intrinsic semiconductor is insufficient for most practical applications. This means we cannot get enough useful current with an intrinsic semiconductor. One way to increase conductivity is by *doping*. This means deliberately adding impurity atoms to an intrinsic semiconductor to alter its electrical conductivity. The doped semiconductor is then known as an *extrinsic* semiconductor.

Increasing the Free Electrons

To begin with, the manufacturer melts a pure silicon crystal. This breaks the covalent bonds and changes the silicon from a solid to a liquid. To increase the number of free electrons, *pentavalent* atoms (also called *donor atoms*) can be added to the molten silicon. Pentavalent atoms have five electrons in the valence orbit. Examples of pentavalent impurities include arsenic, antimony, and phosphorus.

Assume that a small amount of arsenic has been added. The arsenic atoms will diffuse uniformly throughout the molten silicon. After the silicon has cooled, a solid crystal forms. Once more, we find a crystalline structure where each atom has four neighboring atoms that share valence electrons. Most of the atoms in the crystal are still silicon, but occasionally we will find an arsenic atom, as shown in Fig. 1-11a. This arsenic atom has taken the place of a silicon atom, and therefore it has eight electrons in its valence orbit. Originally, this donor atom had five valence electrons. With each neighbor now sharing one of its valence electrons with the arsenic atom, an extra electron is left over. Since the valence orbit can hold no more than eight electrons, the extra electron is a free electron traveling in a

Fig. 1-11.
Pentavalent doping.
(a) Donor atom produces a
free electron.
(b) Conduction band has
many free electrons.

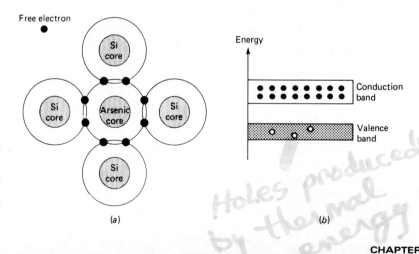

conduction-band orbit. In effect, the pentavalent or donor atom has produced one free electron.

Figure 1-11*b* shows the energy bands of a doped crystal. The conduction band now contains many free electrons. Since each arsenic atom contributes one free electron, a manufacturer can control the number of free electrons by controlling the amount of donor impurity. Furthermore, notice that the valence band has a few holes. These are the holes produced by thermal energy.

Silicon that has been doped by a pentavalent impurity is called an *n-type* semiconductor, where the *n* stands for negative. Since the free electrons outnumber the holes in an *n*-type semiconductor, the free electrons are referred to as *majority carriers* and the holes are known as *minority carriers*.

Increasing the Number of Holes

How can we dope a pure silicon crystal to get an excess of holes? By using a *trivalent* impurity, one whose atoms have only three valence electrons. Examples are aluminum, boron, and gallium. Suppose aluminum has been used to dope pure silicon. After the molten silicon has cooled and formed a crystal, we would find that the aluminum atom has four neighbors as shown in Fig. 1-12*a*. Since the aluminum atom originally had only three valence electrons and each neighbor shares one electron, only seven electrons are in the valence orbit. This means a hole appears in the valence orbit of each aluminum atom. A trivalent atom is also called an *acceptor* atom because each hole it contributes may accept an electron during recombination.

Figure 1-12*b* shows the energy bands. Notice the extra holes in the valence band, a direct result of trivalent doping. There are a few free electrons in the conduction band because thermal energy still produces some electron-hole pairs. This type of doped silicon is known as a *p-type* semiconductor, where *p* stands for positive. In a *p*-type semiconductor, the holes are the majority carriers and the free electrons are the minority carriers.

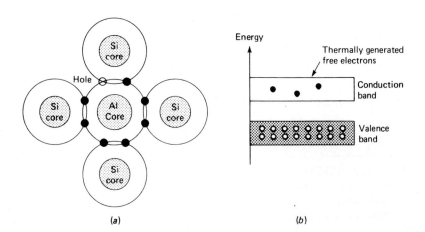

Fig. 1-12.
Trivalent doping.
(a) Acceptor atom produces a hole. (b) Valence band has many holes.

Fig. 1-13.

Two types of extrinsic
semiconductors. (a) *n*-type
has an excess of free
electrons. (b) *p*-type has an
excess of holes.

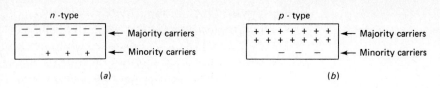

(a)

(b)

Points to
Remember

Before a manufacturer can dope a semiconductor, it must be
produced as a pure crystal. Then by controlling the amount of impurity, the manufacturer can precisely control its conductivity. Historically, pure germanium crystals were easier to produce than pure
silicon crystals. This is why the earliest transistors were made of
germanium. Eventually, manufacturing techniques improved and
pure silicon crystals became available. Because of its advantages (discussed later), silicon has emerged as the most popular and useful
semiconductor material. As described earlier, extrinsic silicon comes
in two varieties: *n*-type or *p*-type as shown in Fig. 1-13. In *n*-type
silicon the free electrons are the majority carriers. In *p*-type silicon
the holes are the majority carriers.

1-8
SEMICONDUCTOR
DEVICES

By combining *p* and *n* materials in different ways, we get the semiconductor devices used in modern electronics. The remaining chapters discuss these devices and their applications. For now, here is a
brief description of the most basic semiconductor devices.

Figure 1-14*a* shows a *diode*. This device has two doped regions;
one is *n*-type and the other is *p*-type. The diode is the most basic
solid-state device. It allows majority carriers to flow easily in one
direction but not in the other. In other words, it is like a one-way
street. Because it is a one-way conductor, the diode is used in power
supplies, circuits that convert ac line voltage to a dc voltage suitable
for electronics equipment.

Figure 1-14*b* is a *transistor,* a device with three doped regions. This
basic device can amplify (increase) a weak signal from an antenna
until it is large enough to be useful. The transistor can also act like
an electronic switch. For this reason, it has had a major impact on the
construction of computers. The invention of the transistor was the
beginning of a tidal wave of related inventions such as integrated
circuits, optoelectronic devices, and microprocessors.

Fig. 1-14.
Basic semiconductor
devices. (a) Diode.
(b) Transistor. (c) Chip.

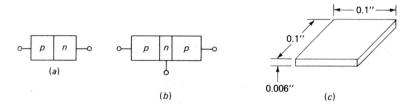

Figure 1-14c shows a *chip,* a small piece of semiconductor material. The dimensions are representative. Chips are often smaller than this, sometimes larger. By advanced photographic techniques, a manufacturer can produce circuits on the surface of this chip, circuits with many diodes, transistors, etc. The finished network is so small you need a microscope to see the connections. We call a circuit like this an *integrated circuit* (IC).

A *discrete circuit* is different. It is the kind of circuit you build when you connect separate resistors, capacitors, transistors, etc. Each component or device is discrete, or distinct from the others. This differs from an integrated circuit where the components are atomically part of the chip.

Summary

Germanium and silicon are examples of semiconductors, materials whose electrical conductivity is between that of a conductor and an insulator. When isolated, germanium and silicon atoms have four valence electrons.

Each orbit is equivalent to an energy level; the larger the orbit, the greater the energy of the electron. If an outside source of energy lifts an electron into a larger orbit or higher energy level, this electron will radiate energy when it falls back to its original energy level.

When silicon atoms combine into a crystal, they share valence electrons and produce eight electrons in each valence orbit. The shared electrons create covalent bonds between the atoms. Above absolute zero temperature, thermal energy breaks some of the covalent bonds, creating free electrons and holes. These free electrons and holes recombine after a period of time called the lifetime.

Because no two electrons in a silicon crystal can have the same energy, we must visualize the energy levels grouped into energy bands. Forbidden gaps separate these energy bands. The conduction band lies above the valence band. Above absolute zero, the conduction band of an intrinsic semiconductor has a few electrons and the valence band has an equal number of holes.

A semiconductor has two atomic paths for charge flow: the conduction band and the valence band. Free electrons travel in the conduction band and holes travel in the valence band. By doping with pentavalent impurities, we can produce an *n*-type semiconductor, one whose majority carriers are free electrons. Doping with trivalent impurities results in a *p*-type semiconductor, one whose majority carriers are holes.

Glossary

ambient temperature The temperature of the surrounding air or environment.

conduction band The group of permitted energy levels above the valence band.

conductor A material (usually a metal) with many free electrons.

covalent bond A valence electron being shared by two adjacent atoms.

crystal The solid structure that silicon or germanium atoms form by covalent bonding.

doping Adding impurity atoms to an intrinsic semiconductor to produce either an excess of free electrons or an excess of holes.

free electron An electron traveling in a conduction-band orbit. The attraction of the nucleus is so weak that the electron can move easily from one atom to the next.

hole A vacancy in the valence orbit of a semiconductor atom.

hole flow The movement of holes through the valence band produced by the flow of valence electrons in the opposite direction.

lifetime The average amount of time between the creation and recombination of a free electron and a hole.

***n*-type semiconductor** A semiconductor with an excess of free electrons produced by doping with pentavalent or donor atoms.

***p*-type semiconductor** A semiconductor with an excess of holes produced by doping with trivalent or acceptor atoms.

recombination The merging of a free electron and a hole.

semiconductor A material like silicon or germanium whose electrical properties lie between those of a conductor and an insulator.

valence orbit This is the outer orbit of an isolated atom.

Review Questions

1. Name three conductive materials.
2. Why is the valence electron of a conductor called a free electron?
3. Name two semiconductor materials. How many valence electrons does an isolated semiconductor atom have?
4. As the orbit increases in size, what happens to the energy level?
5. What happens when an electron falls from a higher energy level to a lower one?
6. Explain what a covalent bond is.
7. Define ambient temperature. What does room temperature equal?
8. What is a hole? What do recombination and lifetime mean?
9. What is the band of energy above the valence band?
10. Does an intrinsic semiconductor act like a conductor or an insulator at absolute zero temperature? Explain why.
11. Describe the two types of charge flow in an intrinsic semiconductor at room temperature.
12. Explain the two types of doping and the effect they have on a semiconductor.

Problems

1-1. Draw a core diagram similar to Fig. 1-3*d* for an isolated atom of gold.

1-2. Sketch an orbit diagram like Fig. 1-5*a* and an energy diagram like Fig. 1-5*b* for a germanium atom.

1-3. Draw a bonding diagram like Fig. 1-6*b* for an aluminum atom in the center.

1-4. An intrinsic silicon crystal has one million thermally produced free electrons at room temperature. If the temperature increases to 75°C, will the number of holes be less than, equal to, or greater than one million?

1-5. Sketch an energy diagram like Fig. 1-8*b* for *n*-type silicon with 12 majority carriers and 3 minority carriers.

1-6. Draw an energy diagram like Fig. 1-8*b* for *p*-type germanium with 3 minority carriers and 9 majority carriers.

1-7. An external voltage source is applied to a *p*-type semiconductor. If the left end of the crystal is positive and the right end is negative, do the majority carriers flow to the left or to the right? Which way do the minority carriers flow?

1-8. An intrinsic semiconductor has been doped with one billion acceptor atoms. At absolute zero temperature, how many majority carriers are there? Are these free electrons or holes?

1-9. Match each of the 12 entries in column A as closely as possible with one entry in column B:

Column A	Column B
conductor	four valence electrons
semiconductor	surrounding
hole	energy levels of free electrons
ambient	free electron falls into hole
covalent bond	one valence electron
forbidden gap	valence band
recombination	acceptor atom
silicon	doped
larger orbit	widely used semiconductor
conduction band	shared electron
atomic path for holes	unstable orbits
extrinsic	higher energy level

Rectifier Diodes

Faraday: Known as the electrical wizard of the nineteenth century, his discoveries have had a major impact on world evolution. For advice on how to do it, listen to his words: "Let your imagination go, guiding it by judgment and principle, but holding it in and directing it by experiment. Nature is your best friend and critic in experimental science if you only allow her intimations to fall unbiased on your mind. Nothing is so good as an experiment which, while it sets an error right, gives you as a reward for your humility an absolute advance in knowledge."

In 1883, Edison put an extra electrode inside a light bulb in an attempt to remove gases emitted by the incandescent filament. When this electrode was made positive with respect to the filament, a current flowed. When the electrode was negative, the current stopped. Edison dismissed this phenomenon (called the Edison effect) as having no practical value. Without knowing it, Edison had built the first *diode,* a device historically recognized as the starting point of modern electronics.

The word "diode" is a contraction of two (*di*) and electrode (*ode*). A diode is a two-electrode device with a preferred direction of current. In other words, a diode acts like a conductor when the applied voltage is in one direction but like an insulator when it is in the opposite direction. The earliest diodes were vacuum-tube devices with a hot filament (called the cathode) that emitted free electrons and a positive plate (called the anode) that collected the free electrons. The modern diode is a semiconductor device with *n*-type material supplying the free electrons and *p*-type material collecting them.

2-1 THE pn JUNCTION

At room temperature, a *p*-type semiconductor has mostly holes produced by doping and a few free electrons produced by thermal energy. On the other hand, an *n*-type semiconductor has mainly free electrons and only a few holes. By itself, an *n*-type semiconductor is about as useful as a carbon resistor; the same can be said about *p*-type material. But dope a crystal so half is *n*-type and the other half *p*-type, and you have something that acts like a one-way conductor. The following discussion tells you why.

The Carriers

Imagine a neutral atom. It has the same number of electrons as protons. Suppose one of the electrons is removed. Then the atom has a positive charge and is called a *positive ion*. Conversely, if an electron is added to a neutral atom, the atom becomes negatively charged and is known as a *negative ion*.

Figure 2-1*a* shows a *p*-type semiconductor. Each plus sign represents a hole, and each circled minus sign is the acceptor atom that the hole is in. Together, the hole and acceptor atom represent a neutral charge. When a hole disappears by recombining with an electron, however, the acceptor atom then contains an extra negative charge and becomes a negative ion. As it now stands, the *p*-type material of Fig. 2-1*a* is electrically neutral because the number of plus signs equals the number of minus signs.

Similarly, Fig. 2-1*b* shows an *n*-type semiconductor. Here, the minus signs symbolize the free electrons, and the circled plus signs represent the donor atoms these free electrons are orbiting. Each free electron and its donor atom represent a neutral charge. If a free electron leaves its orbit around the donor atom and moves to another atom, the donor atom becomes a positive ion. Unlike the free electrons, these positive ions cannot move around because they are embedded in the crystal structure. In Fig. 2-1*b,* the *n*-type material is electrically neutral because the number of minus signs equals the number of plus signs.

Depletion Layer

A manufacturer can produce a crystal with *p*-type material on one side and *n*-type on the other as shown in Fig. 2-2*a*. This is not two separate pieces of semiconductor material somehow glued together, but rather a single continuous crystal with two doped areas. The border where the *p* material meets the *n* material is called the *junction*. Because of their mutual repulsion, the free electrons on the *n* side of the junction tend to *diffuse* (spread) in all directions. Some diffuse across the junction (see Fig. 2-2*a*). When a free electron enters the *p* region, it becomes a minority carrier. With so many holes around it, this minority carrier has a short lifetime and soon falls into

Fig. 2-1.
Majority carriers and ions.
(a) Holes and negative ions.
(b) Free electrons and
positive ions.

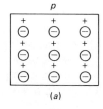

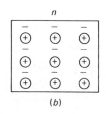

(a) (b)

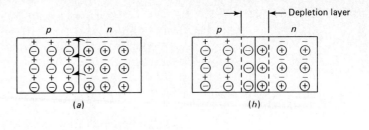

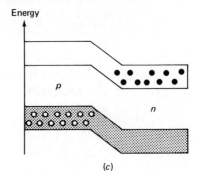

Fig. 2-2.
pn junction. (a) At the
instant of formation.
(b) Depletion layer.
(c) Energy bands.

a hole. When this happens, the hole disappears and the free electron becomes a valence electron.

As the free electrons cross the junction and recombine, they produce negative ions on the left and leave positive ions behind on the right (see Fig. 2-2*b*). As the number of ions increases, more free electrons and holes near the junction disapper. The region containing the positive and negative ions is called the *depletion layer* because it is depleted or emptied of charge carriers.

Barrier Potential As the depletion layer builds up, a difference of potential appears across the junction because of the negative ions on the left and the positive ions on the right. Eventually, this difference of potential becomes large enough to prevent continued diffusion of free electrons across the junction.

To understand why, look at Fig. 2-2*b*. If a free electron enters the depletion layer, it encounters a wall of negative ions that repels it back into the *n* region. At first, the wall of negative ions is small and free electrons have enough energy to get over it. Each crossing electron, however, creates an additional pair of negative and positive ions. Therefore, when the wall of negative ions is large enough, the free electrons no longer have enough energy to overcome the repulsion of the negative ions.

The difference of potential produced by the negative and positive ions is called the *barrier potential*. At room temperature the barrier potential is approximately 0.7 V for a silicon diode and 0.3 V for a germanium diode.

Energy Levels

When we say a free electron cannot break through the barrier potential, we are saying it does not have enough energy to enter the *p* region. The barrier potential represents a difference in energy levels as shown in Fig. 2-2c. Because of the barrier potential, the bottom of each *p* band is at the same height as the top of the corresponding *n* band. This means free electrons on the *n* side no longer have enough energy to get across the junction. Stated another way, free electrons on the *n* side are traveling in orbits that are not large enough to match the available orbits on the *p* side. Unless we increase the energy levels of the free electrons, the diode will not conduct.

Points to Remember

1. At the instant the *pn* junction is formed, free electrons diffuse across the junction and fall into holes.
2. The recombination of free electrons and holes near the junction produces a region of negative and positive ions called the depletion layer.
3. Because of the barrier potential, the diffusion of free electrons across the junction eventually stops.
4. At room temperature a silicon diode has a barrier potential of approximately 0.7 V (a Ge diode has about 0.3 V).

2-2
FORWARD BIAS

A *pn* crystal can act like a diode because it allows current to flow in one direction but not the other. To understand why, look at Fig. 2-3a. Notice that the negative terminal of the battery is connected to the *n* side of the crystal. Because of this, free electrons on the *n* side are repelled toward the junction. This connection is called *forward bias*.

Large Forward Current

Forward bias causes the free electrons on the *n* side to move toward the junction. This leaves positive ions at the right end of the crystal. The positive ions then attract free electrons from the battery. These free electrons flow from the negative battery terminal through the wire into the right end of the crystal.

Since the positive battery terminal is connected to the *p* side, all the holes in the *p* region are repelled toward the junction. As the holes move to the right, they leave negative ions at the left end of the crystal. Valence electrons then flow from these negative ions into the wire connected to the positive battery terminal. As these valence electrons leave, new holes are created at the left end of the crystal.

Here is what we have so far. Inside the crystal, free electrons and holes are moving toward the junction. As soon as they move, new free electrons enter the right end of the crystal and new holes are created at the left end. Therefore, the *n* side remains full of free electrons and the *p* side full of holes. The free electrons that cross the junction recombine with the holes arriving at the junction. As a result, a con-

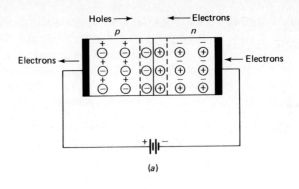

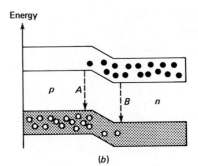

Fig. 2-3.
Forward bias. (a) Holes and free electrons move toward the junction. (b) Free electrons recombine with holes near the junction.

tinuous current is set up in the crystal and the wires connected to it. If you connect an ammeter in series with the battery and diode, it will indicate a current flows in the circuit.

Energy Levels
It improves your understanding if you can describe the same phenomenon in different ways. To deepen your understanding of diode action, look at Fig. 2-3*b*. Applying an external voltage is equivalent to raising the energy level of the free electrons. When the applied voltage is approximately 0.7 V, the free electrons on the *n* side of the junction have enough energy to enter the *p* side. After entering the *p* region, a free electron is a minority carrier with a lifetime typically in nanoseconds. Very quickly, therefore, the free electron falls into a hole (path *A*). Then, as a valence electron, it moves through the holes and arrives at the left end of the crystal.

Sometimes, a free electron falls into a hole even before crossing the junction. Here's how. In Fig. 2-3*b*, a hole may cross the junction, where it becomes a minority carrier on the *n* side. Here it will have a short lifetime, and it soon disappears when a free electron falls into it (path *B*).

Regardless of where the recombination takes place, the result is the same. A steady stream of free electrons flows toward the junction and falls into holes. The captured electrons (now valence electrons) move left in a steady stream through the holes in the *p* region. In this way, we get a continuous current through the diode.

CHAPTER 2

2-3
REVERSE BIAS

What happens if we reverse the polarity of the applied voltage as shown in Fig. 2-4a? In this case, free electrons and holes move away from the junction. This connection is called *reverse bias*. With reverse bias, the diode no longer acts like a conductor. Let's find out why.

Small Reverse Current

Since the positive battery terminal is connected to the *n* side and the negative battery terminal to the *p* side, free electrons and holes temporarily flow away from the junction. This increases the width of the depletion layer until its potential equals the applied voltage; then the majority carriers stop flowing. Within nanoseconds the flow drops to approximately zero.

Why is there no current after a few nanoseconds? Because the applied voltage now adds to the barrier potential, preventing the flow and recombination of majority carriers at the junction. For this reason, a dc ammeter connected in series with the battery and diode will indicate approximately zero current.

Energy Levels

Another way to understand what is going on is by looking at the energy levels of the majority carriers (see Fig. 2-4b). The applied voltage reduces the energy levels of the free electrons on the *n* side of the junction. This is why the *n* bands have dropped well below the *p* bands. Now, there is no way the free electrons can cross the junction because their orbits are too small to match the much larger orbits on the *p* side.

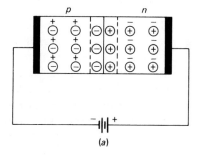

(a)

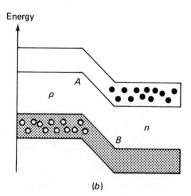

(b)

Fig. 2-4.
Reverse bias. (a) Majority carriers cannot cross junction. (b) Free electrons do not have enough energy to cross junction.

Minority Carriers

At absolute zero temperature only free electrons exist in the n material and only holes in the p material. For this reason, forward bias produces a large dc current and reverse bias produces zero dc current. Therefore, the diode is a conductor in the forward direction and an insulator in the reverse direction.

Above absolute zero temperature, thermal energy produces a few free electrons on the p side and a few holes on the n side. When the diode is reverse-biased, these minority carriers flow toward the junction and recombine. As each thermally produced free electron and hole recombine at the junction, one free electron leaves the negative battery terminal and enters the left end of the crystal; simultaneously, a valence electron leaves the right end of the crystal and enters the positive battery terminal. Since thermal energy continuously produces minority carriers, a continuous flow exists. If you connect an extremely sensitive dc ammeter in series with the diode and battery, it will indicate a very small dc current flows in the external circuit.

Surface Leakage

Besides the minority-carrier current just described, it is possible for a small reverse current to flow along the surface of a crystal. Since the atoms on the surface of a crystal have broken covalent bonds, the skin of a crystal is full of holes and provides a high-resistance path for current. Surface-leakage current is independent of temperature but not voltage. The greater the reverse voltage, the larger the surface-leakage current.

Reverse Current

The total reverse current is the sum of minority-carrier current and surface-leakage current. At room temperature the reverse current is extremely small compared with the forward current. As an example, a 1N914 (a commercially available silicon diode) has a reverse current of 25 nA for a reverse voltage of 20 V. Except for demanding applications, you can ignore the reverse current of a silicon diode because it is too small to matter.

Silicon versus Germanium

Recall the forbidden gap between the valence band and the conduction band. Thermal energy occasionally knocks a valence electron into the conduction band. This creates a free electron and a hole, which can increase the minority-carrier reverse current. The forbidden gap of silicon is wider than the forbidden gap of germanium. For this reason, it is harder for thermal energy to produce minority carriers in a silicon diode than in a germanium diode. In other words, silicon is less sensitive to an increase in temperature than germanium. This is why silicon diodes have much less reverse current than germanium diodes and why silicon has become the industry standard rather than germanium.

Figure 2-5 shows the schematic symbol of a diode. The *p* side is called the *anode* and the *n* side is the *cathode*. When the diode is forward-biased, a large electron flow exists between the cathode and the anode, equivalent to a conventional flow from the anode to the cathode.

**Conventional
versus Electron
Flow**

A word or two about conventional flow and electron flow might be helpful. In 1750, Franklin visualized electricity as an invisible fluid. If a body had more than its normal share of this fluid, he said it had a positive charge; if the body had less than its normal share, its charge was considered negative. On the basis of this theory, Franklin concluded that this "electric fluid" flowed from positive (excess) to negative (deficiency). This imaginary flow from positive to negative is now called *conventional flow*. Between 1750 and 1897, a large number of concepts and formulas based on conventional flow came into existence.

In 1897, Thomson discovered the electron, an atomic particle with a negative charge. Before long, scientists realized that the only charges flowing in a copper wire were free electrons. In other words, when a battery is connected to a circuit, the only flow that physically exists is the flow of electrons from the negative to the positive terminal. This concept of flow is called *electron flow*.

Conventional flow gives the same answers as electron flow, except for the reversed direction of flow. You will encounter both concepts of flow in industry. Some people prefer electron flow because it's closer to the truth. Other people involved in mathematical analysis prefer conventional flow because it ties in better with the mountain of formulas that came into existence before the discovery of the electron. We will use electron flow in explaining how a device works. When analyzing circuits mathematically, you may use either flow as long as you are consistent.

Diode Arrow

Most scientists and engineers prefer conventional flow. When they invent a new semiconductor device, they use a schematic symbol that indicates the forward direction of conventional current. This is why the diode triangle or arrow (Fig. 2-5) points in the easy direction of conventional flow. If you prefer electron flow, just remember that electrons flow easily against the diode arrow. It may help to notice that the diode arrow points in the direction from which the free electrons come.

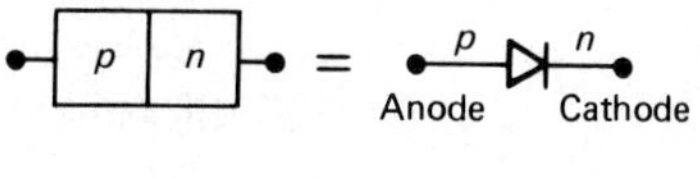

Fig. 2-5.
Schematic symbol of a
diode.

Knee Voltage

Suppose an adjustable dc voltage source is connected in series with a resistor and diode as shown in Fig. 2-6a. When the applied voltage is zero, there is no current. As we increase the voltage, electrons begin to flow. Because of the series connection, the current is the same in all parts of the circuit. This current increases slowly at first. When the applied voltage is approximately equal to the barrier potential, the current increases rapidly. In other words, when the applied voltage overcomes the barrier potential, the forward current is large.

Figure 2-6b summarizes the relation between diode current and voltage. Notice how small the current is until the diode voltage equals V_K. Voltage V_K is called the *knee voltage*. It is the voltage that separates low forward current from high forward current. The knee voltage approximately equals the barrier potential. Therefore, with a silicon diode the current starts to increase rapidly somewhere between 0.6 and 0.7 V. As a conservative estimate, we will use 0.7 V in this book. (Note: for a germanium diode, V_K is approximately 0.3 V.)

Maximum Forward Current

The product of voltage and current equals power. For instance, if the diode current equals 50 mA when the diode voltage is 0.75 V, the power dissipated by the diode is

$$P = VI = (0.75 \text{ V})(50 \text{ mA}) = 37.5 \text{ mW}$$

Because this power produces heat, it raises the temperature of the diode.

Since power is directly proportional to current, there is a limit to how much current a diode can safely conduct. For instance, the 1N4001 is a silicon diode that can tolerate up to 1 A of steady current. If the diode current is greater than 1 A, the diode will be destroyed because it can no longer dissipate the power and its temperature becomes too high.

Small Reverse Current

Figure 2-7a shows reverse bias. If the applied voltage is zero, no current flows. When we increase the voltage, a very small current

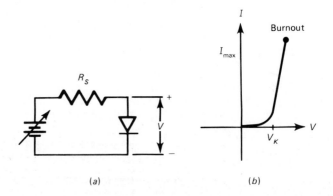

Fig. 2-6.
Forward bias. (a) Circuit.
(b) Knee voltage and
maximum forward current.

(a)

(b)

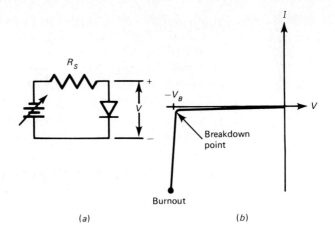

Fig. 2-7.
Reverse bias. (a) Circuit.
(b) Small reverse current
and avalanche.

(a)   *(b)*

appears. This reverse current consists of the minority-carrier current and the surface-leakage current. Therefore, the reverse current is partially temperature-dependent. For instance, with a reverse voltage of 50 V the 1N4001 has a reverse current of 10 μA for a junction temperature of 25°C. The reverse current increases to 50 μA when the junction temperature is 75°C.

Breakdown Voltage

In the forward direction too much current eventually destroys a diode because the power dissipation becomes too large. In the reverse direction, too much voltage will produce electrical breakdown. This breakdown is similar to an avalanche where a falling rock dislodges other rocks, which then knock still more rocks free until the side of the mountain comes down.

Here is how the avalanche of charges may occur in a reverse-biased diode. When the reverse voltage is excessive, a few thermally produced free electrons can reach high speeds and dislodge other electrons from valence orbits. Once free, the valence electrons (now free electrons) accelerate because of the strong electric field and collide with other valence electrons. The process ends in an avalanche with huge numbers of valence electrons breaking loose from their orbits and becoming free electrons. The voltage where the avalanche occurs is called the *breakdown voltage*.

Figure 2-7*b* summarizes the idea. Notice how the reverse current is extremely small until we reach the breakdown voltage V_B. Beyond this point, the reverse current suddenly increases until the diode is destroyed by excessive power dissipation. As an example, the 1N4001 has a breakdown voltage of slightly more than 50 V. When you exceed this voltage, you run the risk of having an avalanche and destroying the diode. Except for special diodes (zener diodes) discussed in the next chapter, the reverse voltage applied to a diode should always be less than the breakdown value specified for the diode. On data sheets this is sometimes called the *PIV rating* (peak inverse voltage).

Figure 2-8 is a diode graph for the forward and reverse characteristics of a diode. It echoes an idea stated several times earlier. The diode conducts easily in the forward direction and poorly in the reverse direction. Properly used, the diode will act like a one-way conductor. Remember the following key ideas because they will help you with troubleshooting and design:

1. The knee voltage in the forward direction is very small, typically 0.7 V for a silicon diode. Beyond the knee voltage the diode current increases rapidly.
2. Unless the forward current is limited in some way, the power dissipation becomes excessive and the diode is destroyed.
3. The breakdown voltage in the reverse direction is usually large, typically greater than 50 V. Reverse current is extremely small below the breakdown voltage and it is temperature-dependent.
4. Beyond the breakdown voltage, the avalanche of valence electrons becoming free electrons will usually destroy the diode.

2-5
THE IDEAL DIODE

When you are troubleshooting a diode circuit, all you need is a basic idea of what the circuit is supposed to do. Then you can estimate and check the voltages in different parts of the circuit. This type of analysis does not require exact answers; all you need are ballpark estimates of the circuit voltages. For this reason, many people use *approximations,* diode models that simplify analysis.

Basic Idea

The *ideal diode* is the first and simplest approximation of a real diode. It has no forward voltage drop, no reverse current, and no breakdown (see Fig. 2–9a). Such a diode cannot be manufactured. It is only a theoretical approximation of a real diode. However, in a well-designed circuit a real diode behaves almost like an ideal diode because the forward voltage across the diode is small compared with the input and output voltages. This is why you can analyze many diode circuits by treating all diodes as ideal. In using the ideal-diode ap-

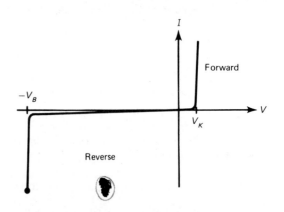

Fig. 2-8.
Complete diode curve.

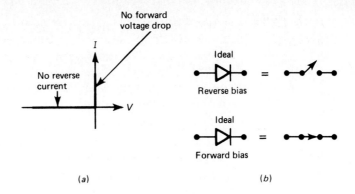

No forward
voltage drop

I

No reverse
current

V

Ideal

Reverse bias

Ideal

Forward bias

Fig. 2-9.
Ideal diode. (a) Graph.
(b) Equivalent circuit.

(a)

(b)

proximation, it helps to visualize the ideal diode as a switch (Fig. 2-9*b*). The switch is closed when the diode is forward-biased and open when the diode is reverse-biased.

Does Not Exist To repeat, the ideal diode does not exist as a manufactured device; it exists only in our minds as a convenient model for quick solutions of diode currents and voltages. The situation is similar to the way we treat the conducting wires in a circuit. Usually, we idealize these wires as perfect conductors by ignoring their resistance. The answers are not exact, but they are reasonably accurate because the designer normally keeps the wire resistance small compared with the resistances in the circuit. Occasionally, the answers we get with the ideal diode have too much error for the application. In this case, we will use better approximations to be discussed in later sections.

EXAMPLE 2-1 In Fig. 2-10*a*, the resistor is called the *load resistor*, designated R_L. Use the ideal-diode approximation to calculate the output voltage and current in Fig. 2-10*a*.

SOLUTION The diode is forward-biased. Therefore, we can visualize it like the closed switch of Fig. 2-10*b*. Since the switch is closed,

$$V_{out} = V_{in} = 15 \text{ V}$$

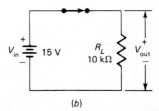

V_{in} 15 V R_L 10 kΩ V_{out}

V_{in} 15 V R_L 10 kΩ V_{out}

Fig. 2-10.
Forward-biased circuit. (a)
Original. (b) Equivalent.

(a)

(b)

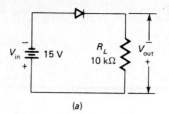

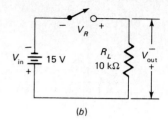

Fig. 2-11.
Reverse-biased circuit.
(a) Original. (b) Equivalent.

(a)
(b)

With Ohm's law, we can calculate the current through the load resistor:

$$I = \frac{V_{out}}{R_L} = \frac{15\ V}{10\ k\Omega} = 1.5\ mA$$

Because of the series connection, the current equals 1.5 mA in any part of the circuit. So, the diode current also equals 1.5 mA.

EXAMPLE 2-2 What is the output voltage in Fig. 2-11a assuming an ideal diode? How much voltage is there across the diode?

SOLUTION Because the diode is reverse-biased, the diode appears like the open switch of Fig. 2-11b. Since no current flows through the resistor, Ohm's law tells us

$$V_{out} = IR_L = 0(10\ k\Omega) = 0$$

Kirchhoff's voltage law says the sum of voltages around a closed circuit is zero. In Fig. 2-11b, this means that all of the source voltage appears across the open diode. In other words, the reverse voltage is

$$V_R = V_{in} - V_{out} = 15\ V - 0 = 15\ V$$

To avoid damage, the diode must have a PIV rating greater than 15 V.

EXAMPLE 2-3 What is the output voltage in Fig. 2-12a if the ideal-diode approximation is used?

SOLUTION The input is an ac voltage (sine wave) with a positive peak of 15 V and a negative peak of −15 V. During the positive half cycle, the diode is forward-biased. Figure 2-12b shows the ideal circuit at the positive peak. Because the switch is closed, the output has a positive peak of 15 V.

During the negative half cycle of input voltage, the diode is reverse-biased. Figure 2-12c shows the ideal circuit at the negative peak. Because the switch is open, no voltage reaches the output. To satisfy Kirchhoff's voltage law, 15 V appears across the open diode.

We can summarize the analysis like this. Each positive half cycle of input voltage appears across the output because the diode acts like a closed switch. Each negative half cycle of input voltage is blocked from the output because the diode is open. This is why the output voltage is

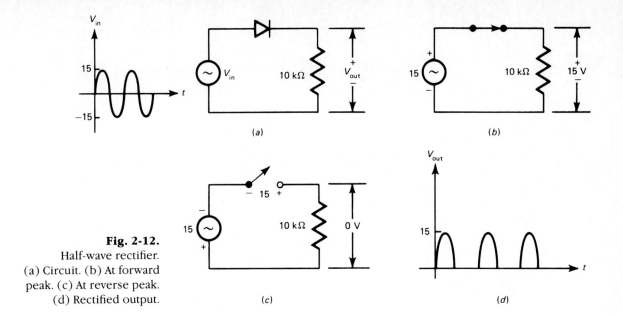

Fig. 2-12.
Half-wave rectifier.
(a) Circuit. (b) At forward
peak. (c) At reverse peak.
(d) Rectified output.

the *half-wave* signal shown in Fig. 2-12d. The circuit of Fig. 2-12a is called a *half-wave rectifier.*

EXAMPLE 2-4 Using the ideal-diode approximation, calculate the peak current through the diode of Fig. 2-12a. Also, what is the maximum reverse voltage the diode must be able to withstand?

SOLUTION The peak current through the diode occurs at the positive peak (see Fig. 2-12b) and equals

$$I_p = \frac{V_p}{R_L} = \frac{15 \text{ V}}{10 \text{ k}\Omega} = 1.5 \text{ mA}$$

As shown in Fig. 2-12c, the maximum reverse voltage occurs at the negative peak and equals

$$V_R = V_{in} - V_{out} = 15 \text{ V} - 0 = 15 \text{ V}$$

For more insight into how a half-wave rectifier works, look at Fig. 2-13. The top waveform is the input voltage, followed by the output voltage, the diode voltage, and the circuit current. Because of Kirchhoff's voltage law, the sum of voltages around the loop must equal zero at each instant. When the diode is forward-biased, it ideally has zero voltage across it. This is why the positive half cycles appear across the output. On the other hand, when the diode is reverse-biased, it acts like an open switch. The lack of current through the resistor implies zero voltage across the output. For this reason, the negative half cycles appear across the diode.

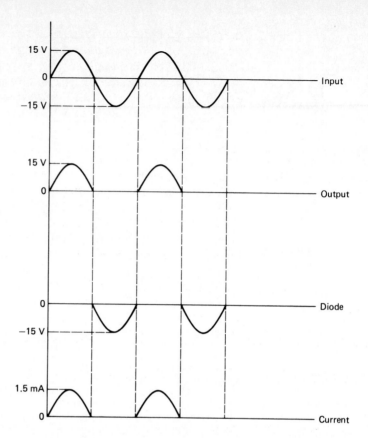

Fig. 2-13.
Timing diagram for half-wave rectifier.

2-6
THE SECOND APPROXIMATION

The ideal diode is the simplest but crudest approximation of a diode. The answers we get with this approximation give us an initial idea of how diode circuits work. To improve the accuracy, we can include the forward voltage drop across the diode. A simple way to do this is to use the knee voltage.

Includes Knee Voltage

Figure 2-14a shows the graph of current versus voltage for the *second approximation*. Notice that the diode does not conduct until the diode voltage equals V_K. Thereafter, the diode voltage remains at V_K

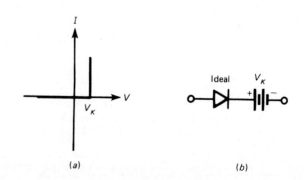

Fig. 2-14.
The second approximation.
(a) Graph. (b) Equivalent circuit.

(a)   *(b)*

even though the diode current may increase. The second approximation is equivalent to an ideal diode in series with a battery (see Fig. 2-14b). For silicon diodes, V_K is approximately 0.7 V. When using the second approximation, we assume the diode is open until the external circuit produces a diode drop of 0.7 V. Thereafter, the current increases but the diode voltage remains at 0.7 V.

Using the Second Approximation

When would you use the second approximation? Whenever 0.7 V is significant compared with the input voltage. For instance, suppose errors of less than 5 percent are acceptable. If the input voltage is more than 14 V, you can use the ideal diode because ignoring the 0.7 V produces an error of less than 5 percent. But if the input voltage is less than 14 V, you have to take the 0.7 V into account because it no longer is less than 5 percent of the input voltage. A lot will depend on what you are trying to do. If you are troubleshooting, the ideal diode may be adequate. But if you are designing, you probably will need the improved accuracy that comes with the second approximation.

EXAMPLE 2-5 Use the second approximation to calculate the output voltage and current in Fig. 2-15a. Also calculate the power dissipation of the diode.

SOLUTION Since the diode is forward-biased, we can visualize the equivalent circuit of Fig. 2-15b. The output voltage equals

$$V_{out} = V_{in} - V_K = 15 \text{ V} - 0.7 \text{ V} = 14.3 \text{ V}$$

The output current is

$$I = \frac{V_{out}}{R_L} = \frac{14.3 \text{ V}}{10 \text{ k}\Omega} = 1.43 \text{ mA}$$

The power dissipated by the diode is

$$P = VI = (0.7 \text{ V})(1.43 \text{ mA}) = 1 \text{ mW}$$

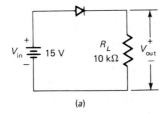

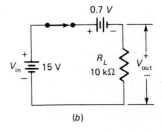

Fig. 2-15.
Example of using second approximation.

(a)   (b)

EXAMPLE 2-6 Use the second approximation to find the output waveform in Fig. 2-16a. Also, work out the peak forward current and the *peak inverse voltage* (abbreviated PIV and equivalent to the peak reverse voltage).

SOLUTION During the positive half cycle of input voltage, the first 0.7 V is wasted in overcoming the barrier potential. Thereafter, the diode can conduct. Figure 2-16b shows the circuit at the positive peak. Kirchhoff's voltage law tells us the peak output voltage equals the peak source voltage minus the drop across the diode:

$$V_p = V_{in} - V_K = 15 \text{ V} - 0.7 \text{ V} = 14.3 \text{ V}$$

During the negative half cycle, the diode is open. Therefore, no voltage reaches the output. For this reason, the final output waveform is a half-wave signal with a peak of 14.3 V, as shown in Fig. 2-16c.

In Fig. 2-16b, the maximum forward current occurs at the positive peak of input voltage and equals

$$I_p = \frac{V_p}{R_L} = \frac{14.3 \text{ V}}{10 \text{ k}\Omega} = 1.43 \text{ mA}$$

What about the peak inverse voltage across the diode? This is the same as the maximum reverse voltage across the diode. When the diode is reverse-biased, it is open and no voltage appears across the output. Because of this, all of the source voltage is across the diode. Therefore,

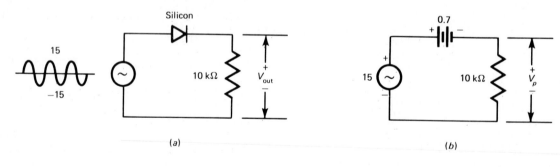

(a)   (b)

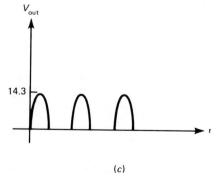

Fig. 2-16.
Using the second approximation. (a) Circuit. (b) At forward peak. (c) Rectified output.

(c)

the peak inverse voltage equals the negative peak of the source voltage. In symbols,

$$PIV = V_{in} - V_{out} = 15\ V - 0 = 15\ V$$

2-7
THE THIRD APPROXIMATION

Still another approximation can be made for a diode to account more accurately for its forward voltage drop. The forward diode curve is not vertical above the knee. Rather, it has a slight upward slope to the right. This means slightly more voltage is dropped across the diode as the current increases.

Graph

Figure 2-17*a* shows the *third approximation* of a diode. This is more accurate than the previous approximations. The diode starts conducting above V_K. Once conducting, it acts like a small resistance because the change in current is directly proportional to the change in voltage. This resistance is called the *bulk resistance* of the diode and is symbolized by r_B.

Equivalent Circuit

Figure 2-17*b* is the equivalent circuit for the third approximation, an ideal diode in series with a battery and the bulk resistance. When the diode is conducting, the voltage across the third approximation is the sum of the knee voltage plus the drop across the bulk resistance. In other words, it takes at least 0.7 V to start conduction in a silicon diode. The diode then drops additional voltage across r_B. The total voltage across the diode therefore is

$$V_F = V_K + I_F r_B \tag{2-1}$$

where V_K is approximately 0.7 V for a silicon diode (0.3 V for a Ge diode).

More on Bulk Resistance

A word or two on finding bulk resistance. If a *curve tracer* is available, you can display the forward diode curve. Above the knee, this curve

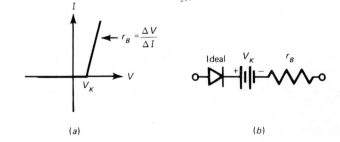

Fig. 2-17.
The third approximation.
(a) Graph. (b) Equivalent circuit.

(a)

(b)

will appear almost linear, as shown in Fig. 2-18. By selecting two points well above the knee, you can calculate the bulk resistance by using

$$r_B = \frac{\Delta V}{\Delta I} \qquad\qquad (2\text{-}2)$$

where ΔV and ΔI are the changes in voltage and current between the two points. For example, we calculate the bulk resistance of Fig. 2-18 as follows:

$$r_B = \frac{\Delta V}{\Delta I} = \frac{0.8\text{ V} - 0.7\text{ V}}{50\text{ mA} - 10\text{ mA}} = \frac{0.1\text{ V}}{40\text{ mA}} = 2.5\ \Omega$$

Which Approximation to Use   When you want a fast and rough idea of how a diode circuit works, the ideal diode is adequate. On the other hand, in critical designs where high accuracy is required, you will need the third approximation because it includes knee voltage and bulk resistance. For everyday work, however, the second approximation is usually the best compromise. It is the one we will use in this book unless otherwise indicated. When you do problems at the end of each chapter, use the second approximation of a silicon diode unless you are specifically told otherwise.

EXAMPLE 2-7   In a half-wave circuit like Fig. 2-19a, the bulk resistance of the silicon diode is 20 Ω. Use the third approximation to calculate the peak load voltage and the peak inverse voltage across the diode.

SOLUTION   Figure 2-19b shows the circuit at the positive peak of source voltage. Since the total series resistance is 1020 Ω, the peak forward current is

$$I_p = \frac{V_{in} - V_K}{R_L + r_B} = \frac{10\text{ V} - 0.7\text{ V}}{1000\ \Omega + 20\ \Omega} = \frac{9.3\text{ V}}{1020\ \Omega} = 9.12\text{ mA}$$

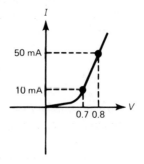

Fig. 2-18.
Estimating the bulk resistance.

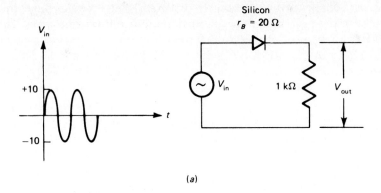

(a)

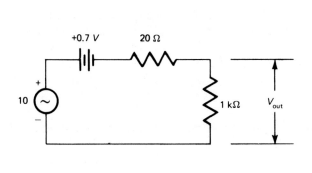

Fig. 2-19. (b)

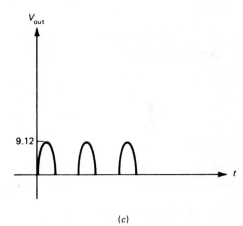

(c)

The peak voltage across the output is given by the peak current times the load resistance:

$$V_p = I_p R_L = (9.12 \text{ mA})(1 \text{ k}\Omega) = 9.12 \text{ V}$$

Figure 2-19c shows the rectified voltage across the load resistor.

During the negative half cycle the diode is open and all of the source voltage appears across it because the load voltage is zero. Therefore, the peak inverse voltage across the diode is

$$\text{PIV} = V_{in} - V_{out} = 10 \text{ V} - 0 = 10 \text{ V}$$

The diode must be able to withstand this peak inverse voltage. For this reason, it must have a PIV rating (breakdown voltage) greater than 10 V.

DC RESISTANCE OF A DIODE

An ordinary resistor is an example of a *linear* device because the total current through it is proportional to the total voltage across it. A diode is different. It is a *nonlinear* device because the total current through it is not proportional to the total voltage across it. Stated another way, the dc resistance of a diode (the ratio of total voltage to total current) is not constant because it is low in the forward direction and high in the reverse direction. Even in the forward direction, the dc resistance will decrease with an increase in current.

Forward Resistance

For example, a 1N914 is a silicon diode with a forward current of 10 mA at 0.65 V, 30 mA at 0.75 V, and 50 mA at 0.85 V. At the first point, the dc resistance is

$$R_F = \frac{V_F}{I_F} = \frac{0.65 \text{ V}}{10 \text{ mA}} = 65 \text{ }\Omega$$

At the second point,

$$R_F = \frac{0.75 \text{ V}}{30 \text{ mA}} = 25 \text{ }\Omega$$

And at the third point,

$$R_F = \frac{0.85 \text{ V}}{50 \text{ mA}} = 17 \text{ }\Omega$$

Notice how the dc resistance decreases as the current increases. In all cases, the forward resistance is relatively low.

Reverse Resistance

In the reverse direction, the 1N914 has 25 nA at 20 V and 5 μA at 75 V. At the first point, the dc resistance is

$$R_R = \frac{V_R}{I_R} = \frac{20 \text{ V}}{25 \text{ nA}} = 800 \text{ M}\Omega$$

At the second point,

$$R_R = \frac{75 \text{ V}}{5 \text{ }\mu\text{A}} = 15 \text{ M}\Omega$$

Notice how the dc resistance decreases as we approach the breakdown voltage (listed as 75 V on the manufacturer's data sheet). Nevertheless, the reverse resistance of the diode is still high, well into the megohms.

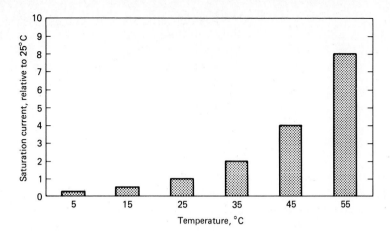

Fig. 2-21.
Saturation current doubles
for each 10°C rise.

devices will encounter. Extremes in temperature are one of the main reasons electronic equipment may fail to perform well. Later chapters will continue to point out the effect of temperature on device operation.

2-10
DATA
SHEETS

Manufacturer's data sheets list absolute maximum ratings and other specifications for diodes. Do not take the absolute maximum ratings lightly; if you exceed these ratings, you almost always destroy the diode or seriously degrade its performance. This section briefly discusses the maximum ratings and other diode specifications.

Breakdown Ratings

Most diodes should not be allowed to break down, because this may destroy the diode or seriously degrade its performance. There is no standard symbol for breakdown voltage. It has been symbolized on data sheets as follows:

$V_{R(\text{max})}$ maximum reverse voltage
$V_{(BR)}$ voltage breakdown
BV   breakdown voltage
PRV   peak reverse voltage
PIV   peak inverse voltage
V_{RWM} voltage reverse working maximum
V_{RM} voltage reverse maximum

Some of these are dc ratings and some are ac ratings. You will have to consult individual data sheets for the conditions attached to these breakdown ratings.

Power and Current Ratings

One way to destroy a diode is to exceed its reverse breakdown voltage. Another way to ruin a diode is to exceed its power rating, symbolized P_D. In the forward direction the voltage at the destruct point is well above the knee voltage, a volt or more. The product of this voltage and the current produces so much heat that the silicon melts, ruining the diode.

Manufacturers sometimes specify the power rating of a diode on data sheets. For instance, the 1N914 has a maximum power rating of

$$P_D = 250 \text{ mW}$$

More often, the data sheet lists only the *current ratings,* the maximum current a diode can handle. This current rating, often symbolized by I_0, represents the maximum dc current the diode can safely conduct. Specifying the maximum current rather than the maximum power is more convenient because current is easily measured.

As an example, the data sheet of a 1N4001 does not include a power rating, but it does list the following:

$$I_0 = 1 \text{ A}$$

This tells us that if we allow more than 1 A of steady current through a 1N4001, it may be destroyed or its life shortened.

Current-Limiting Resistor

This brings us to why a current-limiting resistor is almost always used in series with a diode. Back in Fig. 2-6a, R_S is called a *current-limiting* resistor. The larger we make R_S, the smaller the diode current for a given source voltage. Diode circuits always include a current-limiting resistance to keep the maximum forward current below the maximum current rating of the diode. Even in circuits where you cannot see a resistor (like a black box driving a diode), the Thevenin resistance facing the diode may be enough to keep the current at safe levels. The point is there must always be enough resistance in series with the diode to limit the current to less than the current rating.

Small-Signal versus Power Diodes

Incidentally, data sheets define two classes of diodes: *small-signal* diodes and *power* diodes (also called rectifiers). Small-signal diodes have a power rating less than 0.5 W, while rectifiers have power rating more than 0.5 W. The 1N914 is a small-signal diode because its power rating is 0.25 W; the 1N4001 is a rectifier because its power rating is 1 W.

Surge-Current Rating

In power-supply circuits, to be discussed later, the current through a diode may be temporarily higher when the power is first applied. This initial current is called the *surge current;* it usually exists for less than 50 ms. Data sheets list the surge-current rating as I_{surge}, $I_{FM(\text{surge})}$, I_{FSM}, etc. The surge-current rating of a diode is larger than the dc-current rating because the diode cannot reach destructive temperatures during the brief time that surge current exists. As an example, the data sheet of a 1N4001 lists an I_{surge} of 30 A and an I_0 of 1 A. This means a 1N4001 can conduct up to 30 A on a temporary basis and up to 1 A on a steady basis.

Forward Characteristics

A data sheet normally includes at least one value of forward current for the corresponding forward voltage. For instance, a 1N4001 has an I_F of 1 A for a V_F of 1.1 V. One use for this data is to calculate the approximate bulk resistance. With the foregoing values,

$$r_B = \frac{V_F - V_K}{I_F} = \frac{1.1 \text{ V} - 0.7 \text{ V}}{1 \text{ A}} = 0.4 \text{ }\Omega$$

Reverse Characteristics

Also included on diode data sheets is at least one value of reverse current for the corresponding reverse voltage. As an example, the data sheet of the 1N4001 specifies an I_R of 10 μA for a V_R of 50 V and a junction temperature of 25°C. For these values, we can calculate a dc reverse resistance of

$$R_R = \frac{V_R}{I_R} = \frac{50 \text{ V}}{10 \text{ }\mu\text{A}} = 5 \text{ M}\Omega$$

Summary

The barrier potential of a silicon diode is approximately 0.7 V. This barrier potential is equivalent to a difference in the energy levels of electrons on each side of the junction. For a diode to conduct, an external voltage must overcome the barrier potential, equivalent to increasing the energy of free electrons on the n side of the junction.

A diode is a one-way conductor. When forward-biased, it conducts well. When reverse-biased, it conducts poorly. Below the breakdown voltage, the only current in the reverse direction consists of thermally produced minority carriers and surface-leakage current. Above the breakdown voltage, an avalanche of minority carriers appears and the diode conducts heavily. Normally, a diode is operated below the breakdown voltage.

We can use three approximations to simplify diode-circuit analysis. The ideal diode treats a diode as a closed switch when forward-biased and an open switch when reverse-biased. The second approximation includes the knee voltage. The third approximation includes knee voltage and bulk resistance.

The total diode voltage divided by the total diode current equals the dc resistance of the diode. We can measure the dc resistance of a diode with an ohmmeter. For a silicon diode, the ratio of reverse to forward resistance is typically greater than 1000.

Temperature changes can alter diode characteristics. An increase in temperature will reduce the knee voltage approximately 2 mV for each degree rise in Celsius temperature. Furthermore, the reverse minority-carrier current doubles every time the Celsius temperature increases 10°.

You should be aware of the diode's ratings. These include the breakdown voltage, the maximum power dissipation, the maximum dc current, and the maximum surge current. Data sheets also list at least one pair of forward current and voltage, plus one pair of reverse current and voltage.

Glossary

avalanche When the reverse voltage is excessive, a thermally produced free electron can gain enough speed to dislodge a valence electron. The two free electrons then can dislodge two more valence electrons. The process continues until a huge number of free electrons exists. With ordinary diodes, you should avoid avalanche.

barrier potential The voltage across the depletion layer, typically 0.7 V in a silicon diode and 0.3 V in a germanium diode.

breakdown voltage The reverse voltage where avalanche occurs, normally avoided in typical diode circuits.

bulk resistance The ohmic resistance of the p and n regions of a diode. Since it does not include the voltage drop across the junction, bulk resistance is different from dc resistance.

curve tracer An instrument with an oscilloscope display that can graph device current versus device voltage.

dc resistance of diode The ratio of total diode voltage to total diode current. This resistance includes the effect of the junction and the bulk resistance.

depletion layer The region near the junction where the positive and negative ions produce the barrier potential.

forward bias Applying an external voltage to a diode that produces an electron flow from the n side to the p side.

ideal diode This first approximation of a diode treats the diode as a closed switch when forward-biased and an open switch when reverse-biased.

junction This is where the p-type and n-type materials join in a pn crystal.

knee voltage Also called the offset voltage, this is approximately equal to the barrier potential. For a silicon diode, it is between 0.6 and 0.7 V.

linear device One whose current is directly proportional to its voltage. A resistor is an example.

load resistor The resistor that the output voltage appears across.

nonlinear device A device like a diode whose current is not directly proportional to its voltage.

reverse bias Applying an external voltage that tries to produce an electron flow from the *p* side to the *n* side.

second approximation This widely used approximation assumes a voltage drop of 0.7 V across a forward silicon diode.

surge current This is the large current in some diode circuits that appears immediately following the application of power to the circuit. The surge current usually lasts for only a few cycles.

third approximation This approximation includes the diode's bulk resistance in series with an ideal diode and a knee voltage.

Review Questions

Pg.

19. **1.** Define a positive ion and negative ion.
20. **2.** Describe the creation of the depletion layer.
21. **3.** Explain how forward bias can produce a large current in a diode.
23 **4.** Explain why reverse bias results in very little current.
26 **5.** Why does excessive forward current destroy a diode?
27. **6.** Describe the avalanche process that occurs with too much reverse voltage across a diode.
   7. Sketch the complete graph of current versus voltage for a silicon diode with an offset of 0.7 V and a breakdown voltage of 75 V. Explain each part of the graph.
   8. Name the three diode approximations. What is the equivalent circuit for each of these approximations?
39 **9.** How can you test a diode with an ohmmeter? What are some precautions with this test?
39/40 **10.** When the temperature increases, what happens to the knee voltage? The minority-carrier reverse current?
43 **11.** Name three diode ratings for the forward direction and one for the reverse direction.

Problems

2-1. Use the ideal-diode approximation to calculate the current through the diode in Fig. 2-22a.

2-2. In Fig. 2-22b, a sinusoidal source of 50 V rms drives the circuit. (The peak voltage equals 1.414 times the rms voltage.) If the diode is ideal, what is the peak forward current? The peak inverse voltage?

RECTIFIER DIODES

45

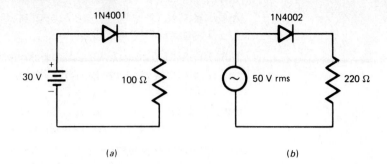

Fig. 2-22.

(a)   (b)

2-3. Use the second approximation to find the peak current in Fig. 2-22a and b.

2-4. In Fig. 2-23a, use the ideal-diode approximation to find the current through the diode.

2-5. Calculate the current through the diode of Fig. 2-23b using the ideal-diode approximation.

2-6. In Fig. 2-23c, what is the maximum current through the diode? The peak inverse voltage? What is the minimum PIV rating needed for the diode?

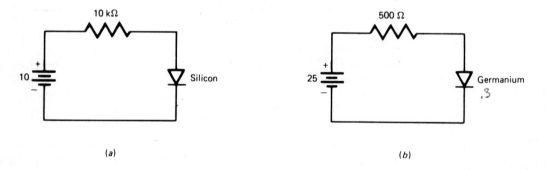

(a)   (b)

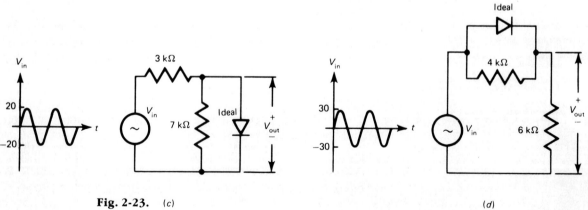

Fig. 2-23. (c)   (d)

2-7. What is the positive peak output voltage in Fig. 2-23c? The negative peak output voltage?

2-8. In Fig. 2-23d, what is the maximum current through the diode? 5ma. The positive peak output voltage? The negative peak output voltage? 30✓ 18v

2-9. Use the second approximation to calculate the current in Fig. 2-23a. What is the power dissipation in the diode?

2-10. What is the current in Fig. 2-23b using the second approximation? The power dissipation of the diode?

2-11. In the worst case, a 1N4002 has a reverse current of 50 μA for a reverse voltage of 100 V and a temperature of 100°C. In Fig. 2-23b, what is the maximum possible reverse current?

2-12. A silicon diode has a forward of current of 50 mA at 1 V and a reverse current of 20 nA at 50 V. Calculate the dc resistance in the forward direction and the reverse direction. What does the ratio of reverse to forward resistance equal?

2-13. A silicon diode has an I_S of 5 nA and a V_K of 0.7 V at 25°C. What does I_S equal at 65°C? IN 1185

2-14. Here are some diodes and their specifications:

Diode	I_o	PIV rating	I_F		I_R	
1N914	200 mA	75 V	10 mA at 1 V	100~	25 nA at 20 V	800 m~
1N4001	1 A	50 V	1 A at 1.1 V	1.1~	10 μA at 50 V	5m~
1N1185	35 A	120 V	10 A at 0.95 V	.095~	4.6 mA at 100 V	21.7K~

Which of these diodes is all right to use in the circuit of Fig. 2-24a?

2-15. Use the data of Prob. 2-14. Which of the diodes breaks down when used in the circuit of Fig. 2-24b?

2-16. Calculate the forward and reverse dc resistances for the diodes of Prob. 2-14.

2-17. You are troubleshooting the circuit of Fig. 2-25. When you look at the input voltage with an oscilloscope, you see a sinusoidal voltage with a peak of 15 V. On the other hand, you see no voltage at all

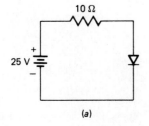

(a)

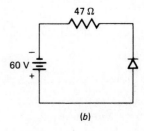

(b)

Fig. 2-24.

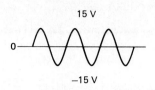

15 V

0

−15 V

Fig. 2-25.

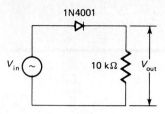

1N4001

V_{in} ~ 10 kΩ V_{out}

across the output. Which of the following (one or more) is a possible trouble:

a. Diode shorted
b. Diode open
c. Load resistor shorted
d. Load resistor open

Note: Excessive power dissipation usually causes resistors to open rather than short. But you can get a *solder bridge* (a splash of solder) on a circuit board that shorts out a resistor. This is why choice c is included in the foregoing list of troubles.

2-18. The output voltage of Fig. 2-25 is a complete sine wave with a peak of approximately 15 V. Which of the following is a possible trouble:

a. Diode shorted
b. Diode open
c. Load resistor shorted

Special Diodes

Labs 5,6,7.

Gibbs: Many historians rate Josiah Willard Gibbs as the greatest American scientist, placing him close to Newton, Einstein, and other towering geniuses. In spite of his greatness, Gibbs remains unknown to most people. Perhaps, the reason is his incredibly obscure style of writing. It took years for scientists to understand what he was talking about, not because of technical difficulties but rather because of language difficulties. As many joked, "It was easier to rediscover Gibbs than to read him."

Chapter 2 discussed the rectifier diode, a device that is optimized for one-way conduction. The main application of rectifier diodes is in power supplies, circuits that convert ac line voltage to the dc supply voltage needed by most electronic circuits. But one-way conduction is not all that a diode can do. It is possible to optimize the characteristics of a semiconductor diode for other applications. This chapter is about special-purpose diodes such as LEDs, Schottky diodes, varactors, and zener diodes.

3-1

THE LIGHT-EMITTING DIODE

Optoelectronics is the technology that combines optics and electronics. This exciting field includes many devices based on the action of a *pn* junction. The most widely used optoelectronic device is the *light-emitting* diode (LED), which is discussed in this section.

Basic Idea

By using elements like gallium, arsenic, and phosphorus, a manufacturer can produce LEDs that emit red, green, yellow, blue, orange, and infrared (invisible) light. LEDs that produce visible radiation are useful with test instruments, pocket calculators, etc. The infrared LED finds applications in burglar-alarm systems and other areas requiring invisible radiation.

Why does a LED radiate light? When a diode is forward-biased, free electrons recombine with holes near the junction, as shown in Fig. 3-1*a*. As the free electrons fall from higher energy levels to lower

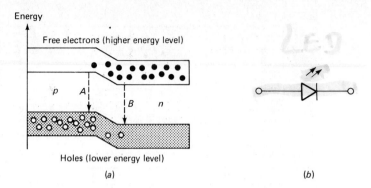

Energy

Free electrons (higher energy level)

p A

B n

Holes (lower energy level)

(a)

(b)

Fig. 3-1.
LED. (a) Energy diagram.
(b) Schematic symbol.

ones, they radiate energy. In rectifier diodes, this energy goes off almost totally in the form of heat. But in a LED, some of the energy radiates as light. Figure 3-1*b* shows the schematic symbol for a LED. The arrows are a reminder of the emitted light. LEDs have replaced incandescent lamps in many applications because of the following advantages:

1. Low operating voltage
2. Long life
3. Fast on-off switching

A Typical Red LED The TIL221 is a commercially available LED made of gallium arsenide phosphide. This LED emits red light when forward-biased. Figure 3-2*a* shows how the current varies with voltage for an ambient temperature of 25°C. The knee voltage of a TIL221 is approximately 1.5 V, considerably higher than the knee voltage of an ordinary silicon diode. As the LED current increases, the LED voltage increases. When the LED current is 20 mA, the LED voltage is approximately 1.6 V.

Figure 3-2*b* shows how the emitted light varies with forward current. The graph shows the light intensity relative to its value at 20 mA. As an example, when the current decreases to 4 mA, the relative light intensity decreases to approximately 0.2; this means the light has only 0.2 of the intensity it had at 20 mA. When the current increases to 40 mA, the light intensity increases to 1.8 times the reference level.

Changes in ambient temperature will affect the amount of light. Given a constant current through a TIL221, an increase in temperature produces a decrease in emitted light as shown in Fig. 3-2*c*. The light intensity at any ambient temperature T_A is compared with its value at a reference temperature of 25°C. For instance, when the ambient temperature increases to 80°C, the light intensity decreases to approximately 0.62 of its value at 25°C.

A Green LED The TIL222 is a gallium-phosphide diode. This LED emits green light. Figure 3-3 shows how the forward current varies with forward voltage. Notice that the knee voltage is approximately 2.1 V. When the

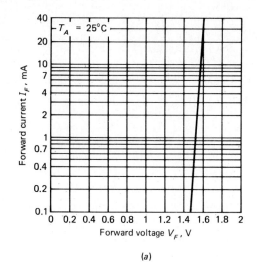

(a)

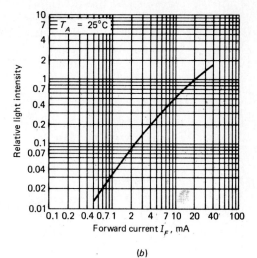

(b)

Fig. 3-2.
Data for a TIL221, a red
LED.

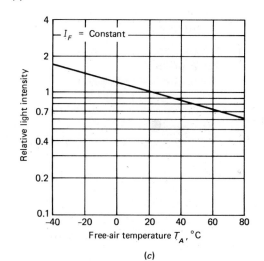

(c)

Fig. 3-3.
Data for a TIL222, a green
LED.

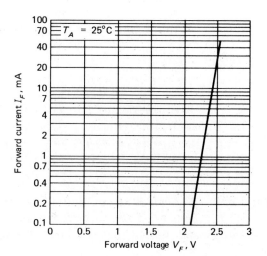

current is 20 mA, the forward voltage drop is approximately 2.5 V. This means that a green LED drops more voltage than a red LED. The graphs for light intensity versus forward current and ambient temperature are similar to Fig. 3-2*b* and *c*.

LED Voltage and Current

The TIL221 and 222 are representative of commercial LEDs. As you can see from Figs. 3-2*a* and 3-3, LEDS have a forward voltage drop from approximately 1.5 to 2.6 V for currents between 1 and 50 mA. The exact voltage depends on the LED current, color, tolerance, and other factors. Unless otherwise specified, use a nominal voltage drop of 2 V when troubleshooting or analyzing the LED circuits in this book. Typically, a LED current of 10 to 20 mA produces adequate light for most applications. Also, the maximum LED current is usually around 50 mA. You can consult a manufacturer's data sheet for the maximum ratings, voltage drops, and tolerances of a particular LED.

Seven-Segment Display

Figure 3-4*a* shows a *seven-segment* display. It contains seven rectangular LEDs (*a* through *g*). Each LED is called a *segment* because it forms part of the character being displayed. Figure 3-4*c* is the schematic diagram of a seven-segment display. External current-limiting resistors are included to keep the LED currents at safe levels. By grounding one or more resistors, we can form any digit from 0 through 9. For instance, by grounding *a, b,* and *c,* we get a 7 like Fig. 3-4*b*. Grounding *a, b, c, d,* and *g* produces a 3.

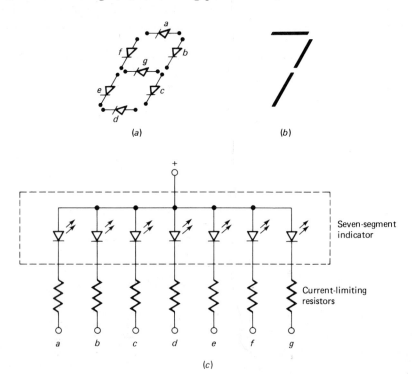

(a) (b)

(c)

Fig. 3-4.
Seven-segment indicator.
(a) Individual LEDs.
(b) Displaying a seven.
(c) With current-limiting
resistors.

A seven-segment display can also indicate the capital letters A, C, E, and F, plus lowercase letters b and d. Microprocessor trainers often use seven-segment displays that show all digits from 0 through 9, plus A, b, C, d, E, and F. (Besides 0 through 9 and A through F, a seven-segment indicator can display other letters of the alphabet.)

EXAMPLE 3-1   Calculate the LED current in Fig. 3-5a.

SOLUTION   The current through the series-limiting resistor equals the voltage across the resistor divided by the resistance:

$$I = \frac{V_S}{R_S} = \frac{V_{in} - V_{LED}}{R_S} = \frac{12\text{ V} - 2\text{ V}}{470\ \Omega} = 21.3\text{ mA}$$

Unless you have a more accurate value for the LED voltage drop, use an approximate drop of 2 V for all troubleshooting and design.

Industrial schematics don't usually show complete circuits like Fig. 3-5a. Instead, you get an abbreviated drawing like Fig. 3-5b. When the dc source is grounded on one side, the usual practice is to show only the potential of the other side with respect to ground. In Fig. 3-5a, the potential of the positive side with respect to ground is +12 V. When you see an industrial schematic like Fig. 3-5b with voltage at a point, always remember that the potential is with respect to ground.

EXAMPLE 3-2   Suppose the dc source of Fig. 3-5a is replaced by a variable dc source. Calculate the LED current for each of the following source voltages: 5 V, 10 V, 15 V, and 20 V.

SOLUTION   When the supply voltage is 5 V, the LED current is

$$I = \frac{V_{in} - V_{LED}}{R_S} = \frac{5\text{ V} - 2\text{ V}}{470\ \Omega} = 6.38\text{ mA}$$

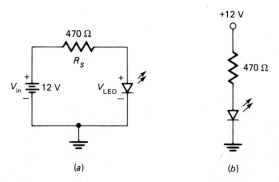

Fig. 3-5.

(a)   (b)

When V_{in} = 10 V, the current is

$$I = \frac{10\ V - 2\ V}{470\ \Omega} = 17\ mA$$

For V_{in} = 15 V,

$$I = \frac{15\ V - 2\ V}{470\ \Omega} = 27.7\ mA$$

And when V_{in} = 20 V,

$$I = \frac{20\ V - 2\ V}{470\ \Omega} = 38.3\ mA$$

Figure 3-6 shows a graph of the LED current versus the different supply voltages. As you see, the LED current is linearly related to the supply voltage. In other words, changes in LED current are directly proportional to changes in supply voltage.

3-2 SCHOTTKY DIODES

At lower frequencies a rectifier diode can easily turn off when the operation changes from forward bias to reverse bias. But as the frequency increases, we reach a point where the ordinary diode cannot turn off fast enough to prevent significant reverse current. This means an ordinary diode no longer is an efficient rectifier at high frequencies because it remains closed during the initial part of the reverse cycle.

Charge Storage

Figure 3-7a shows a forward-biased diode with a current I_F. Figure 3-7b illustrates the corresponding energy bands. Free electrons in the

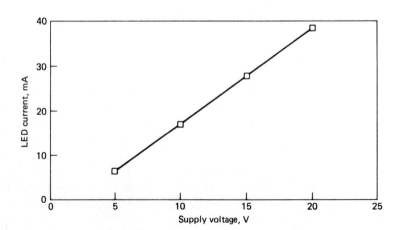

Fig. 3-6.
LED current versus supply voltage.

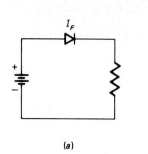
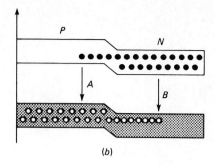

Fig. 3-7.
Charge storage.
(a) Forward-biased diode.
(b) Energy diagram.

(a)   *(b)*

conduction band (the upper band) cross the junction and travel into the *p* side. Once on the *p* side, these free electrons become minority carriers with a lifetime that depends on the doping level. The more lightly doped the *p* material is, the longer the lifetime of the free electrons. Eventually, the free electrons recombine by falling into holes (path *A*).

Similarly, holes in the valence band (the lower band) cross the junction and travel into the *n* side before recombination occurs (path *B*). Again, the recombination is not instantaneous but rather takes place after a brief period called the lifetime. As an example, if the lifetime of the holes equals 1 μs, holes can cross the junction and exist as minority carriers on the *n* side for approximately 1 μs before recombination takes place.

Because of lifetime, minority carriers are continuously stored near the junction of a forward-biased diode. The greater the lifetime, the longer these stored charges can exist before recombining. The concentration of minority carriers near the junction of a forward-biased diode is called *charge storage*.

Reverse-Recovery Time

Charge storage is important when you try to switch a diode from on to off. Why? Because if you suddenly reverse-bias a diode, the stored charges will reverse direction and set up a temporary current in the opposite direction. The greater the lifetime, the longer the stored charges can contribute to reverse current.

For example, suppose a forward-biased diode is suddenly reverse-biased, as shown in Fig. 3-8*a*. Because of the stored charges shown in Fig. 3-8*b,* a reverse current can exist temporarily. This reverse current I_R has the same magnitude as the forward current I_F. The longer the lifetime, the longer the reverse current will persist.

Manufacturers define the *reverse-recovery time* t_{rr} as the time it takes for the reverse current to drop to 10 percent of the foward current. Figure 3-8*c* illustrates reverse-recovery time. When the source voltage is reversed, the current through the diode suddenly changes from a forward value of I_F to a reverse value of I_R. Because the stored charges have reversed direction, I_R initially has the same magnitude as I_F. The reverse current I_R is constant for a while as

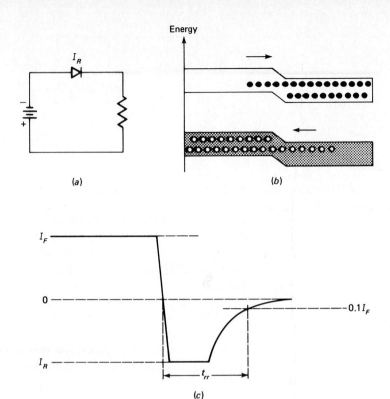

Fig. 3-8.
Charge storage.
(a) Reverse-biased diode.
(b) Energy diagram.
(c) Reverse-recovery time.

shown in Fig. 3-8c, but eventually it begins to decrease. The time interval t_{rr} is the time it takes for the reverse current to decrease to $0.1I_F$.

As an example, the 1N4148 has a t_{rr} of 4 ns. Suppose this diode has a forward current of 10 mA. If you suddenly reverse-bias the diode, a current of 10 mA will flow in the opposite direction. Because of the lifetime of minority carriers, this current remains at 10 mA for a while. Then as recombinations take place, the current decreases. Since the 1N4148 has t_{rr} of 4 ns, it will take approximately 4 ns for the reverse current to decrease to 1 mA.

Reverse-recovery time is so short in small-signal diodes that you don't even notice its effect at frequencies below 10 MHz. It's only when you get well above 10 MHz that you have to take t_{rr} into account.

Effect on Rectification

What effect does reverse-recovery time have on rectification? Take a look at the half-wave rectifier shown in Fig. 3-9a. At low frequencies the output is well behaved because it is the classic half-wave rectified signal shown in Fig. 3-9b. As the frequency increases into the mega-hertz region, however, the output signal begins to deviate from its normal shape as shown in Fig. 3-9c. As you see, some conduction is noticeable near the beginning of the reverse half cycle. The reverse-recovery time is now becoming a significant part of the period. For instance, if $t_{rr} = 4$ ns and the period is 50 ns, then the early part of the reverse half cycle will have a wiggle in it similar to Fig. 3-9c.

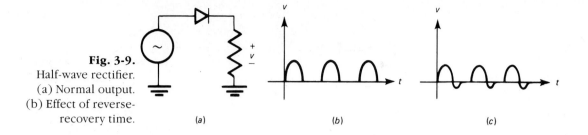

Fig. 3-9.
Half-wave rectifier.
(a) Normal output.
(b) Effect of reverse-
recovery time.

(a)   (b)   (c)

Eliminating Charge Storage

What is the solution to reverse-recovery time? A *Schottky* diode. This special-purpose diode uses a metal like gold, silver, or platinum on one side of the junction and doped silicon (usually n-type) on the other side, as shown in Fig. 3-10a. When a Schottky diode is unbiased, free electrons on the n side are in smaller orbits (lower energy level) than free electrons on the metal side. This difference in orbit size or energy level is called the Schottky barrier.

When the diode is forward-biased, free electrons on the n side can gain enough energy to travel in larger orbits. Because of this, free electrons can cross the junction and enter the metal, producing a large forward current. Since the metal has no minority carriers, there is no charge storage and almost no reverse-recovery time.

Figure 3-10b shows the schematic symbol of a Schottky diode. As a memory aid, notice that the lines look almost like a rectangular S. The lack of charge storage means that the Schottky diode can switch off faster than an ordinary diode. In fact, a Schottky diode can easily rectify frequencies above 300 MHz. As an example, the signal source

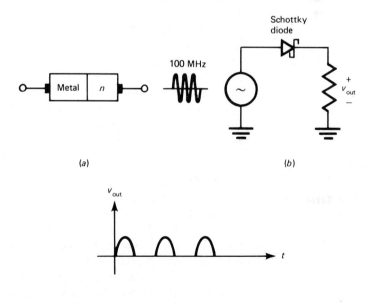

Fig. 3-10.
Schottky diode.
(a) Structure. (b) Half-wave
rectifier. (c) Rectified
output.

SPECIAL DIODES

57

of Fig. 3-10*b* has a frequency of 100 MHz. Despite this, the rectified output appears like the half-wave signal of Fig. 3-10*c*.

One important application of Schottky diodes is in digital computers. The speed of computers depends on how fast their diodes and other devices can turn on and off. This is where the Schottky diode is useful. Because it has no charge storage, the Schottky diode is the key device in *low-power Schottky TTL,* a group of widely used digital devices.

Forward Voltage Drop

One final point: In the forward direction, the Schottky diode has an offset voltage of approximately 0.25 V. Therefore, when using the second approximation of a Schottky diode, you need only subtract 0.25 V instead of 0.7 V. In some applications such as low-voltage power supplies you may see Schottky diodes used instead of silicon diodes because of their lower voltage drop. Typical Schottky diodes are low-power devices; so they are not widely used in power supplies.

3-3
THE VARACTOR

The *varactor* (also called a voltage-variable capacitance, epicap, and tuning diode) is widely used in television receivers, FM receivers, and other communications equipment. Let's find out more about this special-purpose silicon diode.

Diode Capacitance

When reverse-biased, a diode has a large resistance R_R. This resistance depends on the minority carriers and the surface leakage. Typically, R_R is well into megohms. Besides this resistance, a diode also has a built-in capacitance C_T. To understand why this capacitance exists, look at Fig. 3-11*a*. When reverse-biased, a silicon diode resembles a capacitor: the *p* and *n* regions are like the plates of a capacitor, and the depletion layer is like the dielectric.

Transition Capacitance

The foregoing capacitance is called *transition* capacitance. The word "transition" refers to the transition from *p*-type to *n*-type material. Transition capacitance is also known as depletion-layer capacitance,

Fig. 3-11.
Transition capacitance.
(a) *p* and *n* regions appear as charged plates.
(b) Equivalent circuit.
(c) Capacitance decreases when reverse voltage increases.

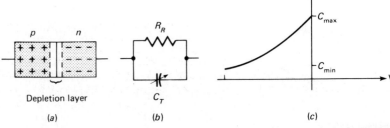

barrier capacitance, and junction capacitance. The thing that makes transition capacitance useful is this: Since the depletion layer gets wider with more reverse voltage, the transition capacitance becomes smaller. It's as though you moved the plates of a capacitor apart. The key idea is this: capacitance is controlled by voltage.

Equivalent Circuit

Figure 3-11b shows the equivalent circuit for a reverse-biased diode. A large reverse resistance R_R is in parallel with the transition capacitance C_T. Recall that capacitive reactance is given by

$$X_C = \frac{1}{2\pi f C}$$

At low frequencies, the capacitive reactance is very high. Therefore, the diode appears open because both X_C and R_R approach infinity.

At higher frequencies, however, X_C decreases and the equivalent circuit of Fig. 3-11b reduces to a capacitance C_T. Figure 3-11c shows how this capacitance varies with reverse voltage. Notice that the capacitance decreases nonlinearly for larger reverse voltages. In other words, at higher frequencies a varactor is equivalent to a voltage-controlled capacitance.

Tuning Range

Silicon diodes optimized for their variable capacitance are called *varactors*. Data sheets for varactors usually list a reference value of capacitance measured at a specific reverse voltage, typically -4 V. For instance, the data sheet of a 1N5142 lists a reference capacitance of 15 pF at -4 V.

Data sheets include a tuning range TR, defined as

$$TR = \frac{C_{max}}{C_{min}}$$

over the permitted voltage range of the varactor. For example, the data sheet of a 1N5142 lists a reference capacitance of 15 pF for a reverse voltage of -4 V and a tuning range of 3:1 for a voltage range of -4 to -60 V. This means that the capacitance decreases from 15 to 5 pF when the voltage varies from -4 to -60 V.

Abrupt and Hyperabrupt Junctions

The tuning range of a varactor depends on the doping level. Figure 3-12a shows the doping profile for an *abrupt-junction* diode (the ordinary type of diode). Notice that the doping is uniform on both sides of the junction. This means that the holes and free electrons are equally distributed. The tuning range of an abrupt-junction diode is between 3:1 and 4:1.

To get larger tuning ranges, some varactors have a *hyperabrupt* junction, one whose doping profile looks like Fig. 3-12b. This profile tells us that the density of charges increases near the junction. The heavier concentration leads to a narrower depletion layer and a

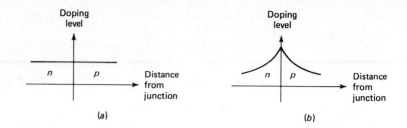

Fig. 3-12.
Doping levels. (a) Abrupt
junction. (b) Hyperabrupt
junction.

larger capacitance. Furthermore, changes in reverse voltage have more pronounced effects on capacitance. A hyperabrupt varactor has a tuning range of more than 10:1, enough to tune an AM radio through its frequency range (535 to 1605 kHz).

Tuning with a Varactor

Figure 3-13*a* shows the schematic symbol for a varactor. Because it acts like a variable capacitance, the varactor is replacing the mechanically tuned capacitor in many applications. For instance, an inductor in parallel with a capacitor has a resonant frequency of

$$ f = \frac{1}{2\pi\sqrt{LC}} \qquad (3\text{-}1) $$

If we use a varactor for the capacitor, we can change the resonant frequency by varying the reverse voltage rather than by turning the shaft of a movable-plate capacitor.

Figure 3-13*b* illustrates this idea of electronic tuning. Notice that the varactor is reverse-biased. As the tuning control is varied, the reverse voltage changes. This changes the capacitance of the varactor and affects the rest of the circuit. If the varactor is part of a resonant circuit, the resonant frequency changes. This is how we can tune in different radio or television stations.

EXAMPLE 3-3

Figure 3-14 shows the variation of capacitance with reverse voltage for some hyperabrupt varactors. What is the capacitance of an MV1403 for a reverse voltage of 1 V? For 2 V, 4 V, and 8 V?

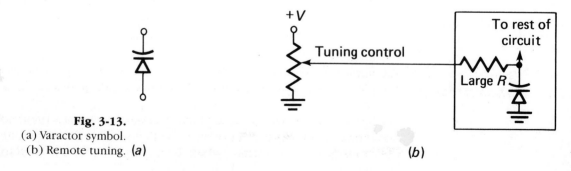

Fig. 3-13.
(a) Varactor symbol.
(b) Remote tuning. *(a)*

(b)

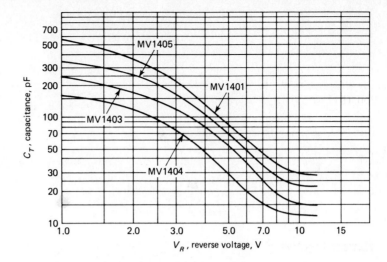

Fig. 3-14.
Diode capacitance versus
reverse voltage.

SOLUTION   Read the third graph from the top. When the reverse voltage is 1 V, the transition capacitance is approximately 250 pF. Similarly, C_T is approximately 175 pF at 2 V, 80 pF at 4 V, and 19 pF at 8 V.

EXAMPLE 3-4   What is the tuning range of an MV1401 over a voltage range of 1 to 10 V?

SOLUTION   Read the top curve of Fig. 3-14. As you see, the transition capacitance is approximately 560 pF at 1 V and 30 pF at 10 V. Therefore,

$$\text{Tuning range} = \frac{560 \text{ pF}}{30 \text{ pF}} = 18.7$$

3-4
ZENER DIODES

Small-signal and rectifier diodes are never intentionally operated in the breakdown region because of possible damage. A *zener* diode is different; it has been optimized for breakdown operation. In other words, zener diodes are normally used in the breakdown region. The main application of zener diodes is in *voltage regulators*, circuits that hold dc load voltage approximately constant in spite of changes in line voltage or load resistance.

Avalanche and
Zener Effects

Breakdown is caused by either of two effects: avalanche or zener. First, consider the *avalanche effect*. When a diode is reverse-biased, minority carriers produce a very small reverse current below the breakdown point. When the reverse voltage exceeds the breakdown voltage, the minority carriers have enough energy to dislodge valence electrons from their normal orbits. These liberated valence electrons are now free electrons and can dislodge other valence electrons (similar to a rock or snow slide). The resulting avalanche of free electrons produces a large reverse current.

Another way a diode can break down is the *zener effect*. The electric field across the junction of a reverse-biased diode becomes very intense for heavily doped diodes. If the field is strong enough, it can pull valence electrons out of their shells. This produces a large reverse current. Zener effect is sometimes called high-field emission because it is the electric field that dislodges the valence electrons.

For breakdown voltages less than 5 V, the zener effect causes the breakdown. Above 6 V, the avalanche effect dominates. Between 5 and 6 V, both effects are present. Strictly speaking, zener diodes with breakdown voltages greater than 6 V should be called avalanche diodes, but the general practice in industry is to refer to diodes exhibiting either effect as zener diodes. (Zener, an early researcher in this field, proposed the zener effect long before the avalanche theory was advanced. The name "zener diode" caught on and prevails today.)

I-V Graph

Figure 3-15*a* shows the schematic symbol of a zener diode; Fig. 3-15*b* is an alternative symbol. In either symbol, the lines resemble a *z*, which reminds us of zener. By varying the doping level, a manufacturer can produce zener diodes with breakdown voltages from about 2 to 200 V. These diodes can operate in any of three regions: forward, leakage, or breakdown. In a voltage regulator, the diodes always operate in the breakdown region. In other circuits, they may operate in all three regions during an ac cycle.

Figure 3-16 shows the *I-V* graph of a zener diode. In the forward region, it starts conducting around 0.7 V, just like an ordinary silicon diode. In the leakage region (between zero and breakdown), it has only a small leakage or reverse current. In a zener diode, the breakdown has a very sharp knee, followed by an almost vertical increase in current. Note that the voltage is almost constant, approximately equal to V_z over most of the breakdown region. Data sheets usually specify the value of V_z at a particular test current I_{zr}.

Maximum Ratings

The power dissipation of a zener diode equals the product of its voltage and current. In symbols,

$$P_z = V_z I_z \qquad (3\text{-}2)$$

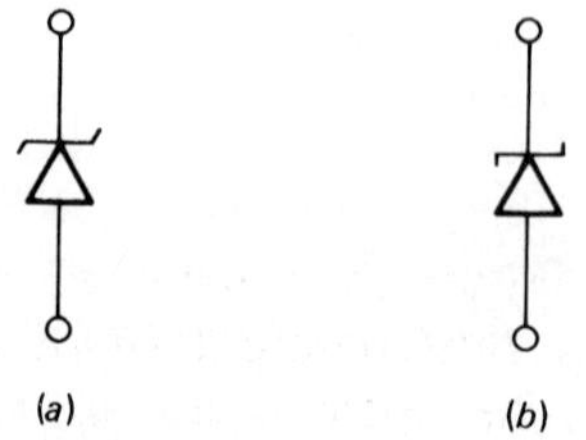

Fig. 3-15.
Schematic symbols for
zener diode.

(a)   (b)

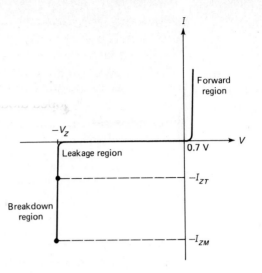

Fig. 3-16.
Graph for zener diode.

For example, if $V_Z = 12$ V and $I_Z = 10$ mA, then

$$P_Z = (12 \text{ V})(10 \text{ mA}) = 120 \text{ mW}$$

As long as P_Z is less than the power rating, the zener diode can operate in the breakdown region without being destroyed. Commercially available zener diodes have power ratings from 1/4 to more than 50 W.

Data sheets often include the maximum current a zener diode can handle without exceeding its power rating. This maximum current is related to the power rating as follows:

$$I_{ZM} = \frac{P_{ZM}}{V_Z} \qquad\qquad (3\text{-}3)$$

where I_{ZM} = maximum rated zener current
P_{ZM} = power rating
V_Z = zener voltage

In Fig. 3-16, I_{ZM} represents the zener current at the burn-out point. (Note: I_{ZM} is the magnitude of reverse current. A minus sign is included in the graph to indicate reverse bias.)

For instance, a 12-V zener diode with a power rating of 400 mW has a current rating of

$$I_{ZM} = \frac{P_{ZM}}{V_Z} = \frac{400 \text{ mW}}{12 \text{ V}} = 33.3 \text{ mA}$$

Therefore, if there's enough current-limiting resistance to keep the zener current less than 33.3 mA, the zener diode can operate in the breakdown region without being destroyed. (Some people use a safety factor of 80 percent, which means they limit the maximum current to $0.8I_{ZM}$.)

Voltage Regulation

A zener diode is sometimes called a *voltage-regulator* diode because it maintains a constant output voltage even though the current through it changes. For normal operation of a zener diode, you have to reverse-bias it as shown in Fig. 3-17a or the equivalent circuit of Fig. 3-17b. Furthermore, to produce breakdown operation the source voltage V_{in} must be greater than the zener breakdown voltage V_Z. A series resistor R_S is always used to limit the zener current to less than its current rating. Otherwise, the zener diode will burn out like any device with too much power dissipation. Once operating in the breakdown region, the zener diode will produce an almost constant output voltage, even though its current changes.

In Fig. 3-17a or b, no load resistor is connected across the zener diode. Therefore, the current through the zener diode equals the current through the series resistor. Since the voltage across the resistor is $V_{in} - V_Z$, the current through the resistor is given by

$$I_S = \frac{V_{in} - V_Z}{R_S} \qquad (3\text{-}4)$$

where I_S = current through series resistor
V_{in} = input voltage
V_Z = zener breakdown voltage
R_S = series resistance

This current depends on the source voltage, zener voltage, and series resistance.

Here is a numerical example. Suppose the source voltage is 30 V, the zener voltage is 12 V, and the series resistance is 1.8 kΩ, as shown in Fig. 3-17c. Because the input voltage is greater than the zener voltage, the zener diode is operating in the breakdown region. Therefore, it produces an output voltage of approximately 12 V.

With Eq. (3-4), we can calculate a zener current of

$$I_S = \frac{V_{in} - V_Z}{R_S} = \frac{30 \text{ V} - 12 \text{ V}}{1.8 \text{ k}\Omega} = 10 \text{ mA}$$

Figure 3-17d shows the operating point, which is designated Q_1. As indicated, the zener voltage is 12 V and the zener current is 10 mA. (Note: the minus signs are used in the graph to indicate reverse bias.)

If the input voltage is changed from 30 V to 48 V as shown in Fig. 3-17e, the zener current changes to

$$I_S = \frac{48 \text{ V} - 12 \text{ V}}{1.8 \text{ k}\Omega} = 20 \text{ mA}$$

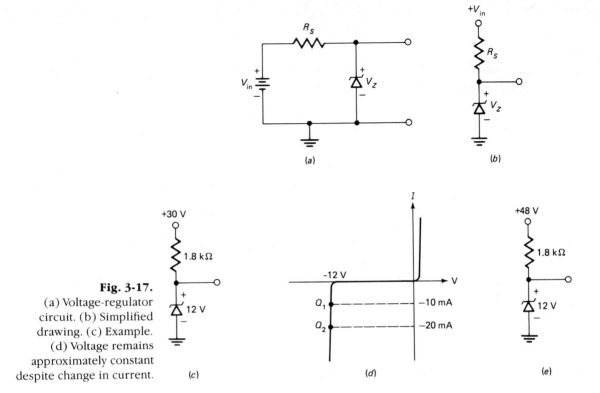

Fig. 3-17.
(a) Voltage-regulator circuit. (b) Simplified drawing. (c) Example. (d) Voltage remains approximately constant despite change in current.

Because the breakdown region is almost vertical in Fig. 3-17d, the zener voltage is still approximately 12 V even though the operating point has shifted to Q_2. This is what voltage regulation is all about. Even though the input voltage has changed from 30 V to 48 V, the output voltage across the zener diode remains fixed at approximately 12 V.

Ideal Zener Diode

For troubleshooting and preliminary design, we can approximate the breakdown region as vertical. This means we can treat the zener voltage as absolutely constant. Figure 3-18a shows the ideal approximation of a zener diode. This equivalent circuit means that a zener diode operating in the breakdown region is equivalent to a constant voltage source like an ideal battery with no internal resistance. In circuit analysis it means that you can replace a zener diode by an ideal battery, provided the zener diode is operating in the breakdown region.

Be sure to notice the reverse polarity of the battery. The ideal battery representing the zener diode does not supply current to the external circuit. Rather, the external circuit is forcing current through the battery in reverse, similar to a battery under charge. So, remember this idea when you replace a zener diode by an ideal battery.

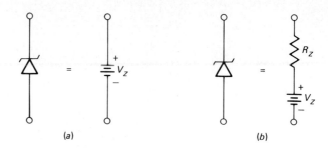

Fig. 3-18.
Equivalent circuits for
zener diode. (a) Ideal.
(b) Second approximation.

(a)

(b)

Second Approximation

When a zener diode is operating in the breakdown region, an increase in zener current produces a slight increase in zener voltage. This implies that a zener diode has a small resistance. Data sheets specify this *zener resistance* R_z (often called zener impedance) at the same test current I_{ZT} used to measure V_z. The zener resistance at this test current is designated R_{ZT} (or Z_{ZT}). For instance, the data sheet of a 1N3020 lists the following: $V_{ZT} = 10$ V, $I_{ZT} = 25$ mA, and $Z_{ZT} = 7$ Ω. This means the 1N3020 has a zener voltage of 10 V and a zener resistance of 7 Ω when the zener current is 25 mA.

The second approximation of a zener diode includes the zener resistance in series with a battery (see Fig. 3-18b). Because R_z is very small, it causes the total voltage across the zener diode to increase only slightly when I_z increases. Later examples will show you how to use the ideal and second approximations of a zener diode.

The 1N746 Series

Types 1N746 through 1N759 are a group of zener diodes with a power rating of 400 mW for an ambient temperature up to 75°C. Table 3-1 lists the data for an ambient temperature of 25°C and a test current of 20 mA. If the zener current does not equal 20 mA, the listed values

TABLE 3-1. Types 1N746 through 1N759 with $I_{ZT} = 20$ mA

Type	V_z	R_z	I_{ZM}	T_c
1N746	3.3 V	28 Ω	121 mA	−0.062%/°C
1N747	3.6 V	24 Ω	111 mA	−0.055%/°C
1N748	3.9 V	23 Ω	103 mA	−0.049%/°C
1N749	4.3 V	22 Ω	93 mA	−0.036%/°C
1N750	4.7 V	19 Ω	85 mA	−0.018%/°C
1N751	5.1 V	17 Ω	78 mA	−0.008%/°C
1N752	5.6 V	11 Ω	71 mA	+0.006%/°C
1N753	6.2 V	7 Ω	65 mA	+0.022%/°C
1N754	6.8 V	5 Ω	59 mA	+0.035%/°C
1N755	7.5 V	6 Ω	53 mA	+0.045%/°C
1N756	8.2 V	8 Ω	49 mA	+0.052%/°C
1N757	9.1 V	10 Ω	44 mA	+0.056%/°C
1N758	10 V	17 Ω	40 mA	+0.060%/°C
1N759	12 V	30 Ω	33 mA	+0.060%/°C

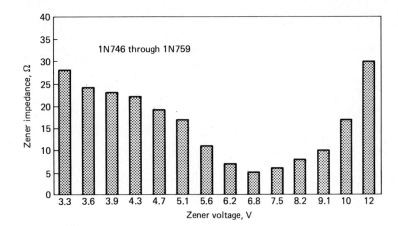

Fig. 3-19.
Zener impedance reaches a
minimum value around
6.8 V.

may be used as approximations at other zener currents. The zener voltages of the 1N746 series have a tolerance of ±10 percent. A tolerance of ±5 percent is available for devices with the suffix *A*. For instance, the 1N763A has a zener voltage of 6.2 V with a tolerance of ±5 percent. In Fig. 3-19, notice that the zener impedance reaches a minimum value when the zener voltage is 6.8 V.

**Temperature
Coefficient**

What is the temperature coefficient? Raising the ambient temperature changes the zener voltage slightly. On data sheets the effect of temperature is listed under *temperature coefficient,* defined as the percent change per degree change. The zener effect produces a negative temperature coefficient, meaning the breakdown voltage decreases with temperature. On the other hand, the avalanche effect produces a positive temperature coefficient; so the zener voltage increases with temperature.

Figure 3-20 shows the variation of temperature coefficient with zener voltage. For breakdown voltages less than 5 V, the zener effect is dominant and the temperature coefficient is negative. For break-

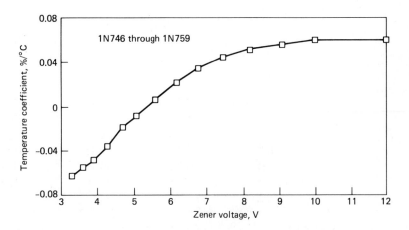

Fig. 3-20.
Temperature coefficient
equals zero between 5
and 6 V.

down voltages greater than 6 V, the avalanche effect is dominant and the temperature coefficient is positive. Between 5 and 6 V, the temperature coefficient passes through zero because the breakdown phenomenon is changing from zener effect to avalanche effect.

Having a zero temperature coefficient is important in some applications where a solid zener voltage is needed over a large temperature range. In this case, a designer will select a zener diode with a breakdown voltage between 5 and 6 V.

EXAMPLE 3-5   In Fig. 3-21a, V_{in} may vary from 20 to 40 V. Use the ideal approximation to calculate the minimum and maximum zener currents. What is the output voltage?

SOLUTION   Look up a 1N758 in Table 3-1 and read a zener voltage of 10 V. Since the source voltage varies from 20 to 40 V, the zener diode always operates in the breakdown region. This allows us to replace the zener diode by a battery of 10 V, as shown in Fig. 3-21b. The minimum zener current occurs for a source voltage of 20 V. With Eq. (3-4), the zener current equals

$$I_z = \frac{V_{in} - V_z}{R_S} = \frac{20\ V - 10\ V}{820\ \Omega} = 12.2\ mA$$

The maximum zener current exists when the source voltage is 40 V and equals

$$I_z = \frac{40\ V - 10\ V}{820\ \Omega} = 36.6\ mA$$

Ideally, the output voltage is

$$V_{out} = V_Z = 10\ V$$

for any source voltage between 20 and 40 V.

EXAMPLE 3-6   Use the second approximation in Fig. 3-21a to calculate the minimum and maximum zener current. Also work out the minimum and maximum output voltage.

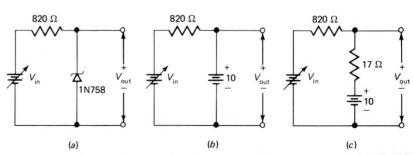

Fig. 3-21.   (a)   (b)   (c)

SOLUTION Table 3-1 lists a zener resistance of 17 Ω. Although measured for a test current of 20 mA, this value of zener resistance can be used as an approximation for any zener current. Figure 3-21c shows the zener diode replaced by its second approximation. Now, the total series resistance is

$$R_T = R_S + R_Z = 820 \ \Omega + 17 \ \Omega = 837 \ \Omega$$

When $V_{in} = 20$ V, the zener current equals

$$I_Z = \frac{V_{in} - V_Z}{R_S + R_Z} = \frac{20 \ \text{V} - 10 \ \text{V}}{837 \ \Omega} = 11.9 \ \text{mA}$$

When $V_{in} = 40$ V, the zener current increases to

$$I_Z = \frac{40 \ \text{V} - 10 \ \text{V}}{837 \ \Omega} = 35.8 \ \text{mA}$$

Because of the zener resistance in Fig. 3-21c, the output voltage will be slightly higher than 10 V. The minimum output voltage is

$$V_{out} = V_Z + I_Z R_Z = 10 \ \text{V} + (11.9 \ \text{mA})(17 \ \Omega) = 10.2 \ \text{V}$$

and the maximum output voltage is

$$V_{out} = 10 \ \text{V} + (35.8 \ \text{mA})(17 \ \Omega) = 10.6 \ \text{V}$$

This example illustrates how effective the voltage regulation is. The source voltage varies from 20 to 40 V, a change of 100 percent. Ideally, the load voltage remains fixed at 10 V. To a second approximation, it varies from 10.2 to 10.6 V, a change of only 3.9 percent. Therefore, the zener diode holds the load voltage approximately constant as shown in Fig. 3-22.

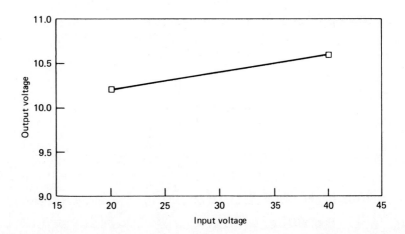

Fig. 3-22.
Output voltage is almost constant over large range of input voltage.

EXAMPLE 3-7   As the ambient temperature rises, the internal temperature of a device increases, which implies a decrease in the power rating. Figure 3-23 shows how the power rating of any zener diode in the 1N746–759 series varies with ambient temperature. What is the maximum current rating of a 1N753 for an ambient temperature of 150°C?

SOLUTION   In Fig. 3-23, the power rating is 100 mW at 150°C. Therefore, the current rating decreases to

$$I_{ZM} = \frac{100 \text{ mW}}{6.2 \text{ V}} = 16.1 \text{ mA}$$

3-5

THE ZENER REGULATOR

In the preceding section we discussed a zener circuit with no load resistor. Because of this, the current through the zener diode equaled the current through the series resistor. An unloaded zener circuit is rare. Usually, a load resistor is in parallel with the zener diode. As a result, the zener current no longer equals the series current.

Basic Idea

Figure 3-24 shows a zener regulator with a load resistor connected across the output. The input voltage V_{in} typically comes from a *power supply,* a circuit that converts ac line voltage into dc voltage. The only problem with this dc voltage is that it is *unregulated,* meaning that it can vary when line voltage or load current changes. All you have to know for now is that V_{in} is a variable dc voltage. What we are trying to do with the overall circuit of Fig. 3-24 is produce a constant dc output voltage V_{out}.

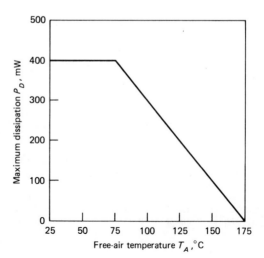

Fig. 3-23.
Derating curve.

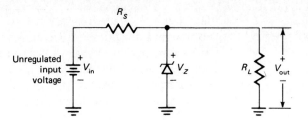

Fig. 3-24.
Load resistor connected in
parallel with zener diode.

A series-limiting resistor R_S prevents the zener current from exceeding its rated maximum value of I_{ZM}. Ideally, the zener diode acts like a battery; therefore, the load voltage is constant. For instance, if line voltage increases, V_{in} also increases. The zener voltage, however, remains constant so that V_{out} equals V_Z. To a second approximation, V_{out} is not perfectly constant because the zener resistance produces a small additional voltage drop depending on the value of zener current.

Fundamental Relations In Fig. 3-24, the voltage across the series-limiting resistor equals $V_{in} - V_{out}$. Therefore, the current through the resistor is

$$I_S = \frac{V_{in} - V_{out}}{R_S}$$ (3-5)

The current through the load resistor is

$$I_L = \frac{V_{out}}{R_L}$$ (3-6)

Because the load resistor is in parallel with the zener diode, the series current equals the sum of the zener current and load current:

$$I_S = I_Z + I_L$$

which can be rearranged as

$$I_Z = I_S - I_L$$ (3-7)

When you analyze a zener regulator, you calculate the series current and the load current. Then with Eq. (3-7) you can calculate the zener current. This zener current must be greater than zero; otherwise, the zener diode is not operating in the breakdown region.

In most troubleshooting and design you can use the ideal zener approximation:

$$V_{out} \cong V_Z$$ (3-8)

The symbol $\cong$ means "approximately equals." This is a reasonable approximation because the zener impedance has only a minor effect on the output voltage. Furthermore, the tolerance of a zener diode is at least 5 percent, large enough to justify using the ideal approximation in most analysis.

Occasionally, you may want to include the effect of zener resistance as follows. Any increase in zener current produces an additional voltage drop across the zener resistance. Therefore, the change in output voltage is given by

$$\Delta V_{out} = (\Delta I_Z)R_Z \qquad \textbf{(3-9)}$$

where ΔV_{out} = change in output voltage
ΔI_Z = change in zener current
R_Z = zener resistance

After you calculate the foregoing change in output voltage, you can add it to the ideal zener voltage to get a more accurate value of output voltage (see Example 3-10).

EXAMPLE 3-8   Use the ideal zener approximation to calculate I_S, I_L, and I_Z in Fig. 3-25 for an R_L of 200 Ω.

SOLUTION   The input voltage is 25 V. Ideally,

$$V_{out} = V_Z = 12 \text{ V}$$

With Eq. (3-5), the series current is

$$I_S \cong \frac{V_{in} - V_{out}}{R_S} = \frac{25 \text{ V} - 12 \text{ V}}{180 \text{ }\Omega} = 72.2 \text{ mA}$$

With Eq. (3-6), the load current is

$$I_L = \frac{V_{out}}{R_L} \cong \frac{12 \text{ V}}{200 \text{ }\Omega} = 60 \text{ mA}$$

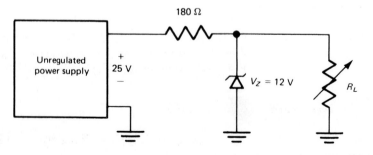

Fig. 3-25.

With Eq. (3-7), the zener current equals the difference of series current and load current:

$$I_z = I_s - I_s \cong 72.2 \text{ mA} - 60 \text{ mA} = 12.2 \text{ mA}$$

EXAMPLE 3-9 Use the ideal zener approximation to calculate all currents in Fig. 3-25 for the following load resistances: 200 Ω, 400 Ω, 600 Ω, 800 Ω, and 1 k Ω.

SOLUTION We already worked out the currents for 200 Ω in the preceding example and found that

$$I_s = 72.2 \text{ mA}, \quad I_L = 60 \text{ mA}, \quad I_z = 12.2 \text{ mA}$$

When the load resistance is 400 Ω, Eqs. (3-5) through (3-7) give

$$I_s \cong \frac{V_{in} - V_{out}}{R_S} = \frac{25 \text{ V} - 12 \text{ V}}{180 \text{ }\Omega} = 72.2 \text{ mA}$$

$$I_L = \frac{V_{out}}{R_L} \cong \frac{12 \text{ V}}{400 \text{ }\Omega} = 30 \text{ mA}$$

$$I_z = I_s - I_L \cong 72.7 \text{ mA} - 30 \text{ mA} = 42.2 \text{ mA}$$

Similar calculations produce the following results:

When R_L = 600 Ω, I_s = 72.2 mA, I_L = 20 mA, I_z = 52.2 mA
When R_L = 800 Ω, I_s = 72.2 mA, I_L = 15 mA, I_z = 57.2 mA
When R_L = 1 kΩ, I_s = 72.2 mA, I_L = 12 mA, I_z = 60.2 mA

Figure 3-26 summarizes these results graphically. As you see, the series current (left bar) remains constant when the load resistance increases. As expected, the load current (middle bar) decreases and the zener current (right bar) increases. In any case, the sum of load current and zener current equals the series current.

EXAMPLE 3-10 In the preceding example, the zener current varies from 12.2 to 60.2 mA. For this current range, how much does the output voltage change when the zener resistance is 7 Ω?

SOLUTION With Eq. (3-9),

$$\Delta V_{out} = (\Delta I_z)R_z = (60.2 \text{ mA} - 12.2 \text{ mA})(7\Omega) = 0.336 \text{ V}$$

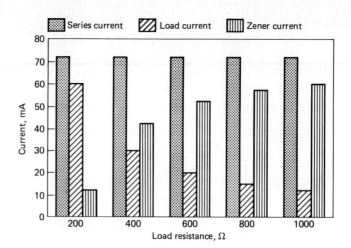

Fig. 3-26.
Comparison of currents for
different load resistances.

So, the output voltage increases 0.336 V when the zener current increases from 12.2 to 60.2 mA. The zener voltage is 12 V at the test current with a tolerance of 5 to 10 percent, depending on the particular zener diode. If a zener current of 12.2 mA happens to produce an output voltage of 12 V, then a zener current of 60.2 mA will produce an output voltage of

$$V_{out} = 12 \text{ V} + 0.336 \text{ V} = 12.336 \text{ V}$$

EXAMPLE 3-11   Calculate I_S, I_L, and I_Z for a V_{in} of 15 V in Fig. 3-27.

SOLUTION   The series current equals

$$I_S = \frac{15 \text{ V} - 12 \text{ V}}{200 \text{ }\Omega} = 15 \text{ mA}$$

The load current is

$$I_L = \frac{12 \text{ V}}{1 \text{ k}\Omega} = 12 \text{ mA}$$

The zener current is the difference of the series and load currents:

$$I_Z = 15 \text{ mA} - 12 \text{ mA} = 3 \text{ mA}$$

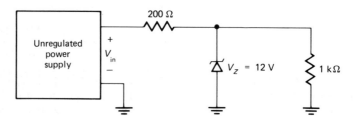

Fig. 3-27.

EXAMPLE 3-12   Calculate the I_S, I_L, and I_Z in Figure 3-27 for the following input voltages: 15 V, 20 V, 25 V, 30 V, and 35 V.

SOLUTION   We have already found the currents for 15 V in the preceding example. When V_{in} = 20 V, Eqs. (3-5) through (3-7) give

$$I_S = \frac{V_{in} - V_{out}}{R_S} \cong \frac{20 \text{ V} - 12 \text{ V}}{200 \text{ }\Omega} = 40 \text{ mA}$$

$$I_L = \frac{V_{out}}{R_L} \cong \frac{12 \text{ V}}{1 \text{ k}\Omega} = 12 \text{ mA}$$

$$I_Z = I_S - I_L \cong 40 \text{ mA} - 12 \text{ mA} = 28 \text{ mA}$$

Similar calculations give the following:

When V_{in} = 25 V, I_S = 65 mA, I_L = 12 mA, and I_Z = 53 mA
When V_{in} = 30 V, I_S = 90 mA, I_L = 12 mA, and I_Z = 78 mA
When V_{in} = 35 V, I_S = 115 mA, I_L = 12 mA, and I_Z = 103 mA

Figure 3-28 summarizes the results. Notice that load current (middle bar) is a constant 12 mA at all input voltages because the load resistance is fixed at 1 kΩ. When the input voltage increases, the series current (left bar) increases. Likewise, the zener current (right bar) increases and remains 12 mA less than the series current. As always, the series current equals the sum of load current and zener current.

Summary   When a LED is forward-biased, free electrons recombine with holes near the junction. As they fall from a higher to a lower energy level, the free

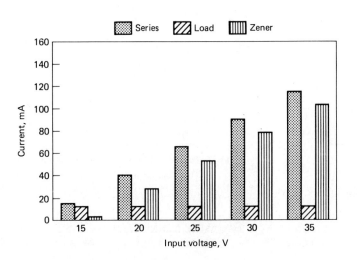

Fig. 3-28.
Comparison of currents for different input voltages.

electrons radiate energy in the form of light. As an approximation, the typical LED has a drop of about 2 V. A seven-segment LED indicator has seven LEDs that can display any digit from 0 through 9.

Because of the lifetime, stored charges exist near the junction of a forward-biased diode. If the diode is suddenly reverse-biased, these stored charges can temporarily flow in the reverse direction. The reverse-recovery time is the time it takes for the reverse current to drop to 10 percent of the forward current. Because a Schottky diode has almost no charge storage, its reverse-recovery time is much shorter than that of an ordinary silicon diode.

The varactor has been optimized for its variable capacitance when reverse-biased. By varying the reverse voltage, we can vary the capacitance. With a varactor, we can tune resonant circuits electronically rather than mechanically.

The zener diode is normally used in the breakdown region. Also known as a voltage-regulator diode, the zener diode can hold the output voltage almost constant despite changes in line voltage and load current. Ideally, a reverse-biased zener diode acts like a battery. To a second approximation, it is like a battery in series with a small zener resistance.

Glossary

avalanche effect At higher reverse voltages, minority carriers reach such high speeds that they can dislodge valence electrons from their orbits. The valence electrons become free electrons which then attain sufficient speeds to dislodge other valence electrons. The process continues until an all-out avalanche produces enormous numbers of free electrons. The avalanche effect dominates in zener diodes whose breakdown voltage is greater than approximately 6 V.

charge storage In a forward-biased diode, minority carriers are temporarily stored near the junction for a brief period known as the lifetime. These stored charges can temporarily sustain current flow when the diode is reverse-biased.

light-emitting diode Known as a LED, this diode emits light when forward-biased.

power supply Most electronic devices need a dc voltage to work properly. For this reason, electronic equipment usually contains a power supply, a circuit that converts ac line voltage into dc output voltage. An exception is battery-powered equipment.

reverse-recovery time Assume a forward-biased diode is suddenly reverse-biased. The reverse recovery time is the time it takes for the reverse current to decrease to 10 percent of the forward current.

Schottky diode Because it has almost no charge storage, this diode can switch off much faster than an ordinary silicon diode.

seven-segment display An indicator with seven rectangular sections that can be individually activated to display any digit from 0 through 9, plus the letters A, b, C, d, E, and F.

temperature coefficient The percent change in zener voltage per degree rise in temperature. The temperature coefficient of a zener diode equals zero somewhere between 5 and 6 V, depending on the zener current.

varactor A silicon diode optimized for its variable capacitance when reverse-biased. The equivalent circuit of a varactor is a voltage-controlled capacitance.

voltage regulator A circuit that converts a varying dc input voltage into a constant output dc voltage.

zener diode A diode designed to operate in the breakdown region. Because its voltage is almost constant, it is ideal for voltage regulation.

zener effect For heavily doped diodes, reverse bias produces an intense electric field across the junction. Zener effect occurs when the field is intense enough to strip valence electrons from their orbits, producing diode breakdown. The zener effect dominates in zener diodes whose breakdown voltage is less than approximately 5 V.

zener resistance The small series resistance of a zener diode when it operates in the breakdown region. This resistance depends on the zener voltage and the current.

Review Questions

1. Why does a LED radiate light?
2. Name three advantages of a LED over an incandescent lamp.
3. What is the current rating of a typical LED? Approximately how much voltage is there across a LED?
4. Which digits and letters can a seven-segment indicator display?
5. How is lifetime related to charge storage? What is reverse-recovery time?
6. What is the main advantage of a Schottky diode?
7. What is a varactor?
8. How does the capacitance of a varactor change when the reverse voltage increases?
9. What is the difference between the avalanche and zener effects?
10. Explain how a zener diode regulates the dc output voltage.
11. What effect does zener resistance have on the regulated output voltage?
12. If the dc input voltage to a zener regulator increases, what happens to the series current, load current, and zener current?

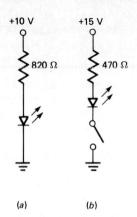

Fig. 3-29.

(a)   (b)

Problems

3-1. What is the approximate current through the LED of Fig. 3-29a?

3-2. In Fig. 3-29b, what is the approximate LED current when the switch is closed?

3-3. If a TIL221 is used in Fig. 3-29a, the actual voltage drop across the LED will be approximately 1.6 V. Calculate the LED current.

3-4. Suppose the LED of Fig. 3-29b is a TIL222. If the LED voltage drop is 2.5 V, what is the LED current?

3-5. The input signal of Fig. 3-30a has a peak of 10 V. Ideally, what does the output voltage waveform look like? If the Schottky diode has a knee voltage of 0.25 V, what does the output voltage waveform look like using the second approximation?

3-6. In Fig. 3-30b, what is the minimum reverse voltage across the varactor? The maximum?

3-7. The varactor of Fig. 3-30b is in parallel with 100 μH because the capacitors labeled "Couple" appear as ac shorts. If the varactor's capacitance can change from 10 to 40 pF, what is the minimum resonant frequency? The maximum?

3-8. Suppose we want to tune through the AM radio band (535 to 1605 kHz). What range of capacitance does the varactor of Fig. 3-30b need?

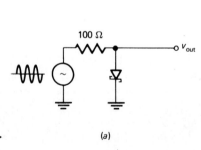

Fig. 3-30.   (a)   (b)

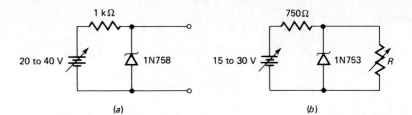

Fig. 3-31.

(a)   (b)

3-9. Use the ideal approximation in Fig. 3-31*a* to calculate the output voltage. What is the minimum zener current? The maximum?

3-10. In Fig. 3-31*a,* use the second approximation to calculate the minimum and maximum output voltage. *11.7ma*

3-11. Ideally, what is the minimum series current in Fig. 3-31*b* for a load resistance of 820 Ω? The load current? The zener current? *4.1 ma* *7.6 ma.*

3-12. Use the ideal approximation to calculate the maximum I_S, I_L, and I_Z in Fig. 3-31*b* for an R_L of 1 kΩ.

3-13. The output voltage is 10 V in Fig. 3-31*a* when the input voltage is 20 V. Suppose you increase the input voltage to 40 V and measure an output voltage of 40 V. Which of the following is a possible trouble?

a. 1N753 used instead of 1N758.
b. Short across the output terminals.
c. Series resistor is shorted.
d. 470 Ω used instead of 1 kΩ.

3-14. Suppose you measure zero output voltage in Fig. 3-31*b*. Name three possible troubles.

Diode Applications

Tesla: He invented the induction motor before working for Edison. After the two grew to hate each other, Tesla left to start his own business. He fanatically supported ac power, while Edison stubbornly championed dc power. For years the two fought bitterly. Tesla won, and ac power became the standard throughout the world. In 1912, the Nobel prize was to be awarded jointly to Tesla and Edison, but Tesla refused to be associated with Edison, and the prize went to a third party.

Since most electronic equipment requires dc voltage, we need to rectify or change the ac voltage from the power line to dc voltage. The circuit that accomplishes this conversion is called the *power supply*. The key devices in a power supply are the rectifier diodes that allow current to flow in one direction only. By including a filter, we can produce a dc output voltage that is approximately constant in value.

4-1
THE HALF-WAVE RECTIFIER

Power companies in the United States supply a nominal line voltage of 115 V rms at 60 Hz. (In Europe, it's 230 V at 50 Hz.) To get a dc voltage from ac line voltage, we can use a *half-wave rectifier* like Fig. 4-1a. When connected to a power outlet, the three-wire plug grounds the chassis (middle prong) and delivers 115 V rms to the circuit. Figure 4-1b shows how the ac line voltage looks.

RMS and Peak Values

As you know from basic circuit theory, the relation between the rms value and the peak value is

$$V_{rms} = 0.707V_p \qquad (4\text{-}1)$$

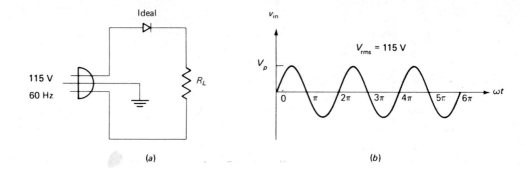

Fig. 4-1.
Half-wave rectifier.
(a) Circuit. (b) Input.
(c) Output.

If you want the peak value in terms of the rms value, then rearrange Eq. (4-1) to get

$$V_p = \frac{V_{rms}}{0.707} \tag{4-2}$$

For instance, if the line voltage has an rms value of 115 V, then its peak value is

$$V_p = \frac{115\ V}{0.707} = 163\ V$$

Average Value and Output Frequency As you saw in Chap. 2, the half-wave rectifier produces a half-wave output like Fig. 4-1c. With calculus, it can be shown that the average value of this half-wave voltage is

$$V_{dc} = 0.318 V_p \tag{4-3}$$

If we ignore the diode voltage drop in Fig. 4-1a, the half-wave voltage of Fig. 4-1c has a peak value of 163 V. Therefore, the average value is

$$V_{dc} = 0.318(163\ V) = 51.8\ V$$

The average value is also known as the *dc value*. Why? Because if you use a dc voltmeter to measure the half-wave voltage of Fig. 4-1c, it will indicate the average value of the waveform, which is 51.8 V.

The period of a half-wave signal is the same as the input. Because of this, the output frequency equals the input frequency. If the sinusoidal voltage of Fig. 4-1b has a frequency of 60 Hz, the rectified voltage of Fig. 4-1c also has a frequency of 60 Hz. In symbols, a half-wave rectifier has this property:

$$\frac{1}{2} \text{ WAVE} \rightarrow \quad f_{out} = f_{in} \tag{4-4}$$

Transformer and Peak Inverse Voltage

Most electronics equipment is transformer-coupled at the input, as shown in Fig. 4-2a. This has two important advantages. First, the transformer steps the voltage up or down, as needed in the application. Second, the transformer electrically isolates the line voltage from the rest of the electronics equipment. This isolation reduces the risk of electrical shock.

In Fig. 4-2a, the secondary voltage drives a diode in series with a load resistor. If we ignore the small diode voltage drop, the full secondary voltage appears across the load resistor during the positive half cycle. Because of this, the peak output voltage equals the peak secondary voltage:

$$V_{out(peak)} = V_{2(peak)} \tag{4-5}$$

The dc load voltage is

$$V_{dc} = 0.318V_{out(peak)} \tag{4-6}$$

and the dc load current is

$$I_{dc} = \frac{V_{dc}}{R_L} \tag{4-7}$$

Figure 4-2b shows the half-wave rectifier at the peak of the negative half cycle of secondary voltage. The diode is shaded or dark to indicate that is it off at this time. Since there is no current through the

Fig. 4-2.
(a) Transformer steps input voltage up or down.
(b) Peak secondary voltage appears across open diode.

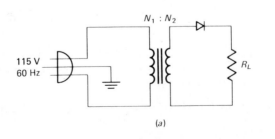
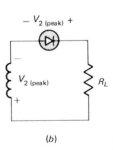

diode, there is no voltage drop across the resistor. This means that all of the secondary voltage appears across the open diode. In a rectifier circuit, the maximum reverse voltage that appears across a diode is called the *peak inverse voltage* (PIV). So, Fig. 4-2*b* tells us that a half-wave rectifier has a peak inverse voltage of

$$\text{PIV} = V_{2(\text{peak})} \qquad (4\text{-}8)$$

where $V_{2(\text{peak})}$ is the peak secondary voltage. To avoid damage to the diode, the PIV must be less than the diode breakdown voltage. Some designers include a safety factor of 2 or more, because this improves diode lifetime.

Current Rating of Diode

With Eq. (4-6), we can calculate the average or dc load voltage. If the load resistance is known, we then can calculate the load current I_{dc}. Because the half-wave rectifier is a single-loop circuit, the dc diode current is equal to the dc load current:

$$I_{\text{diode}} = I_{\text{dc}} \qquad (4\text{-}9)$$

On data sheets, I_{diode} is usually listed as I_0. Therefore, one of the things designers must look for is the I_0 rating of the diode. This rating tells them how much direct current the diode can handle.

For example, the data sheet of a 1N4001 gives an I_0 rating of 1 A. If the dc voltage is 10 V and the load resistance is 20 Ω, then the dc load current is 0.5 A. The 1N4001 would be all right to use in a half-wave rectifier because its current rating (1 A) is greater than the dc current through it (0.5 A). Often, designers include a safety factor of 2 or more, because this lengthens diode lifetime.

EXAMPLE 4-1

In Fig. 4-3*a*, what is the dc voltage across the load resistor? The peak inverse voltage across the diode? The dc current through the diode?

SOLUTION

Since the secondary voltage has an rms value of 40 V, the peak secondary voltage is

$$V_{2(\text{peak})} = \frac{V_{2(\text{rms})}}{0.707} = \frac{40\ \text{V}}{0.707} = 56.6\ \text{V}$$

Almost always, we can ignore the diode voltage drop when analyzing rectifier circuits. Therefore, the load voltage is a half-wave rectified sine wave with a peak of

$$V_{\text{out(peak)}} = V_{2(\text{peak})} = 56.6\ \text{V}$$

DIODE APPLICATIONS

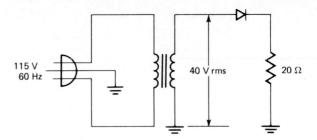

Fig. 4-3.

The dc load voltage is

$$V_{dc} = 0.318V_{out(peak)} = 0.318(56.6 \text{ V}) = 18 \text{ V}$$

The peak inverse voltage is

$$\text{PIV} = V_{2(peak)} = 56.6 \text{ V}$$

Since the dc load current is

$$I_{dc} = \frac{V_{dc}}{R_L} = \frac{18 \text{ V}}{20 \text{ }\Omega} = 0.9 \text{ A}$$

the dc diode current is

$$I_{diode} = I_{dc} = 0.9 \text{ A}$$

4-2
THE FULL-WAVE RECTIFIER

Figure 4-4a is a circuit called a *full-wave rectifier.* The secondary winding has a center tap connected to one end of the load resistor. The other end of the load resistor is connected to the diodes. This circuit acts like two half-wave rectifiers conducting on alternate half cycles.

Basic Idea   During the positive half cycle of secondary voltage, the upper diode is forward-biased and the lower diode is reverse-biased. Therefore, current flows through the load resistor, the upper diode, and the upper-half winding of Fig. 4-4c. During the negative half cycle, current flows through the load resistor, the lower diode, and the lower-half winding of Fig. 4-4d.

Especially important, the load current is in the same direction during both half cycles. This is why the load voltage has the same plus-minus polarity in Fig. 4-4c and d. For this reason, the load voltage is the *full-wave* signal shown in Fig. 4-4b. If we ignore the small diode voltage drop, the full-wave signal is like a sine wave whose negative half cycle has been inverted.

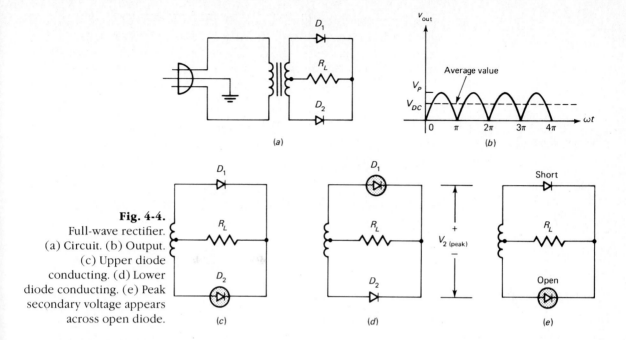

Fig. 4-4.
Full-wave rectifier.
(a) Circuit. (b) Output.
(c) Upper diode
conducting. (d) Lower
diode conducting. (e) Peak
secondary voltage appears
across open diode.

Average Value and Output Frequency

As you would expect, the average value of a full-wave signal is twice that of a half-wave signal, given by

$$V_{dc} = 0.636V_{out(peak)} \qquad (4\text{-}10)$$

where $V_{out(peak)}$ is the peak load voltage. If we ignore the small diode voltage drop, $V_{out(peak)}$ equals the peak voltage across half the secondary winding:

$$V_{out(peak)} = \frac{V_{2(peak)}}{2} \qquad (4\text{-}11)$$

In Fig. 4-4b, two cycles of output voltage appear for each cycle of input voltage. This is why the output frequency of a full-wave rectifier is twice the input frequency:

$$f_{out} = 2f_{in} \qquad (4\text{-}12)$$

If the line frequency is 60 Hz, the output frequency of a full-wave rectifier is

$$f_{out} = 2(60 \text{ Hz}) = 120 \text{ Hz}$$

Peak Inverse Voltage

Figure 4-4e shows the circuit at the instant the secondary voltage reaches its maximum value. $V_{2(peak)}$ is the peak voltage across the secondary winding. Since the upper diode is conducting at this time, it is approximately a short circuit. If you add the voltages around the outside loop, you can see that all of the secondary voltage must appear across the lower diode. Therefore, the peak inverse voltage across the nonconducting diode is

$$PIV = V_{2(peak)} \qquad \textbf{(4-13)}$$

Current Rating of Diodes
The I_0 rating of each diode need only be half of the dc load current. If the dc load current is 1 A, then each diode needs a current rating of only 0.5 A. Why is this true? In Fig. 4-4a, each diode conducts for only half a cycle. This means the current through a diode is a half-wave rectified current. Therefore, the dc current through each diode is half the dc load current. In symbols,

$$I_{diode} = \frac{I_{dc}}{2} \qquad \textbf{(4-14)}$$

The I_0 rating of each diode must be equal to or greater than I_{diode}. Typically, a safety factor of 2 or more is used.

EXAMPLE 4-2
In Fig. 4-5a, the center tap is grounded. Since the bottom of the load resistor is also grounded, the circuit is equivalent to the full-wave rectifier discussed earlier. What is the dc load voltage? The PIV across each diode? The dc current through each diode?

SOLUTION
Work out the peak secondary voltage:

$$V_{2(peak)} = \frac{V_{rms}}{0.707} = \frac{40 \text{ V}}{0.707} = 56.6 \text{ V}$$

The peak output voltage is half of this:

$$V_{out(peak)} = \frac{56.6 \text{ V}}{2} = 28.3 \text{ V}$$

The dc load voltage is

$$V_{dc} = 0.636 V_{out(peak)} = 0.636(28.3 \text{ V}) = 18 \text{ V}$$

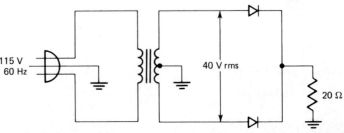

Fig. 4-5.

The peak inverse voltage is

$$PIV = V_{2(peak)} = 56.6 \text{ V}$$

Since the dc load current is

$$I_{dc} = \frac{V_{dc}}{R_L} = \frac{18 \text{ V}}{20 \text{ }\Omega} = 0.9 \text{ A}$$

the dc diode current is half of the dc load current:

$$I_{diode} = \frac{I_{dc}}{2} = \frac{0.9 \text{ A}}{2} = 0.45 \text{ A}$$

4-3
THE BRIDGE RECTIFIER

Figure 4-6a is a *bridge rectifier,* the most widely used of all rectifier circuits. During the positive half cycle of secondary voltage, diodes D_2 and D_3 are forward-biased as shown in Fig. 4-6c while diodes D_1 and D_4 are reverse-biased. Because of this, electrons flow up through D_3, to the right through the load resistance, and up through D_2. (Conventional flow is the opposite way.) During the negative half cycle, diodes D_1 and D_4 are conducting, while diodes D_2 and D_3 are off, as shown in Fig. 4-6d. This time, electrons flow down through D_1, to the right through the load resistance, and down through D_4.

During either half cycle, the electron flow is to the right through the load resistor (conventional flow is to the left). Because of this,

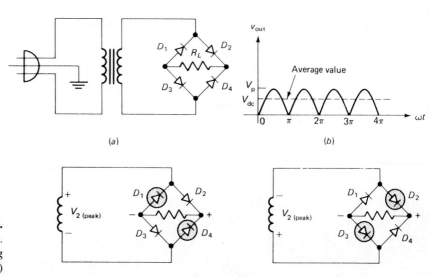

Fig. 4-6.
Bridge rectifier. (a) Circuit.
(b) Output. (c) During
positive half cycle. (d)
During negative half cycle.

DIODE APPLICATIONS

87

the plus-minus polarity of load voltage is the same in Fig. 4-6c and d. This is why the load voltage is the full-wave signal shown in Fig. 4-6b. The dc load voltage is

$$V_{dc} = 0.636V_{out(peak)} \qquad (4\text{-}15)$$

If we neglect the diode voltage drop, the peak output voltage in Fig. 4-6c or d is equal to the peak secondary voltage:

$$V_{out(peak)} = V_{2(peak)} \qquad (4\text{-}16)$$

Because of the full-wave rectification, the output frequency is twice the input frequency:

$$f_{out} = 2f_{in} \qquad (4\text{-}17)$$

Figure 4-6c shows the circuit at the instant of the positive peak in secondary voltage. Since D_2 is approximately a short circuit, all of the secondary voltage appears across D_4. Similarly, since D_3 is approximately a short circuit, all of the secondary voltage appears across D_1. In other words, the full secondary voltage appears across each open diode. This means the peak inverse voltage across each nonconducting diode is

$$PIV = V_{2(peak)} \qquad (4\text{-}18)$$

Since each diode conducts for only half of a cycle, the dc diode current is half of the dc load current:

$$I_{diode} = \frac{I_{dc}}{2} \qquad (4\text{-}19)$$

EXAMPLE 4-3   In Fig. 4-7, the ground on the left side of the bridge and the ground at the bottom of the load resistor make this circuit equivalent to the bridge rectifier discussed earlier. What are the following values: V_{dc}, PIV, and I_{diode}?

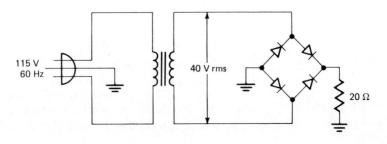

Fig. 4-7.

SOLUTION As calculated in Examples 4-1 and 4-2, the peak secondary voltage is

$$V_{2(peak)} = 56.6 \text{ V}$$

As usual, we ignore the diode voltage drop, which allows us to write

$$V_{out(peak)} = V_{2(peak)} = 56.6 \text{ V}$$

Next, we can calculate the dc load voltage:

$$V_{dc} = 0.636 V_{out(peak)} = 0.636(56.6 \text{ V}) = 36 \text{ V}$$

The peak inverse voltage is

$$\text{PIV} = V_{2(peak)} = 56.6 \text{ V}$$

The dc load current is

$$I_{dc} = \frac{V_{dc}}{R_L} = \frac{36 \text{ V}}{20 \text{ }\Omega} = 1.8 \text{ A}$$

Therefore, the dc diode current is

$$I_{diode} = \frac{I_{dc}}{2} = \frac{1.8 \text{ A}}{2} = 0.9 \text{ A}$$

EXAMPLE 4-4 Compare the dc load voltage, peak inverse voltage, and dc diode currents calculated in Examples 4-1 through 4-3.

SOLUTION In Examples 4-1 through 4-3, we calculated the following values: Half-wave rectifier:

$$V_{dc} = 18 \text{ V}, \text{ PIV} = 56.6 \text{ V}, I_{diode} = 0.9 \text{ A}$$

Full-wave rectifier:

$$V_{dc} = 18 \text{ V}, \text{ PIV} = 56.6 \text{ V}, I_{diode} = 0.45 \text{ A}$$

Bridge rectifier:

$$V_{dc} = 36 \text{ V}, \text{ PIV} = 56.6 \text{ V}, I_{diode} = 0.9 \text{ A}$$

The same secondary voltage (40 V rms) was used for each rectifier. In other words, given the same transformer, all peak inverse voltages are the same, but the bridge rectifier produces the largest dc output. The only disadvantage of a bridge rectifier is that it uses four diodes. But this is a minor disadvantage because diodes are relatively inexpensive. Almost always, the bridge rectifier is the best compromise for a power supply, and you will see it used more than any of the other rectifier circuits.

EXAMPLE 4-5 Summarize the important relations between the voltages and currents of the three basic rectifiers.

SOLUTION First, consider the peak secondary voltage. Transformer manufacturers rarely specify the turns ratio. Instead, they tell you what the rms secondary voltage is. For this reason, it helps to summarize voltages relative to the rms secondary voltage. All rectifiers have the same peak secondary voltage, given by

$$V_{2(peak)} = 1.414V_{2(rms)}$$

Second, consider the dc output voltages. The half-wave rectifier has a dc output voltage of

$$V_{dc} = 0.318V_{2(peak)} = 0.318(1.414)V_{2(rms)} = 0.45V_{2(rms)}$$

The full-wave rectifier has a dc output voltage of

$$V_{dc} = 0.636\frac{V_{2(peak)}}{2} = 0.318(1.414)V_{2(rms)} = 0.45V_{2(rms)}$$

The bridge rectifier has a dc output of

$$V_{dc} = 0.636V_{2(peak)} = 0.636(1.414)V_{2(rms)} = 0.9V_{2(rms)}$$

Third, consider the peak inverse voltages. All rectifiers have the same peak inverse voltage, given by

$$PIV = V_{2(peak)}$$

Fourth, consider the diode and load currents. In a half-wave rectifier the dc diode current equals the dc load current. In the full-wave rectifier and the bridge rectifier, the dc diode current is half the dc load current.

Figure 4-8 shows how the different voltages are related. The key numbers to remember are 1.414, 0.45, and 0.9. All voltages are relative to the rms secondary voltage, which has a value of unity (first bar). In a half-wave rectifier the peak secondary voltage is 1.414 (second bar), the dc load voltage is 0.45 (third bar), and the peak inverse voltage is 1.414 (fourth bar). The full-wave rectifier has the same values. The bridge rectifier has a peak secondary voltage of 1.414, a dc load voltage of 0.9, and a peak inverse voltage of 1.414.

Figure 4-9 shows how the average currents are related. As you can see, the half-wave rectifier has a dc diode current (second bar) equal to the dc load current (first bar). The full-wave rectifier has a dc diode

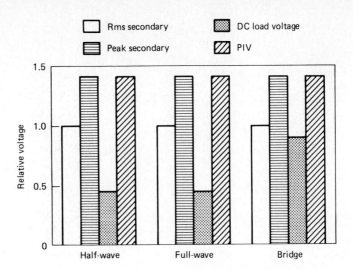

Rms secondary   DC load voltage

Peak secondary   PIV

Fig. 4-8.
Voltage relations in average
rectifiers.

current that is half of its dc load current. Similarly, the bridge rectifier
has a dc diode current equal to half of its dc load current. After you
have calculated the dc load current of any rectifier, you can quickly
work out the dc diode current.

4-4
THE CAPACITOR-
INPUT FILTER

The half-wave and full-wave rectified voltages are pulsating dc voltages. Before they are suitable for driving electronic circuits, these
pulsating dc voltages have to be filtered or smoothed out to get an
output that is almost a constant dc voltage. The most widespread
method currently used for this is the *capacitor-input filter*.

Basic Idea

Figure 4-10*a* shows a capacitor-input filter. When the diode is on, the
source charges the capacitor. Because the diode has a low resistance,

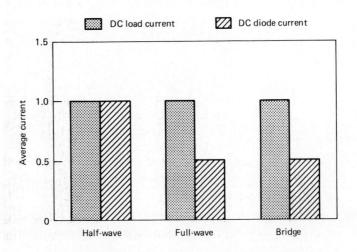

DC load current   DC diode current

Fig. 4-9.
Current relations in average
rectifiers.

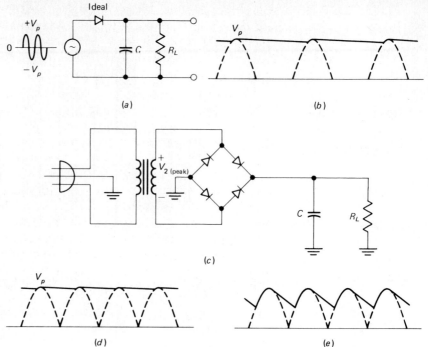

Fig. 4-10.
Capacitor-input filter.
(a) Half-wave circuit.
(b) Output with half-wave
rectifier. (c) Bridge circuit.
(d) Output with bridge
rectifier. (e) Large ripple.

the charging time constant is very short. On the other hand, when the diode is off, the capacitor discharges through the load. By deliberate design, the discharging time constant is made very long. This is why the voltage across the capacitor is almost a constant dc voltage, as shown in Fig. 4-10*b* (solid curve). Without the capacitor, the circuit would be a half-wave rectifier with the half-wave voltage shown (dashed curve).

Here is how the circuit works. On the positive half cycle of input voltage, the diode turns on and charges the capacitor. Because the diode has only a very small resistance, the capacitor charges almost immediately to the peak input voltage. Slightly beyond the positive peak, the diode shuts off. The capacitor then discharges through the load resistor. But the discharging time constant is very long compared with the period of the input signal. This is why the capacitor voltage decreases only slightly before the next peak arrives. At this peak, the diode turns on briefly and recharges the capacitor to the full input voltage (ignore diode voltage drop).

The output voltage of Fig. 4-10*b* is almost a pure dc voltage equal to the peak input voltage. This is why the circuit is sometimes called a *peak rectifier* or *peak detector.* The only deviation from pure dc is the small *ripple* (fluctuation) caused by the slight charging and discharging of the filter capacitor.

Full-Wave Action

Figure 4-10*c* shows a bridge rectifier driving a capacitor-input filter. This results in better peak detection because the capacitor charges twice as often (see Fig. 4-10*d*). As a result, the ripple is cut in half and the dc output voltage more closely approaches the peak voltage. Full-wave peak rectifiers are more common than half-wave rectifiers. From now on, we will emphasize the full-wave rectifier driving a capacitor-input filter.

Lightly Loaded and Heavily Loaded Cases

In Fig. 4-10*d* the peak-to-peak ripple is small because the capacitor discharge is very light. This is accomplished by having a long discharging time constant. This implies either a large capacitance or a large load resistance or both, so long as the product of R_L and C is much longer than the input period. We will refer to this as the *lightly loaded* case. There is no strict definition here, but as a guide in this book we will consider the loading to be light if the peak-to-peak ripple is equal to or less than 10 percent of the input peak voltage.

On the other hand, if either the load resistance or the filter capacitor is too small, the discharging time constant will be short and the ripple will become excessive, as shown in Fig. 4-10*e*. This is known as the *heavily loaded* case. In this book, a capacitor-input filter is heavily loaded when the peak-to-peak ripple is more than 10 percent of the peak input voltage.

Ripple Formula

Whether troubleshooting or designing, you will need a way to estimate the ripple out of a capacitor-input filter. The capacitor-input filter of most power supplies is designed with about 10 percent ripple or less, the lightly loaded case. The Appendix derives this formula for the ripple in the lightly loaded case:

$$V_{rip} = \frac{I_{dc}}{fC} \qquad \textbf{(4-20)}$$

where V_{rip} = peak-to-peak ripple
I_{dc} = dc load current
f = ripple frequency
C = filter capacitance

Memorize this formula. You will find this approximation extremely useful when analyzing power supplies with a capacitor-input filter. The ripple frequency is 60 Hz for the half-wave rectifier and 120 Hz for the full-wave rectifier and bridge rectifier.

DC Load Voltage

If there is no ripple, the dc output voltage of a capacitor-input filter equals the peak input voltage:

$$V_{dc} = V_{2(peak)} \qquad \textbf{(4-21)}$$

This is ideal, what you would use for a first estimate of the dc load voltage.

If you have calculated the peak-to-peak ripple, you can use the following formula to get a more accurate value of the dc load voltage:

$$V_{dc} = V_{2(peak)} - \frac{V_{rip}}{2} \qquad\qquad (4\text{-}22)$$

This should make sense. The dc load voltage is the average voltage between the top of the ripple and the bottom. By subtracting half the peak-to-peak ripple from the peak input voltage, you get the average or dc load voltage.

It is also possible to include the diode voltage drop, which is 0.7 V or less. Unless otherwise indicated, we will ignore the diode voltage drop in a power supply using a capacitor-input filter. The tolerances in the line voltage, the transformer, and the filter capacitor make precise calculations unnecessary for troubleshooting and preliminary design. In other words, the tolerances in the circuit will usually produce errors much larger than the diode voltage drops.

Peak Inverse Voltage

Because the capacitor remains charged to approximately the peak input voltage, the peak inverse voltage of a half-wave rectifier increases. In Fig. 4-10a, the peak inverse voltage across the diode is

$$PIV = 2V_{2(peak)} \qquad \text{(half wave)} \qquad\qquad (4\text{-}23)$$

This occurs when the source voltage is at its negative peak, while the capacitor is still charged to the positive peak.

In Fig. 4-10c, the peak inverse voltage across a nonconducting diode in the bridge rectifier is

$$PIV = V_{2(peak)} \qquad \text{(full wave)} \qquad\qquad (4\text{-}24)$$

This is the same as before and applies to a full-wave rectifier as well as a bridge rectifier. You can see this by looking at Fig. 4-10c. When the secondary voltage reaches a peak, the upper right diode is on, which means the full secondary voltage appears across the lower right diode.

Current Rating

The dc diode current in a half-wave rectifier driving a capacitor-input filter is still given by I_{dc}, which equals V_{dc}/R_L. Similarly, the dc diode current in a full-wave or bridge rectifier is still given by $I_{dc}/2$. The reasons are similar to those given earlier.

EXAMPLE 4-6   In Fig. 4-11, calculate the dc load voltage and the peak-to-peak ripple.

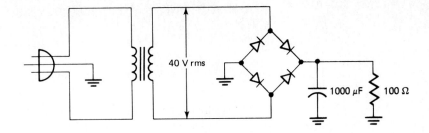

Fig. 4-11.

SOLUTION The peak secondary voltage is 56.6 V (found in Example 4-1). Therefore, a first estimate for the load voltage is 56.6 V. Next, calculate the peak-to-peak ripple. To do this, you have to work out the dc load current:

$$I_{dc} = \frac{V_{dc}}{R_L} = \frac{56.6 \text{ V}}{100 \text{ }\Omega} = 0.566 \text{ A}$$

The ripple is

$$V_{rip} = \frac{I_{dc}}{fC} = \frac{0.566 \text{ A}}{(120 \text{ Hz})(1000 \text{ }\mu\text{F})} = 4.72 \text{ V}$$

With this, we can improve the answer for dc load voltage:

$$V_{dc} = V_{2(peak)} - \frac{V_{rip}}{2} = 56.6 \text{ V} - \frac{4.72 \text{ V}}{2} = 54.2 \text{ V}$$

EXAMPLE 4-7 Summarize the important relations between the voltages and currents of the three basic rectifiers driving capacitor-input filters.

SOLUTION The peak secondary voltage is always given by

$$V_{2(peak)} = 1.414V_{2(rms)}$$

Ideally, the half-wave rectifier and bridge rectifier produce a dc load voltage equal to the peak secondary voltage:

$$V_{dc} = V_{2(peak)} = 1.414V_{(rms)}$$

The full-wave rectifier produces only half of the foregoing:

$$V_{dc} = 0.5V_{2(peak)} = 0.707V_{2(rms)}$$

The half-wave rectifier has a peak inverse voltage of

$$PIV = 2V_{2(peak)} = 2.828V_{2(rms)}$$

while the full-wave and bridge rectifiers have a peak inverse voltage of

$$PIV = V_{2(peak)} = 1.414V_{2(rms)}$$

Finally, for a half-wave rectifier the dc diode current is

$$I_{diode} = I_{dc}$$

With the full-wave and bridge rectifiers,

$$I_{diode} = 0.5I_{dc}$$

Figure 4-12 shows how the voltages are related. All voltages are relative to the rms secondary voltage, which has a value of unity (first bar). As you see, the half-wave rectifier has a peak secondary voltage of 1.414 (second bar), a dc load voltage of 1.414 (third bar), and a peak inverse voltage of 2.828 (fourth bar). The full-wave rectifier has a peak secondary voltage of 1.414, a dc load voltage of 0.707, and a peak inverse voltage of 1.414. The bridge rectifier has a peak secondary voltage of 1.414, a dc load voltage of 1.414, and a peak inverse voltage of 1.414.

Figure 4-13 shows how the average currents are related. With a half-wave rectifier the dc diode current equals the dc load current. With the full-wave and bridge rectifiers the dc diode current is half the dc load current.

4-5
VOLTAGE REGULATION

Line voltage has a tolerance of approximately ± 10 percent. Because of this, the dc load voltage out of a capacitor-input filter will have a tolerance of ± 10 percent. This variation is too large for some electrical circuits. One way to stabilize or regulate the dc load voltage is by

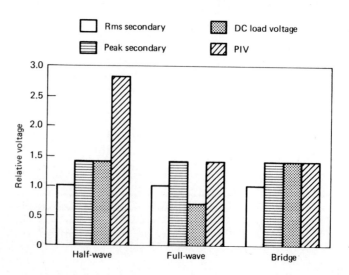

Fig. 4-12.
Voltage relations in peak rectifiers.

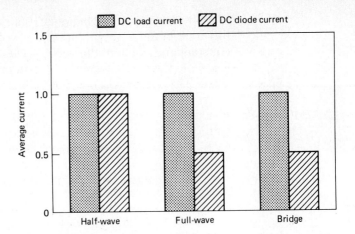

Fig. 4-13.
Current relations in peak
rectifiers.

using a zener diode across the load resistor as shown in Fig. 4-14. The output of a bridge rectifier and capacitor-input filter is the input voltage V_{in} to a zener regulator. As long as the zener diode is operating in the breakdown region, the dc load voltage is approximately equal to V_z.

Section 3-5 analyzed the zener regulator in detail. For your convenience, here are the key formulas derived earlier. The dc current through the series-limiting resistor is

$$I_S = \frac{V_{in} - V_{out}}{R_S} \qquad (4\text{-}25)$$

The dc current through the load resistor is

$$I_L = \frac{V_{out}}{R_L} \qquad (4\text{-}26)$$

The zener current is

$$I_Z = I_S - I_L \qquad (4\text{-}27)$$

The dc output voltage is approximately

$$V_{out} = V_Z$$

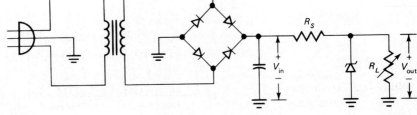

Fig. 4-14.
Power supply with zener
regulator.

DIODE APPLICATIONS

97

With the zener regulator of Fig. 4-14, the line voltage can vary throughout the day, but the dc load voltage will remain approximately constant and equal to the zener voltage. This is one method of voltage regulation. Others are discussed in later chapters.

EXAMPLE 4-8 Assume a secondary voltage of 12.6 V rms in Fig. 4-14. What is the zener current for the following circuit values: $V_Z = 6.8$ V, $R_S = 1$ kΩ, and $R_L = 1.2$ kΩ?

SOLUTION The peak secondary voltage is

$$V_{2(peak)} = \frac{V_{2(rms)}}{0.707} = \frac{12.6 \text{ V}}{0.707} = 17.8 \text{ V}$$

Since the bridge rectifier drives a capacitor-input filter, the approximate dc voltage across the filter capacitor is

$$V_{in} = V_{2(peak)} = 17.8 \text{ V}$$

The current through the series-limiting resistor is

$$I_S = \frac{V_{in} - V_{out}}{R_S} \cong \frac{17.8 \text{ V} - 6.8 \text{ V}}{1 \text{ kΩ}} = 11 \text{ mA}$$

The dc load voltage is

$$V_{out} \cong V_Z = 6.8 \text{ V}$$

and the dc load current is

$$I_L = \frac{V_{out}}{R_L} \cong \frac{6.8 \text{ V}}{1.2 \text{ kΩ}} = 5.67 \text{ mA}$$

Therefore, the zener current is

$$I_Z = I_S - I_L \cong 11 \text{ mA} - 5.67 \text{ mA} = 5.33 \text{ mA}$$

EXAMPLE 4-9 If $C = 100$ μF and $R_Z = 5$ Ω in the preceding example, what is the ripple across the load resistor?

SOLUTION As calculated in Example 4-8, the series-limiting current is 11 mA. Therefore, the filter capacitor has to supply a direct current of 11 mA to the zener regulator. The peak-to-peak ripple across the filter capacitor is

$$V_{rip} = \frac{I_{dc}}{fC} = \frac{11 \text{ mA}}{(120 \text{ Hz})(100 \text{ μF})} = 0.917 \text{ V}$$

The zener resistance of 5 Ω is in parallel with the load resistance of 1.2 kΩ, which means the parallel equivalent resistance is approximately equal to the zener resistance of 5 Ω. The series-limiting resistance forms a voltage divider with this parallel equivalent resistance. Therefore, the ripple is reduced by the voltage-divider factor:

$$V_{out(rip)} = \frac{R_Z}{R_S + R_Z} V_{in(rip)} = \frac{5\ \Omega}{1005\ \Omega} 0.917\ V = 4.56\ mV$$

4-6
L AND π FILTERS

Figure 4-15a shows an inductor in series with the load resistor and a capacitor in parallel with the load resistor. This is called a *choke-input filter* (a choke is an iron-core inductor with a large value of L). The choke has an inductive reactance given by

$$X_L = 2\pi fL$$

The capacitor has a capacitive reactance of

$$X_C = \frac{1}{2\pi fC}$$

At zero frequency, X_L is zero and X_C is infinite. Therefore, the dc voltage out of the bridge rectifier passes through the choke-input filter to the load resistor. Since the output of the bridge rectifier is $0.636V_{2(peak)}$, this ideally is the dc voltage appearing across the load resistor.

On the other hand, the ripple out of the rectifier is greatly attenuated or reduced. Here is the reason. A designer selects values of L and C to produce a large X_L and a small X_C at the ripple frequency

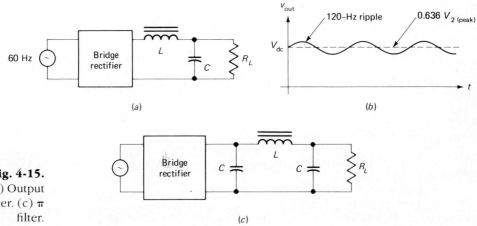

Fig. 4-15.
(a) L filter. (b) Output voltage with L filter. (c) π filter.

of 120 Hz. Therefore, we can approximate the choke as open to the ripple and the capacitor as shorted to the ripple. For this reason, the ripple appearing across the load resistor is small, as shown in Fig. 4-15b.

Incidentally, a choke-input filter is sometimes called an *L filter*. The choke-input filter is excellent for attenuating ripple, but it's rarely used in power supplies because chokes are bulky and expensive.

The π Filter

Figure 4-15c is a π filter. The bridge rectifier drives the capacitor. This produces peak rectification. Then the *L* section attentuates any ripple out of the first capacitor. By deliberate design, X_L is much greater than X_C. Because of this, the ripple is greatly attentuated as it passes through the *L* section. Typically, X_L is at least 10 times greater than X_C. Therefore, the ripple is reduced by at least a factor of 10.

4-7 VOLTAGE MULTIPLIERS

A voltage multiplier is two or more peak rectifiers that produce a dc voltage equal to a multiple of the peak input voltage ($2V_p$, $3V_p$, $4V_p$, and so on). These circuits are used for high-voltage/low-current applications such as cathode-ray tubes (the picture tubes in TV receivers, oscilloscopes, and computer displays).

Voltage Doubler

Figure 4-16a shows a *voltage doubler,* a connection of two peak rectifiers. At the peak of the negative half cycle, D_1 is forward-biased and D_2 is reverse-biased. This charges C_1 to the peak voltage V_p with the polarity shown in Fig. 4-16b. At the peak of the positive half cycle, D_1 is reverse-biased and D_2 is forward-biased. Because the source and C_1 are in series, C_2 will charge toward $2V_p$. After several cycles, the voltage across C_2 equals $2V_p$, as shown in Fig. 4-16c.

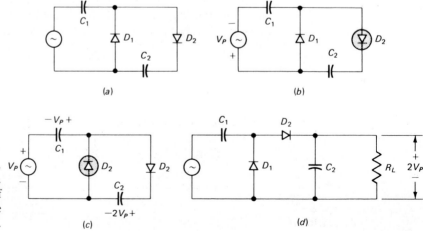

Fig. 4-16. Voltage doubler. (a) Circuit. (b) During negative half cycle. (c) During positive half cycle. (d) Redrawn.

By redrawing the circuit and connecting a load resistor, we get Fig. 4-16d. As long as R_L is large, the discharging time constant is long compared with the period, and the output voltage is approximately equal to $2V_p$. In other words, if the load is light (small current), the dc output voltage is double the peak input voltage.

Voltage Tripler

By connecting another section, we get a *voltage tripler* (Fig. 4-17a). The first two peak rectifiers act like a doubler. At the peak of the negative half cycle of input voltage, D_3 is forward-biased. This charges C_3 to $2V_p$ with the polarity shown in Fig. 4-17b. The tripler output appears across C_1 and C_2.

The load resistance is connected across the tripler output. Again, this load resistance must be large to ensure a long discharging time constant. When this condition is satisfied, the dc output equals approximately $3V_p$.

Theoretically, we could continue adding more peak rectifiers to get higher multiples of peak input voltage. But the output ripple keeps getting worse because the discharge between peaks gets larger. For this reason, the doubler and tripler are the most popular voltage multipliers.

4-8
CLIPPERS AND CLAMPERS

In radar, computers, and other applications, we sometimes want to remove part of a signal with a *clipper* or shift the dc level with a *clamper*.

The Clipper

Figure 4-18a shows a positive clipper, a circuit that removes positive parts of an input signal. The circuit works as follows. During the positive half cycle of input voltage, the diode conducts heavily. To a first approximation, it acts like a closed switch. Therefore, the output voltage equals zero during each positive half cycle.

During the negative half cycle, the diode is reverse-biased and looks like an open switch. As a result, the negative half cycle appears across the output. Ideally, we get the half-wave signal of Fig. 4-18b.

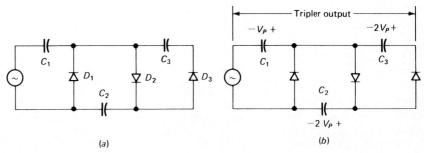

Fig. 4-17.
Voltage tripler.

(a)

(b)

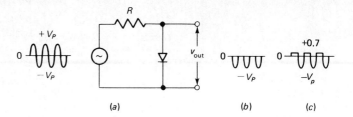

Fig. 4-18.

Positive clipper.

(a) (b) (c)

The clipping is not perfect. To a second approximation, a conducting silicon diode drops 0.7 V. Because of this, the output signal is clipped near +0.7 V rather than 0 V (Fig. 4-18c).

By reversing the polarity of the diode, we get a negative clipper, a circuit that removes the negative half cycles.

The Clamper Figure 4-19 is a positive clamper. Ideally, here is how it works. On the first negative half cycle of input voltage, the diode turns on and charges the capacitor to the peak input voltage. Slightly beyond the negative peak, the diode turns off. Because of the long R_LC time constant, the capacitor remains almost fully charged between negative peaks.

According to Kirchhoff's voltage law, the output voltage is the sum of the source voltage and the capacitor voltage. Because of this, the output is a sine wave that is shifted upward as shown in Fig. 4-19. Ideally, the negative peaks of the output voltage fall at 0 V. To a second approximation, these negative peaks are slightly less than 0 V because of the diode voltage drop.

What happens if we reverse the polarity of the diode? The polarity of capacitor voltage reverses, and the circuit becomes a negative clamper. This means the sine wave is shifted downward until the positive peaks fall at approximately 0 V.

Both positive and negative clampers are widely used. Television receivers, for instance, use a clamper to shift the level of the video signal. In television, the clamper is usually called a *dc restorer*.

4-9
LOGIC GATES

Computers use *logic* circuits. These are circuits that duplicate certain mental processes. With logic circuits, we can electronically add, subtract, and process data to solve problems. A *gate* is a logic circuit with one or more input voltages but only one output voltage.

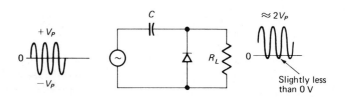

Fig. 4-19.
Positive clamper.

CHAPTER 4

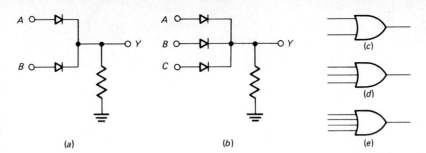

Fig. 4-20.
OR gate. (a) 2-input circuit.
(b) 3-input circuit. (c) 2-
input symbol. (d) 3-input
symbol. (e) 4-input symbol.

The OR Gate

The OR gate has two or more input voltages. If any input voltage is high, the output voltage is high. Figure 4-20a is one way to build an OR gate. The circuit has input voltages A an B, and an output voltage Y. If both input voltages are low, neither diode conducts and the output voltage is low. If either input voltage is high, the diode with the high input conducts and the output voltage is high. Because of the two inputs, we call this circuit a two-input OR gate.

Table 4-1 summarizes the operation. Notice that the output is high if A or B is high. This is why the circuit is known as an OR gate. One or more high inputs produces a high output.

Figure 4-20b is a three-input OR gate. If all input voltages are low, all diodes are off and the output is low. If one or more inputs are high, the output is high. Table 4-2 summarizes the input-output operation. A table like this is called a *truth table*. It lists all the input possibilities and the corresponding outputs.

An OR gate can have as many inputs as desired by adding one diode for each additional input. Six diodes result in a six-input OR gate, nine diodes in a nine-input OR gate. No matter how many inputs, the action of an OR gate can be summarized like this: One or more high inputs produce a high output.

Figure 4-20c, d, and e are the symbols for two-input, three-input, and four-input OR gates. These standard symbols represent OR gates of any design (diode, transistor, etc.)

The AND Gate

The AND gate has two OR input voltages. All input voltages must be high to get a high output voltage. Figure 4-21a shows one way to build an AND gate. In this particular circuit the input voltages can be low (ground) or high (+5 V). when both inputs are low (grounded), both diodes conduct and pull the output down to a low voltage. If

TABLE 4-1. Two-Input OR Gate

A	B	Y
Low	Low	Low
Low	High	High
High	Low	High
High	High	High

TABLE 4-2. Three-Input OR Gate

A	B	C	Y
Low	Low	Low	Low
Low	Low	High	High
Low	High	Low	High
Low	High	High	High
High	Low	Low	High
High	Low	High	High
High	High	Low	High
High	High	High	High

one of the inputs is low and the other high, the diode with the grounded input still conducts and pulls the output voltage down to a low voltage. When both inputs are high, both diodes shut off and the supply voltage pulls the output up to a high voltage (+ 5 V).

Table 4-3 summarizes the operation. As you can see, both *A and B* must be high to get a high output. This is why the circuit is called an AND gate.

Figure 4-21*b* is a three-input AND gate. If all inputs are low, all diodes conduct and the output is low. Even one conducting diode will hold the output low; therefore, the only way to get a high output is to have all inputs high. When all inputs are high, none of the diodes is conducting, and the supply voltage pulls the output up to a high voltage. Table 4-4 summarizes the three-input AND gate.

AND gates can have as many inputs as desired by adding one diode for each additional input. Eight diodes, for instance, give an eight-input AND gate; sixteen diodes, a sixteen-input AND gate. No matter how many inputs an AND gate has, the action can be summarized like this: All inputs must be high to get a high output.

Figure 4-21*c, d,* and *e* are the symbols for two-input, three-input, and four-input AND gates of any design (diode, transistor, etc.).

Summary

The three basic power-supply circuits are the half-wave rectifier, full-wave rectifier, and bridge rectifier. The PIV rating of a diode must be greater

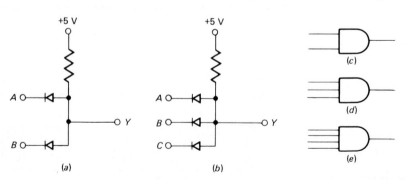

Fig. 4-21.
AND gate. (a) 2-input.
(b) 3-input. (c) 2-input
symbol. (d) 3-input symbol.
(e) 4-input symbol.

TABLE 4-3. Two-Input AND Gate

A	B	Y
Low	Low	Low
Low	High	Low
High	Low	Low
High	High	High

than the peak inverse voltage. The peak inverse voltage is equal to $V_{2(peak)}$ for all three circuits. The I_0 rating must be greater than the dc diode current. For a half-wave rectifier, the dc diode current equals the dc load current; for a full-wave or bridge rectifier, the dc diode current equals half the dc load current. For either the full-wave rectifier or bridge rectifier, the ripple frequency is twice the input frequency; 120 Hz for a line frequency of 60 Hz.

To filter the pulsating dc voltage out of a rectifier circuit, we can use a capacitor-input filter. The filter capacitor charges to approximately the peak of input voltage. Because of the long discharging time constant, the capacitor voltage remains near the peak input voltage throughout the ac cycle. As a result, the output voltage is a dc voltage with a small ripple. For the lightly load case, the peak-to-peak ripple equals I_{dc}/fC. Because of the filter capacitor, the peak inverse voltage of a half-wave rectifier increases to $2V_{2(peak)}$.

Because of the tolerance in line voltage, the dc output voltage of a capacitor-input filter will vary throughout the day. One method of regulating this voltage is by connecting a zener regulator. In this case, the dc load voltage equals the zener voltage, despite variations in line voltage.

Less popular filtering methods include the choke-input filter and the π filter. These are rarely used because inductors are bulky and expensive. Voltage multipliers allow us to double or triple the voltage out of a peak rectifier. Clippers remove part of the input signal, while clampers shift the dc level of the signal. The basic logic gates are the OR gate and the AND gate.

TABLE 4-4. Three-Input AND Gate

A	B	C	Y
Low	Low	Low	Low
Low	Low	High	Low
Low	High	Low	Low
Low	High	High	Low
High	Low	Low	Low
High	Low	High	Low
High	High	Low	Low
High	High	High	High

AND gate A logic gate with two or more inputs and only one output. All inputs must be high to get a high output.

capacitor-input filter The most common type of filter used with rectifiers. Ideally, the capacitor charges to the peak of the input voltage.

clamper In this circuit, a series capacitor charges to the peak input voltage. This produces a shift in the dc level of the output voltage.

clipper A circuit that clips or cuts off certain parts of the input signal.

gate A logic circuit with one or more input voltages but only one output voltage. The output is high only for certain combinations of the input voltages.

OR gate In this logic circuit the output is high when any input is high.

peak detector A circuit whose output is a dc voltage equal to the peak voltage of the input signal.

peak inverse voltage Abbreviated PIV, this is the maximum reverse voltage across the diode during the cycle.

ripple An unwanted fluctuation in the output voltage of a rectifier circuit.

voltage multiplier A cascade of peak rectifiers whose dc output voltage equals a multiple of the peak input voltage.

voltage regulator Any circuit or device that holds the dc output voltage approximately constant in spite of changes in supply voltage and load resistance.

Review Questions

1. Why is the average value of a signal also called its dc value?
2. What is the average output voltage from a half-wave rectifier?
3. What is the average output voltage from a full-wave rectifier?
4. Which of the three basic rectifier circuits is the most widely used? What does dc output voltage equal?
5. How is the output frequency of a bridge rectifier related to the input frequency?
6. A bridge rectifier drives a capacitor-input filter. Ideally, how much dc output voltage is there?
7. How can you calculate the peak-to-peak ripple out of a rectifier circuit that uses a capacitor-input filter?

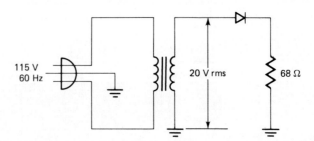

115 V
60 Hz

20 V rms

68 Ω

Fig. 4-22.

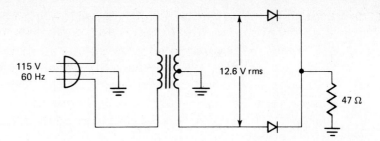

Fig. 4-23.

8. Describe one method for regulating the dc output voltage of a power supply.
9. What is a voltage multiplier?
10. What does a clipper do? A clamper?
11. Name two logic gates.

Problems

4-1. If the line voltage is 115 V with a tolerance of ±10 percent, what are the lowest and highest possible peak line voltages?

4-2. In Fig. 4-22, what is the dc load voltage? The peak inverse voltage across the diode? The dc current through the diode?

4-3. The rms secondary voltage is 12.6 V rms in Fig. 4-22 when the line voltage is 115 V rms. If the line voltage is 10 percent higher than 115 V rms, what are the values of dc load voltage, peak inverse voltage, and dc diode current?

4-4. In Fig. 4-23, work out the values of dc load voltage, peak inverse voltage, and dc diode current.

4-5. Repeat Prob. 4-4 with a load resistance of 22 Ω instead of 47 Ω.

4-6. In Fig. 4-24, what is the dc load voltage, peak inverse voltage, and dc diode current?

4-7. Repeat Prob. 4-6 for an rms secondary voltage of 20 V.

4-8. What is the peak-to-peak ripple in Fig. 4-25? The dc load voltage? The peak inverse voltage across each diode? The dc current through each diode?

4-9. Repeat Prob. 4-8 for an rms secondary voltage of 12.6 V.

4-10. Repeat Prob. 4-8 for a filter capacitance of 2200 μF.

4-11. Repeat Prob. 4-8 for a load resistance of 220Ω.

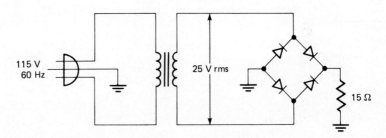

Fig. 4-24.

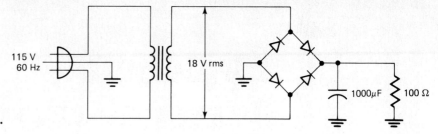

Fig. 4-25.

4-12. In Fig. 4-26, the rms secondary voltage is 24 V. The equivalent load resistance across the filter capacitor is 800 Ω. What is the peak-to-peak ripple across the filter capacitor? The dc voltage across this capacitor?

4-13. The dc voltage across the capacitor of Fig. 4-26 is 30 V. If the LED has 1.75 V across it, what is the LED current? If the zener diode has a zener voltage of 5.1 V, how much current is there through the series-limiting resistor? How much current is there in the 1 kΩ resistor?

4-14. The input to a voltage doubler is 115 V rms. What is the approximate value of dc output voltage?

4-15. The input to a voltage tripler is 360 V rms. What is the ideal dc output voltage?

4-16. One of the diodes in Fig. 4-25 opens. Which of the following happens:
a. Dc load voltage drops in half.
b. Dc load voltage doubles.
c. Ripple frequency drops in half.
d. Rms secondary voltage drops in half.

4-17. In Fig. 4-25, you measure a dc load voltage of 16.2 V. Which of the following is a possible trouble:
a. One of the diodes is open.
b. Open filter capacitor.
c. Shorted filter capacitor.
d. Open load resistor.

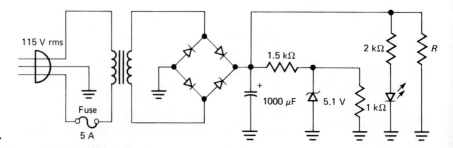

Fig. 4-26.

Bipolar Transistors

Maslow: This inspiring teacher believed that each of us is capable of greatness and excellence. He liked to ask his college classes this question: "Who among you will be the great poet, or painter, or entrepreneur, statesman, senator, president, or surgeon of your generation?" This question was always followed by an embarrassed silence, a shuffling of feet, and a few giggles. Then Maslow would ask, "If not you, then *who?*"

Shockley worked out the theory of the *junction transistor* in 1949, and the first one was produced in 1951. He later received a Nobel prize for his monumental achievement. The transistor's impact on electronics has been enormous. Besides starting the multibillion-dollar semiconductor industry, the transistor has led to many related inventions like integrated circuits, optoelectronic devices, and microprocessors. Most electronic equipment now being designed uses semiconductor devices.

The transistor far outperforms the vacuum tube in most applications. It allows us to do things that were either difficult or impossible with vacuum tubes. This is especially noticeable in the computer industry. Historians agree that the transistor did not revise the computer industry; it created it.

5-1
STRUCTURE

The transistor is a crystal with three doped regions like the *npn* transistor of Fig. 5-1*a*. The *emitter* is heavily doped; its job is to emit, or inject, electrons into the base. The *base* is lightly doped and very thin; it passes most of the emitter-injected electrons on to the *collector*. The doping level of the collector lies between that of the emitter and the base. The collector is largest of the three regions because it must dissipate more heat than the emitter or base. With an *npn* transistor, free electrons are the majority carriers in the emitter and collector.

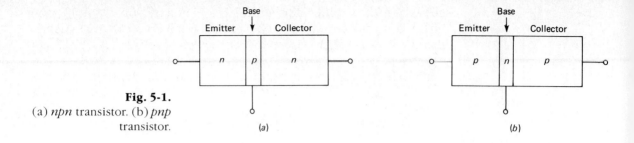

Fig. 5-1.
(a) *npn* transistor. (b) *pnp* transistor.

(a)   (b)

Emitter and Collector Diodes

The transistor of Fig. 5-1*a* has two junctions, one between the emitter and the base, and another between the base and the collector. Because of this, a transistor is similar to two diodes connected back-to-back. We call the diode on the left the emitter-base diode, or simply the *emitter diode*. The diode on the right is the collector-base diode or the *collector diode*.

Figure 5-1*b* shows the other possibility, a *pnp* transistor. The *pnp* transistor is the complement of the *npn* transistor because holes are the majority carriers in the emitter and collector. This means that opposite currents and voltage polarities are involved in the action of a *pnp* transistor compared with an *npn*. To avoid confusion, we will emphasize the *npn* transistor in our early discussions.

Unbiased Transistor

Free electrons in the emitter of an *npn* transistor will diffuse across the emitter junction into the base, producing a depletion layer between the emitter and base, as shown in Fig. 5-2*a* (EB depletion layer). Similarly, free electrons in the collector diffuse across the junction and produce the CB depletion layer. For each of these depletion layers, the barrier potential is approximately 0.7 V at 25°C for a silicon transistor (0.3 V for a germanium transistor). As with diodes, we will emphasize silicon transistors because they are more widely used than germanium transistors. Silicon transistors provide higher voltage ratings, greater current ratings, and less temperature sensitivity. In the discussions that follow, the transistors are silicon unless otherwise indicated.

Because the three regions have different doping levels, the depletion layers do not have the same width. The more heavily doped a region is, the greater the concentration of ions near the junction. This

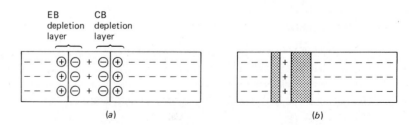

Fig. 5-2.
Depletion layers.

(a)   (b)

means that the emitter depletion layer penetrates only slightly into the emitter region (heavily doped), but deeply into the base (lightly doped). The collector depletion layer extends well into the base and into the collector. Figure 5-2*b* summarizes the different widths. The emitter depletion layer is small and the collector depletion layer is large. Notice how the depletion layers are shaded to indicate the lack of majority carriers.

Schematic Symbols

Figure 5-3*a* is the schematic symbol of an *npn* transistor. An arrowhead is on the emitter but not on the collector. This arrowhead points in the easy direction of conventional flow. If you prefer electron flow, the easy direction of current is against the arrowhead. It may help to remember that the arrowhead points to where the electrons are coming from.

Similarly, Fig. 5-3*b* shows the symbol for a *pnp* transistor. This time, the arrowhead points into the emitter, meaning the easy direction of conventional flow is into the emitter. Therefore, the easy direction for electron flow is out of the emitter.

Biasing a Transistor

Figure 5-4*a* shows an *npn* transistor with a grounded base. Because the base is common to both the emitter and the collector loops, this circuit is called the *common-base* (CB) connection (or configuration). Sometimes, it is referred to as the grounded-base connection. In Fig. 5-4*a*, the dc supply on the left will reverse-bias the emitter diode. Likewise, the dc supply on the right will reverse-bias the collector diode. Since both diodes are reverse-biased, the current in each diode is approximately zero.

Figure 5-4*b* shows another possibility. This time, the dc supply on the left forward-biases the emitter diode, and the dc supply on the right forward-biases the collector diode. Since each diode is forward-biased, each diode has a large current through it.

In Fig. 5-4*a* or *b*, nothing unusual happens. Either we get no current (both diodes reverse-biased) or we get a lot of current (both diodes forward-biased). Transistors are rarely biased like this.

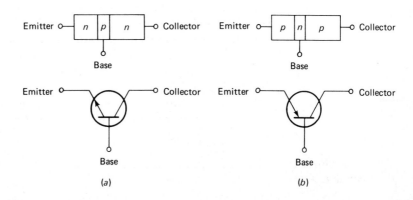

Fig. 5-3.
Schematic symbols.
(a) *npn*. (b) *pnp*.

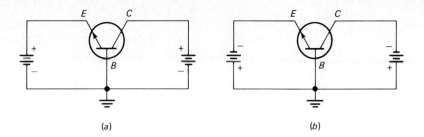

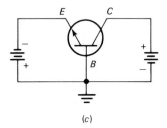

(c)

5-2
FORWARD-
REVERSE BIAS

If we forward-bias the emitter diode and reverse-bias the collector diode as shown in Fig. 5-4c, something very unusual happens. Our initial reaction says there should be a large emitter current and a small collector current. This is not what happens! We do get a large emitter current but we also get a large collector current. This unexpected result is why the transistor has become the important device that it is.

Preliminary Explanation

At the instant the forward bias is applied to the emitter diode, free electrons in the emitter have not yet entered the base region (Fig. 5-5a). If the applied voltage exceeds approximately 0.7 V, many free electrons will enter the base (Fig. 5-5b). The free electrons in the base can now flow in either of two directions: down the thin base into the external base lead, or across the collector junction into the collector region. The downward component of base current is called *recombination* current because the free electrons must fall into holes before they can flow out the base lead. Recombination current is small because the base is lightly doped with a few holes.

Because the base is very thin and lightly doped, most of the free electrons in the base diffuse into the collector, as shown in Fig. 5-5c. Once inside the collector, these free electrons flow into the external collector lead and on to the positive battery terminal.

Here is our final picture of what's going on. In Fig. 5-5c, we visualize a steady stream of electrons leaving the negative source terminal and entering the emitter region. The forward bias forces these free electrons to flow into the base region. The thin and lightly doped base gives almost all of these free electrons enough time to diffuse into the collector before they can recombine with holes in the base. The free electrons then flow out of the collector into the positive

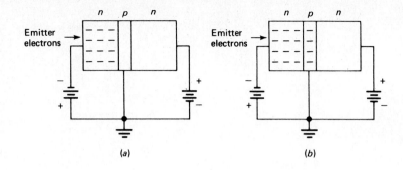

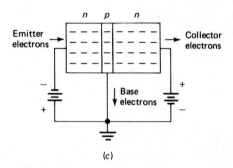

Fig. 5-5.
Flow of free electrons in a
transistor.

terminal of the collector supply. In most transistors, more than 95 percent of the emitter-injected electrons flow to the collector; less than 5 percent fall into base holes and flow out the external base lead.

Here are the three points to remember about transistor action:

1. For normal operation, forward-bias the emitter diode and reverse-bias the collector diode.
2. Collector current is almost equal to emitter current.
3. Base current is very small.

Figure 5-6 shows the relative sizes of the emitter, collector, and base currents. Incidentally, the base current equals the difference of the emitter and collector currents. In symbols,

$$I_B = I_E - I_C \qquad\qquad \textbf{(5-1)}$$

***The Energy
Viewpoint*** An energy diagram is the next step toward a deeper understanding of transistor action. Forward-biasing the emitter diode gives some free electrons in Fig. 5-7 enough energy to move from the emitter into the base. Upon entering the base, the free electrons become minority carriers because they are now inside a *p* region. In nearly all transistors, more than 95 percent of these minority carriers have a long enough lifetime to diffuse into the collector depletion layer and fall into the collector, which is at a lower energy level. As these free electrons fall, they give up energy, mostly in the form of heat.

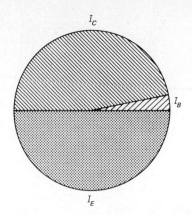

Fig. 5-6.
Relative size of three
transistor currents.

The collector must be able to dissipate this heat, and for this reason, it is physically the largest of the three doped regions. Less than 5 percent of the emitter-injected electrons fall along the recombination path shown in Fig. 5-7. Those that do recombine become valence electrons and flow through base holes into the external base lead.

DC Alpha   As mentioned in the preceding discussion, the collector current almost equals the emitter current. The *dc alpha* of a transistor indicates how close in value the two currents are. It is defined as the ratio of collector current to emitter current:

$$\alpha_{dc} = \frac{I_C}{I_E} \tag{5-2}$$

For example, if we measure an I_C of 9.8 mA and an I_E of 10 mA, then

$$\alpha_{dc} = \frac{9.8 \text{ mA}}{10 \text{ mA}} = 0.98$$

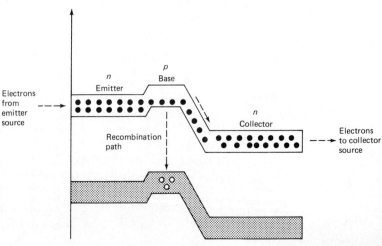

Fig. 5-7.
Energy diagram.

The thinner and more lightly doped the base is, the higher the α_{dc}. Ideally, if all injected electrons go to the collector, α_{dc} would equal unity. Many transistors have α_{dc} greater than 0.99, and almost all have α_{dc} greater than 0.95. Because of this, we can treat α_{dc} as approximately equal to unity in most discussions.

Figure 5-8 graphically illustrates the emitter and collector currents for different values of α_{dc}. In all cases, you can see that collector current almost equals emitter current. Remember this basic idea because it will simplify troubleshooting and design in later discussions.

Base-Spreading Resistance

With two depletion layers penetrating the base, the base holes are confined to a thin channel of p-type semiconductor. The resistance of this thin channel is called the *base spreading resistance* r'_b. The small recombination current I_B flows through r'_b and out the external base lead. When it does, it produces a voltage of $I_B r'_b$. We will discuss the importance of this voltage later. For now, just be aware that r'_b exists. Typically, r'_b is in the range of 50 to 150 Ω. The effects of r'_b are important in high-frequency circuits. At low frequencies, r'_b usually has little effect.

5-3 THE CE CONNECTION

A bipolar transistor can be connected with the emitter grounded instead of the base, as shown in Fig. 5-9a. In this circuit, the emitter is common to the base and collector loops. This is why the circuit is called the *common-emitter* (CE) connection. It is sometimes referred to as a grounded-emitter circuit.

CE Action

Just because you change from a CB connection (Fig. 5-5a) to a CE connection (Fig. 5-9a), you do not change the way a transistor oper-

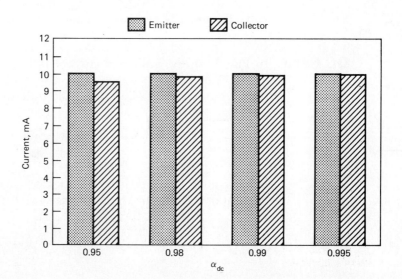

Fig. 5-8.
Collector current almost
equals emitter current.

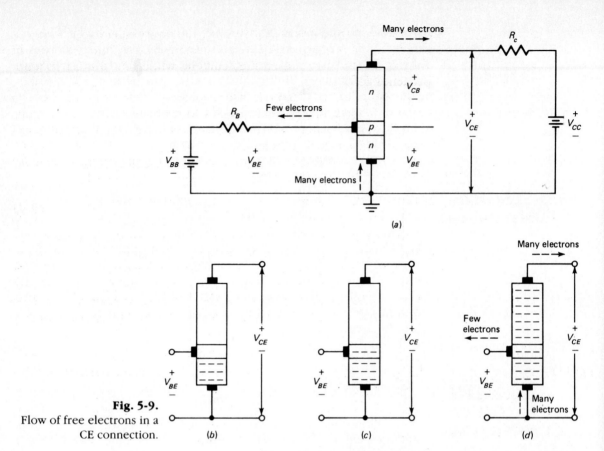

Many electrons

R_c

Few electrons

R_B

n

$+$ V_{CB} $-$

$+$ V_{CE} $-$

$+$ V_{CC} $-$

p

n

$+$ V_{BB} $-$

$+$ V_{BE} $-$

$+$ V_{BE} $-$

Many electrons

(a)

Many electrons

$+$ V_{CE} $-$

$+$ V_{CE} $-$

Few electrons

$+$ V_{CE} $-$

$+$ V_{BE} $-$

$+$ V_{BE} $-$

$+$ V_{BE} $-$

Many electrons

(b)

(c)

(d)

Fig. 5-9.
Flow of free electrons in a CE connection.

ates. Free electrons move exactly the same as before. That is, the emitter is full of free electrons (Fig. 5-9b). When V_{BE} is greater than 0.7 V or so, the emitter injects these free electrons into the base (Fig. 5-9c). As before, the thin and lightly doped base gives almost all of these free electrons enough lifetime to diffuse into the collector. With a reverse-biased collector diode, these free electrons flow out of the collector into the external voltage source (Fig. 5-9d).

DC Beta   In Fig. 5-9a, the collector current is large and the base current is small. The *dc beta* of a transistor, also called the *dc current gain* for the CE connection, is defined as the ratio of collector current to base current. In symbols,

$$\beta_{dc} = \frac{I_C}{I_B} \tag{5-3}$$

For instance, if we measure a collector current of 10 mA and a base current of 0.1 mA, the transistor has a β_{dc} of

$$\beta_{dc} = \frac{10 \text{ mA}}{0.1 \text{ mA}} = 100$$

Figure 5-10 illustrates how the three transistor currents vary for different values of β_{dc}. As β_{dc} increases, the value of collector current approaches the value of emitter current, while the base current approaches zero.

For almost any transistor, less than 5 percent of the emitter-injected electrons recombine with base holes to produce I_B; therefore, β_{dc} is almost always greater than 20. Usually, it is between 50 and 300. And some transistors have β_{dc} as high as 1000.

In another system of analysis called the *hybrid parameters*, h_{FE} rather than β_{dc} is used for the dc current gain. In symbols,

$$h_{FE} = \beta_{dc} \qquad \textbf{(5-4)}$$

Remember this relation because data sheets use the symbol h_{FE} for the dc current gain of a CE connection. For example, the data sheet of a 2N3904 lists a minimum h_{FE} of 100 and a maximum h_{FE} of 300. Given thousands of 2N3904s, this means some have β_{dc} as low as 100, while others may have β_{dc} as high as 300. For mass production, therefore, circuits using a 2N3904 have to be designed to operate for any β_{dc} between 100 and 300.

Relation between α_{dc} and β_{dc}

The emitter current equals the sum of the collector current and the base current:

$$I_E = I_C + I_B$$

Dividing both sides by I_C gives

$$\frac{I_E}{I_C} = 1 + \frac{I_B}{I_C}$$

or

$$\frac{1}{\alpha_{dc}} = 1 + \frac{1}{\beta_{dc}}$$

With algebra, we can rearrange this to get

$$\alpha_{dc} = \frac{\beta_{dc}}{\beta_{dc} + 1} \qquad \textbf{(5-5)}$$

For instance, if a data sheet lists $h_{FE} = 250$, then $\beta_{dc} = 250$ and

$$\alpha_{dc} = \frac{250}{250 + 1} = 0.996$$

Figure 5-11 shows how α_{dc} varies with β_{dc}. As β_{dc} increases from 20 to 300, α_{dc} increases from 0.95 to 1. The important idea here is that α_{dc} is approximately equal to unity for all practical values of β_{dc}. In other words, when troubleshooting or designing transistor circuits, you can almost always treat α_{dc} as equal to 1.

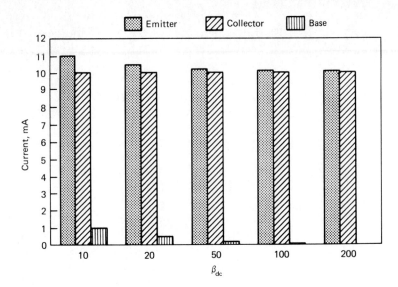

Fig. 5-10.
The three transistor
currents.

5-4
CE COLLECTOR CURVES

One way to visualize how a transistor works is with graphs that relate transistor currents and voltages. These *I-V* curves will be more complicated than those of a diode because we have to include the effect of base current. We can get the data for these curves by using a circuit like Fig. 5-12. The idea is to vary the V_{BB} and V_{CC} supplies to set up different base currents and collector-emitter voltages.

Collector Current Equals β_{dc}Times I_B

Throughout this discussion, assume the transistor has a β_{dc} of 100. When the base current is adjusted to 10 μA, the collector current is

$$I_C = \beta_{dc}I_B = 100(10\ \mu\text{A}) = 1\ \text{mA}$$

As long as the collector diode is reverse-biased, the collector current is approximately 1 mA. In other words, changing V_{CE} has almost no

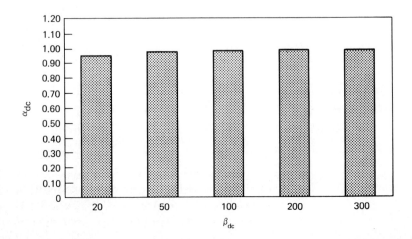

Fig. 5-11.
Dc alpha is approximately
equal to unity.

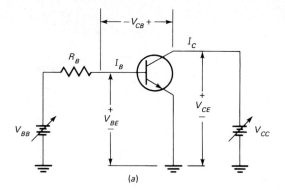

Fig. 5-12.
A CE connection.

(a)

effect on collector current, provided the supply voltage does not exceed the breakdown voltage of the collector diode.

If the V_{BB} supply is increased to get an I_B of 20 μA, then the collector current will increase to

$$I_C = 100(20\ \mu\text{A}) = 2\ \text{mA}$$

Again, the collector current will remain at approximately 2 mA for all values of V_{CE} that reverse-bias the collector diode without exceeding its breakdown voltage.

How large does V_{CE} have to be to reverse-bias the collector diode? In Fig. 5-12,

$$V_{CE} = V_{CB} + V_{BE}$$

With a silicon transistor, this becomes

$$V_{CE} = V_{CB} + 0.7\ \text{V}$$

When the V_{CB} is greater than zero, the collector diode is reverse-biased. Therefore, V_{CE} has to be at least 0.7 V to reverse-bias the collector diode. Furthermore, there may be an additional few tenths of a volt dropped across the bulk resistance of the collector-base region. Therefore, V_{CE} should be at least 1 V or more to reverse-bias the collector diode of most silicon transistors. When the collector diode is reverse-biased, the collector current is approximately 100 times the base current, assuming a β_{dc} of 100.

Four Regions of Operation

Figure 5-13 summarizes transistor action for different base currents. As you see, when V_{CE} is zero, collector current is zero because the collector diode (the collector-base junction) is not reverse-biased. When V_{CE} initially increases, I_C increases rapidly. The almost vertical part of the curves near the origin is called the *saturation region*. In this region the collector diode is slightly in forward bias because V_{CE} is less than 0.7 V. Somewhere beyond a V_{CE} of 0.7 V (the exact value depends on the value of collector current and the bulk resistance),

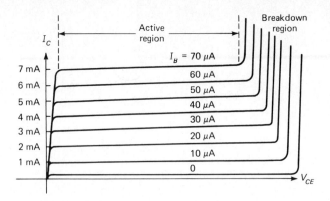

Fig. 5-13.
Collector curves.

the collector diode becomes reverse-biased. When this happens, the collector current levels off and becomes almost constant.

When the collector voltage is too large, the collector diode breaks down, indicated by a rapid increase of collector current in Fig. 5-13. Usually, a designer should avoid operation in the *breakdown region* because the excessive power dissipation may destroy the transistor. For instance, a 2N3904 has a collector breakdown voltage of 40 V. For normal operation, therefore, V_{CE} should be less than 40 V.

The *active region* of a transistor is where the collector curves are almost horizontal or constant. In the active region of Fig. 5-13, the collector current equals approximately 100 times the base current. Stated another way, the active region is all the operating points between the saturation and the breakdown regions. With a 2N3904, the active region occurs for a V_{CE} between approximately 1 and 40 V. In this region, changes in V_{CE} have little effect on I_C. To change I_C, you have to change I_B.

Notice the bottom curve in Fig. 5-13, where $I_B = 0$. This is the *cutoff region*. A small collector current exists here because of the leakage current in the collector diode. For silicon transistors, this leakage current is usually small enough to ignore in most applications. For instance, a 2N3904 has a leakage current of only 50 nA, which is so small that you would not even see the bottom curve of Fig. 5-13.

Curve Tracer

Most transistor data sheets do not show collector curves. If you want to see the collector curves of a particular transistor, use a *curve tracer* (an instrument with a video display that produces the collector curves on its screen). If you try different transistors, you will notice changes in the β_{dc}, breakdown voltage, leakage current, etc. As a rule, transistor curves show a wide variation from one transistor to the next because of manufacturing tolerances.

Ideal CE Approximation

Figure 5-14 is the ideal approximation for an *npn* transistor with a CE connection. The base-emitter part of the transistor acts like a forward-biased diode with a current of I_B. You can use any of the

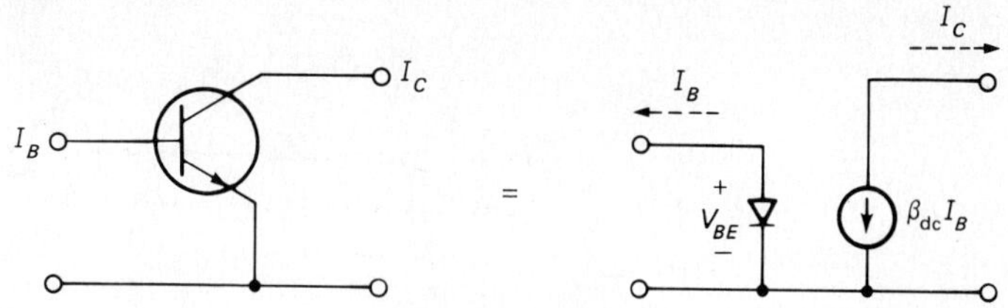

Fig. 5-14.
Dc equivalent circuit of
transistor.

three diode approximations for the base-emitter part of a transistor. In this book we almost always use the second approximation when analyzing the dc operation of transistor circuits.

The collector-emitter part of the transistor acts like a current source with a value of $\beta_{dc}I_B$. If the collector curves were perfectly horizontal in the active region, this current source would have an infinite internal impedance. Since the collector curves have a slight upward slope, the current source has a very high internal impedance. With a CE connection, collector current is controlled by base current. When base current changes, the collector current source automatically changes to a new value equal to $\beta_{dc}I_B$.

Incidentally, notice that the arrow in the current source points in the direction of conventional flow. The electron flow is in the opposite direction, indicated by the dashed arrow.

5-5
BASE BIAS

Biasing a transistor means applying external voltages to produce a desired collector current. Figure 5-15a shows *base bias*, the simplest way to bias a CE connection. Figure 5-15b is an equivalent circuit. In either case, a voltage source V_{BB} forward-biases the emitter diode through a current-limiting resistor R_B. With a silicon transistor, V_{BE} is in the vicinity of 0.6 to 0.7 V. As an approximation, we will use 0.7 V throughout this book unless otherwise indicated. Because of the collector resistor R_C, the value of V_{CE} is less than the supply voltage V_{CC}.

Currents and Voltages

In Fig. 5-15b, the base resistor R_B has a voltage V_{BB} at its left end and a voltage V_{BE} at its right end. Therefore, the voltage across R_B is $V_{BB} - V_{BE}$, and the base current is

$$I_B = \frac{V_{BB} - V_{BE}}{R_B} \qquad (5\text{-}6)$$

In the active region, the collector current is

$$I_C = \beta_{dc}I_B \qquad (5\text{-}7)$$

This collector current produces a voltage drop of $I_C R_C$ across the collector resistor. Therefore, the collector-emitter voltage is

$$V_{CE} = V_{CC} - I_C R_C \qquad (5\text{-}8)$$

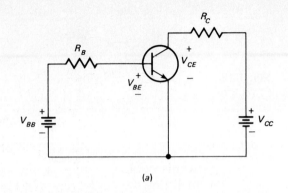

(a)

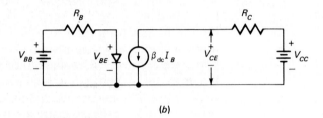

Fig. 5-15.
CE connection and
equivalent circuit.

(b)

With Eqs. (5-6) through (5-8), you can calculate the important currents and voltages in a base-biased circuit. Remember to use 0.7 V for V_{BE} when calculating the base current.

DC Load Line For a particular circuit, V_{CC} and R_C are constants. Therefore, Eq. (5-8) is a linear equation in two variables (V_{CE} and I_C). When graphed, the equation produces a straight line. For example, suppose a circuit has a collector supply voltage of 10 V and a collector resistance of 1 kΩ. Then, Eq. (5-8) gives

$$V_{CE} = 10 - 1000I_C$$

When $I_C = 0$,

$$V_{CE} = 10 - 1000(0) = 10$$

When $I_C = 1$ mA,

$$V_{CE} = 10 - 1000(0.001) = 9$$

When $I_C = 2$ mA,

$$V_{CE} = 10 - 1000(0.002) = 8$$

Continuing in this way, we can calculate values of V_{CE} and I_C that satisfy the equation. Then we can plot the points to get the linear

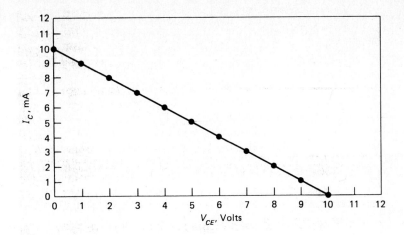

Fig. 5-16.
Dc load line.

graph of Fig. 5-16. A graph like this is called a *dc load line* because it represents all possible dc operating points of the transistor for a specific load resistor (1 kΩ in this example). Notice that the dc load line has a vertical intercept of 10 mA and a horizontal intercept of 10 V.

As you know, two points determine a straight line. Because of this, the simple way to construct a dc load line is to draw a line through the vertical and horizontal intercepts. For example, Fig. 5-17 shows the dc load line superimposed on collector curves. The dc load line intersects the vertical axis at V_{CC}/R_C and horizontal axis at V_{CC}. As a formula,

$$\text{Vertical intercept} = \frac{V_{CC}}{R_C} \qquad \textbf{(5-9)}$$

and
$$\text{Horizontal intercept} = V_{CC} \qquad \textbf{(5-10)}$$

Cutoff and Saturation   The point where the dc load line intersects the $I_B = 0$ curve is known as *cutoff*. At this point, base current is zero and collector current is extremely small. At cutoff, the emitter diode comes out of forward

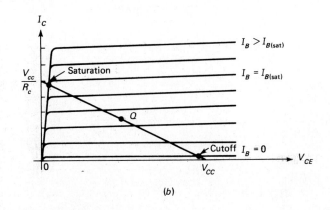

Fig. 5-17.
Saturation and cutoff on dc load line.

(b)

bias and normal transistor action is lost. Notice in Fig. 5-17 that the cutoff point is extremely close to the horizontal intercept of the dc load line. For this reason, Eq. (5-10) is a close approximation to the cutoff voltage.

Saturation is the point of intersection at the upper end of the dc load line. At this point, the collector diode is on the verge of coming out of reverse bias. For convenience, we will designate the coordinates of the saturation point by $I_{C(\text{sat})}$ and $V_{CE(\text{sat})}$. The value of collector current at this point is given by

$$I_{C(\text{sat})} = \frac{V_{CC} - V_{CE(\text{sat})}}{R_C}$$

In Fig. 5-17, you can see that the saturation point is very close to the vertical axis. Because $V_{CE(\text{sat})}$ is typically only a few tenths of a volt, most people approximate the foregoing equation to get

$$I_{C(\text{sat})} \cong \frac{V_{CC}}{R_C} \qquad\qquad (5\text{-}11)$$

The base current that just produces saturation is given by

$$I_{B(\text{sat})} = \frac{I_{C(\text{sat})}}{\beta_{\text{dc}}} \qquad\qquad (5\text{-}12)$$

When the base current is greater than $I_{B(\text{sat})}$, the collector current cannot increase because the collector diode is no longer reverse-biased. In other words, $I_{C(\text{sat})}$ represents the maximum collector current a base-biased circuit can produce for the given supply voltage and collector transistor.

The Q Point   All points on the load line between cutoff and saturation are the active region of the transistor. In the active region the emitter diode is forward-biased and the collector diode is reverse-biased. Usually, a

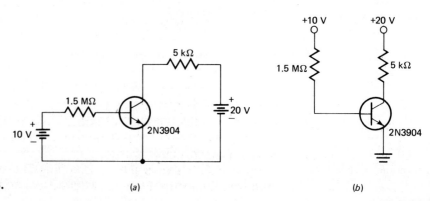

Fig. 5-18.   (a)   (b)

designer will select components to produce an operating point some-where near the middle of the dc load line (designated Q in Fig. 5-17). This point is called the *quiescent* (at rest) point because it is the operating point with no ac input signal.

EXAMPLE 5-1   The 2N3904 of Fig. 5-18a has a β_{dc} of 125. What is the dc voltage between the collector and emitter?

SOLUTION   Figure 5-18b shows the way the circuit is usually drawn in industry. Start by getting the dc base current. It equals the voltage across the base resistor divided by its resistance:

$$I_B = \frac{V_{BB} - V_{BE}}{R_B} = \frac{10\text{ V} - 0.7\text{ V}}{1.5\text{ M}\Omega} = 6.2\ \mu\text{A}$$

The dc collector current equals the dc beta times the base current:

$$I_C = \beta_{dc}I_B = 125(6.2\ \mu\text{A}) = 0.775\text{ mA}$$

The dc collector-emitter voltage equals the collector supply voltage minus the drop across the collector resistor:

$$V_{CE} = V_{CC} - I_C R_C = 20\text{ V} - (0.775\text{ mA})(5\text{ k}\Omega) = 16.1\text{ V}$$

EXAMPLE 5-2   The 2N4401 of Fig. 5-19a has a β_{dc} of 80. Draw the dc load line.

SOLUTION   With Eq. (5-9),

$$\text{Vertical intercept} = \frac{V_{CC}}{R_C} = \frac{30\text{ V}}{1.5\text{ k}\Omega} = 20\text{ mA}$$

To remember Eq. (5-9), notice the following. At the upper end of the dc load line, V_{CE} is zero. This is equivalent to having a short between the collector and emitter of Fig. 5-19a. With a collector-emitter short, all of the supply voltage appears across the collector resistor and the collector current equals V_{CC}/R_C.

Similarly, Eq. (5-10) gives

$$\text{Horizontal intercept} = V_{CC} = 30\text{ V}$$

Here is how to remember Eq. (5-10). At the lower end of the dc load line, the collector current is zero, equivalent to having an open between the collector and the emitter. With no current through the collector resistor of Fig. 5-19a, there is no voltage drop across this resistor: therefore, V_{CE} must equal the supply voltage.

Figure 5-19b shows the dc load line. The ends of the dc load line represent the extremes in transistor current and voltage. In other words,

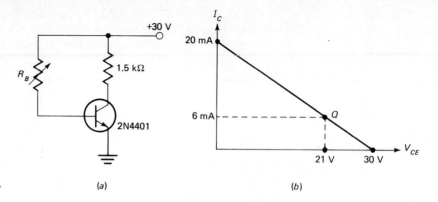

Fig. 5-19.

(a)   (b)

you can get no more than 20 mA of collector current and no more than 30 V of collector-emitter voltage, no matter what the value of R_B. For instance, decreasing R_B eventually saturates the transistor and produces a collector current of approximately 20 mA. On the other hand, increasing R_B toward infinity will cut off the transistor and reduce the collector current to approximately zero. In this case, all of the supply voltage appears across the collector-emitter terminals, so that $V_{CE} = 30$ V.

EXAMPLE 5-3   What are the values of I_C and V_{CE} at the Q point of Fig. 5-19a when $R_B = 390$ kΩ and $\beta_{dc} = 80$?

SOLUTION   The base current is

$$I_B = \frac{V_{CC} - V_{BE}}{R_B} = \frac{30 \text{ V} - 0.7 \text{ V}}{390 \text{ k}\Omega} = 75.1 \ \mu A$$

The collector current is

$$I_C = \beta_{dc} I_B = 80(75.1 \ \mu A) = 6 \text{ mA}$$

The collector-emitter voltage is

$$V_{CE} = V_{CC} - I_C R_C = 30 \text{ V} - (6 \text{ mA})(1.5 \text{ k}\Omega) = 21 \text{ V}$$

Figure 5-19b shows the location of the Q point. It has coordinates of 6 mA and 21 V. The Q point always lies on the dc load line, because the dc load line contains all possible operating points. If we were to decrease R_B, then I_C would increase and V_{CE} would decrease; therefore, the Q point would shift to a higher point on the dc load line. Conversely, an increase in R_B causes I_C to decrease, V_{CE} to increase, and the Q point to move toward the lower end of the load line.

EXAMPLE 5-4   What is the LED current in Fig. 5-20 when the transistor is saturated? If $\beta_{dc} = 150$, what is the value of V_{in} that just saturates the transistor? Assume the LED has a voltage drop of 2 V.

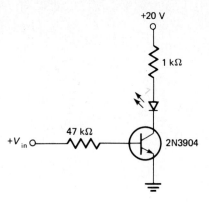

Fig. 5-20.

SOLUTION When the transistor is saturated, its collector is approximately grounded because V_{CE} is almost zero. The voltage across the 1 kΩ equals the supply voltage minus the drop across the LED. We can calculate the LED current as follows:

$$I_{LED} = I_{C(sat)} \cong \frac{V_{CC} - V_{LED}}{R_C} \cong \frac{20 \text{ V} - 2 \text{ V}}{1 \text{ k}\Omega} = 18 \text{ mA}$$

The dc base current that just saturates the transistor is

$$I_{B(sat)} = \frac{I_{C(sat)}}{\beta_{dc}} \cong \frac{18 \text{ mA}}{150} = 0.12 \text{ mA}$$

The corresponding input voltage is

$$V_{IN} = I_B R_B + V_{BE} \cong (0.12 \text{ mA})(47 \text{ k}\Omega) + 0.7 \text{ V} = 6.34 \text{ V}$$

As long as the input voltage is greater than or equal to 6.34 V, the base current is large enough to saturate the transistor and produce a LED current of approximately 18 mA.

5-6

VOLTAGE-DIVIDER BIAS

Linear transistor circuits are those that always operate in the active region. In other words, the transistors in linear circuits are never driven into saturation or cutoff. Typically, the dc source sets up a Q point near the middle of the load line, and a small ac source produces fluctuations in collector current and voltage.

Figure 5-21a shows *voltage-divider bias,* the most widely used bias for linear transistor circuits. Its name comes from the voltage divider (R_1 and R_2) in the base circuit. When the circuit is properly designed,

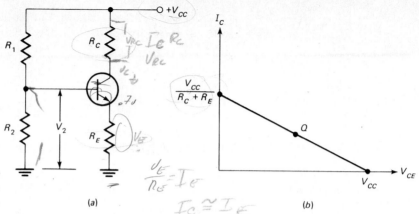

Fig. 5-21.
Voltage-divider bias.
(a) Circuit. (b) Dc load line.

(a)

(b)

the voltage across R_2 forward-biases the emitter diode and produces a collector current that is almost independent of β_{dc}. This is the main reason for the great popularity of voltage-divider bias.

Emitter Current

By deliberate design, the base current in Fig. 5-21a is at least 10 times smaller than the current through R_1 and R_2. Because of this, the voltage divider is lightly loaded. Therefore, the voltage across R_2 is approximately

$$V_2 \cong \frac{R_2}{R_1 + R_2} V_{CC} \qquad (5\text{-}13)$$

The voltage across the emitter resistor equals V_2 minus the V_{BE} drop:

$$V_E = V_2 - V_{BE} \qquad (5\text{-}14)$$

Therefore, the dc emitter current is

$$I_E = \frac{V_E}{R_E} \qquad (5\text{-}15)$$

or

$$I_E = \frac{V_2 - V_{BE}}{R_E}$$

As usual, collector current approximately equals emitter current; so once you have calculated the emitter current, you have the approximate value of the collector current.

DC Load Line

When a transistor is saturated, V_{CE} is approximately zero, which is equivalent to saying the collector and emitter are shorted. In Fig. 5-21a, visualize the collector and emitter shorted. Then you can see that all of the supply voltage is dropped across the series connection of R_C and R_E. Because of this, the saturation current is

$$I_{C(\text{sat})} \cong \frac{V_{CC}}{R_C + R_E} \qquad \textbf{(5-16)}$$

On the other hand, when a transistor is cut off, its collector-emitter terminals appear open. In this case, all the supply voltage of Fig. 5-21a is across the collector-emitter terminals and

$$V_{CE(\text{cutoff})} \cong V_{CC} \qquad \textbf{(5-17)}$$

Figure 5-21b shows the dc load line for voltage-divider bias. Typically, the Q point is near the middle of the dc load line.

Voltages The dc collector-emitter voltage equals the supply voltage minus the drop across the collector and emitter resistors. Because collector and emitter current are approximately equal,

$$V_{CE} = V_{CC} - I_C(R_C + R_E) \qquad \textbf{(5-18)}$$

When troubleshooting, it is convenient to measure voltages with respect to ground. The dc collector-ground voltage V_C equals the supply voltage minus the drop across the collector resistor:

$$V_C = V_{CC} - I_C R_C \qquad \textbf{(5-19)}$$

EXAMPLE 5-5 In Fig. 5-22a, what is the dc voltage from the collector to ground?

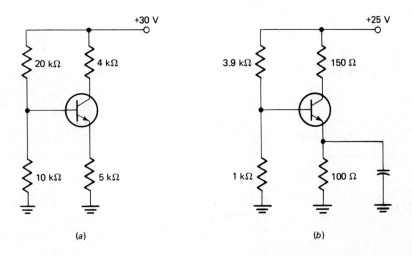

Fig. 5-22. (a) (b)

SOLUTION The voltage across the 10 kΩ is 10 V. The emitter diode drops approximately 0.7 V. Therefore, the voltage from the emitter to ground is

$$V_E = V_2 - V_{BE} \cong 10\text{ V} - 0.7\text{ V} = 9.3\text{ V}$$

Since this voltage is across the emitter resistor, the emitter current is

$$I_E = \frac{V_E}{R_E} \cong \frac{9.3\text{ V}}{5\text{ k}\Omega} = 1.86\text{ mA}$$

Since $I_C \cong I_E$, the collector voltage is

$$V_C = V_{CC} - I_C R_C \cong 30\text{ V} - (1.86\text{ mA})(4\text{ k}\Omega) = 22.6\text{ V}$$

EXAMPLE 5-6 What is the dc collector-to-ground voltage in Fig. 5-22b?

SOLUTION The capacitor is open to direct current. Therefore, it has nothing to do with the problem. With Eq. (5-13), the voltage across the 1-kΩ resistor is

$$V_2 \cong \frac{R_2}{R_1 + R_2} V_{CC} = \frac{1\text{ k}\Omega}{4.9\text{ k}\Omega} 25\text{ V} = 5.1\text{ V}$$

With Eq. (5-14), the dc emitter voltage to ground is

$$V_E = V_2 - V_{BE} \cong 5.1\text{ V} - 0.7\text{ V} = 4.4\text{ V}$$

With Eq. (5-15), the dc emitter current is

$$I_E = \frac{V_E}{R_E} \cong \frac{4.4\text{ V}}{100} = 44\text{ mA}$$

Since $I_C \cong I_E$, the dc collector voltage to ground is

$$V_C = V_{CC} - I_C R_C \cong 25\text{ V} - (44\text{ mA})(150\ \Omega) = 18.4\text{ V}$$

EXAMPLE 5-7 Figure 5-23 shows a two-stage circuit. (A stage is each transistor with its biasing resistors, including R_C and R_E.) Calculate the dc collector voltage to ground for each stage. (Note: *GND* stands for ground.)

SOLUTION The capacitors are open to direct current. Therefore, we can analyze each stage separately because there is no dc interaction between the stages. In the first stage, the dc voltage across the 5 kΩ is 5 V. Subtract 0.7 V for the V_{BE} drop and you have 4.3 V across the emitter resistor. Therefore, the dc emitter current in the first stage is

$$I_E = \frac{V_E}{R_E} \cong \frac{4.3\text{ V}}{2\text{ k}\Omega} = 2.15\text{ mA}$$

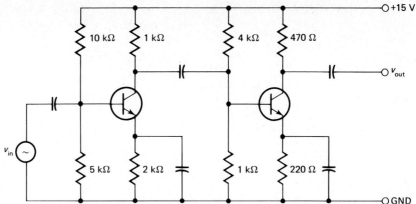

Fig. 5-23.
Two-stage amplifier.

The dc collector voltage to ground is

$$V_C = V_{CC} - I_C R_C \cong 15 \text{ V} - (2.15 \text{ mA})(1 \text{ k}\Omega) = 12.9 \text{ V}$$

In the second stage, the voltage across the 1 kΩ is 3 V. After subtracting 0.7 V, we have 2.3 V across the emitter resistor. The dc emitter current is

$$I_E = \frac{V_E}{R_E} \cong \frac{2.3 \text{ V}}{220 \ \Omega} = 10.5 \text{ mA}$$

and the collector voltage to ground is

$$V_C \cong 15 \text{ V} - (10.5 \text{ mA})(470 \ \Omega) = 10.1 \text{ V}$$

5-7

ACCURATE FORMULA FOR VOLTAGE-DIVIDER BIAS

As long as the base current is at least 10 times smaller than the current through the voltage divider (R_1 and R_2), the approximate analysis of the foregoing section is adequate for most troubleshooting and designing. Occasionally, you may need a more accurate approach.

Here is how to derive an accurate formula for the dc emitter current of a voltage-divider biased circuit. Figure 5-24a shows a voltage-divider biased circuit. Begin by applying Thevenin's theorem to the base circuit as follows: Visualize the base lead open as shown in Fig. 5-24b. The voltage appearing across R_2 is the Thevenin voltage. The voltage-divider theorem gives us

$$V_2 = \frac{R_2}{R_1 + R_2} V_{CC}$$

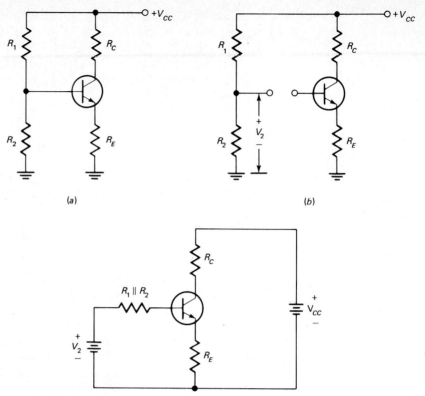

Fig. 5-24.
Voltage-divider bias.
(a) Circuit. (b) With open
base lead. (c) Equivalent
circuit.

Next, visualize the supply voltage reduced to zero. Then you can see
that the Thevenin resistance is R_1 in parallel with R_2:

$$R_{TH} = R_1 \| R_2$$

Figure 5-24c shows the circuit after the base circuit has been theven-
ized. Summing voltages around the base loop gives

$$V_{BE} + I_E R_E - V_2 + I_B(R_1 \| R_2) = 0$$

Since $I_B \cong I_E/\beta_{dc}$, the foregoing equation reduces to

$$I_E \cong \frac{V_2 - V_{BE}}{R_E + (R_1 \| R_2)/\beta_{dc}} \qquad \textbf{(5-20)}$$

In the mass production of transistor circuits, one of the main problems is the variation in β_{dc}. It changes from one transistor to the next.
For instance, a 2N3904 has a minimum β_{dc} of 100 and a maximum β_{dc}
of 300. When using thousands of 2N3904s, you will find some have
β_{dc} as low as 100, while others have β_{dc} as high as 300. Furthermore,
β_{dc} is dependent on temperature and collector current. Because of

the variations in β_{dc}, it is virtually impossible to use base bias in linear circuits, because the Q point is unpredictable. But with voltage-divider bias, we can almost eliminate the effect of β_{dc}.

Here is why voltage-divider bias is a stable way to bias a transistor. When a designer selects values for R_1, R_2, and R_E, he or she can satisfy this condition:

$$R_E \gg \frac{R_1 \parallel R_2}{\beta_{dc}} \qquad\qquad (5\text{-}21)$$

In this expression, the symbol $\gg$ stands for "much greater than." When the foregoing condition is satisfied, Eq. (5-20) simplifies to

$$I_E \cong \frac{V_2 - V_{BE}}{R_E}$$

Since β_{dc} no longer appears in the equation, the emitter current no longer depends on the value of β_{dc}.

EXAMPLE 5-8   Suppose β_{dc} varies from 100 to 300 in Fig. 5-25a and b. Calculate the minimum and maximum collector current for each circuit.

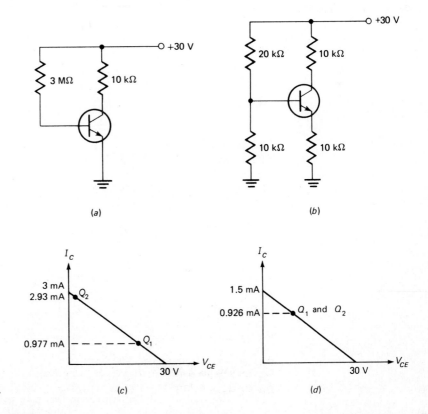

Fig. 5-25.   (a)   (b)   (c)   (d)

SOLUTION In the base-biased circuit of Fig. 5-25a,

$$I_B = \frac{V_{CC} - V_{BE}}{R_B} = \frac{30 \text{ V} - 0.7 \text{ V}}{3 \text{ M}\Omega} = 9.77 \text{ }\mu\text{A}$$

The minimum collector current is

$$I_C = \beta_{dc} I_C = 100(9.77 \text{ }\mu\text{A}) = 0.977 \text{ mA}$$

and the maximum collector current is

$$I_C = 300(9.77 \text{ }\mu\text{A}) = 2.93 \text{ mA}$$

Figure 5-25c shows the Q points for the minimum (Q_1) and maximum (Q_2) cases. As you see, base bias produces an unpredictable Q point that may lie anywhere between the Q_1 and Q_2 extremes. Because of its sensitivity to changes in β_{dc}, base bias is almost worthless for linear transistor circuits.

Voltage-divider bias is different. In Fig. 5-25b,

$$R_1 \parallel R_2 = 20 \text{ k}\Omega \parallel 10 \text{ k}\Omega = 6.67 \text{ k}\Omega$$

With Eq. (5-20), the minimum emitter current is

$$I_E = \frac{10 \text{ V} - 0.7 \text{ V}}{10 \text{ k}\Omega + 6.67 \text{ k}\Omega/100} = \frac{9.3 \text{ V}}{10,067 \text{ }\Omega} = 0.924 \text{ mA}$$

The maximum emitter current is

$$I_E = \frac{10 \text{ V} - 0.7 \text{ V}}{10 \text{ k}\Omega + 6.67 \text{ k}\Omega/300} = \frac{9.3 \text{ V}}{10,022 \text{ }\Omega} = 0.928 \text{ mA}$$

Even though β_{dc} varies from 100 to 300, the emitter current only varies from 0.924 to 0.928 mA, a change of less than half a percent. In other words, the Q point is rock-solid as shown in Fig. 5-25d. Since the variation from minimum to maximum is so small, Q_1 and Q_2 appear to be at the approximately the same point labeled 0.926 mA (the average). This is why voltage-divider bias is used in mass production; it provides a stable Q point for different transistors and changing temperatures.

There's no need to use Eq. (5-20) when troubleshooting or designing circuits. As long as the base current is at least 10 times smaller than the current through the voltage divider, you can use

$$I_E \cong \frac{V_2 - V_{BE}}{R_E}$$

to calculate a fairly accurate value of dc emitter current.

OTHER BIASING CIRCUITS

Up to now, the discussion has been about base bias and voltage-divider bias. Even though voltage-divider bias is the most widely used, other biasing circuits are occasionally used. This section briefly describes these other biasing circuits.

Emitter-Feedback Bias

An increase in temperature typically produces an increase in β_{dc}. In poorly designed circuits, this increases the collector current and forces the Q point to move toward saturation. In some cases the transistor goes into saturation and the circuit is useless. Ideally, we want the Q point to remain fixed in spite of temperature changes.

Figure 5-26a shows *emitter-feedback* bias, an early attempt at compensating for the variations in β_{dc}. The idea is to use the voltage across the emitter resistor to offset the changes in β_{dc}. If β_{dc} increases, the collector current increases. This produces more voltage across the emitter resistor, which decreases the voltage across the base resistor and reduces the base current. The reduced base current results in less collector current, which partially offsets the original increase in β_{dc}.

Incidentally, the word "feedback" is used here to indicate that an output quantity (collector current) produces a change in an input quantity (base current). The emitter resistor is the feedback element because it returns a feedback voltage to the base of the transistor.

As described earlier, you can draw the dc load line through the vertical and horizontal intercepts. To get the vertical intercept, visualize the collector shorted to the emitter. Then, all of the supply voltage is across the series connection of R_C and R_E. This means the collector saturation current is approximately

$$I_{C(\text{sat})} \cong \frac{V_{CC}}{R_C + R_E} \qquad (5\text{-}22)$$

To get the horizontal intercept, visualize the transistor removed from the circuit, equivalent to opening the collector and emitter terminals.

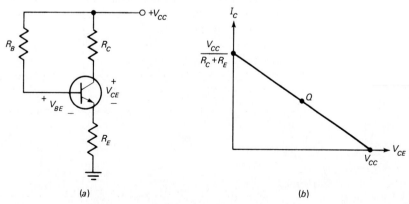

Fig. 5-26.
Emitter-feedback bias.
(a) Circuit. (b) Dc load line.

(a) (b)

BIPOLAR TRANSISTORS

In this case, all of the supply voltage appears between the collector and emitter, which means

$$V_{CE(\text{cutoff})} \cong V_{CC} \qquad\qquad (5\text{-}23)$$

Figure 5-26*b* shows the dc load line.

Here is how to derive a formula for the collector current. If we sum voltages around the base loop, we get

$$V_{BE} + I_E R_E - V_{CC} + I_B R_B = 0$$

Since $I_E \cong I_C$ and $I_B = I_C/\beta_{dc}$, we can rewrite the equation as

$$I_C \cong \frac{V_{CC} - V_{BE}}{R_E + R_B/\beta_{dc}} \qquad\qquad (5\text{-}24)$$

To minimize the variations in β_{dc}, R_E must be much larger than R_B/β_{dc}. This condition is difficult to satisfy in practical circuits. For typical designs where R_E is small compared with the R_C, it turns out that emitter-feedback bias is almost as sensitive as base bias to variations in β_{dc}. For this reason, emitter-feedback bias is rarely used.

Collector-Feedback Bias

Figure 5-27*a* shows *collector-feedback* bias (also called self-bias). The base resistor is returned to the collector rather than to the supply voltage. This is what distinguishes collector-feedback bias from base bias. Here is how the feedback works. Suppose the temperature increases, causing β_{dc} to increase. This produces more collector current. As soon as the collector current increases, the collector-emitter voltage decreases because of the additional voltage drop across R_C. This means there is less voltage across the base resistor, which causes a decrease in base current. The smaller base current then decreases

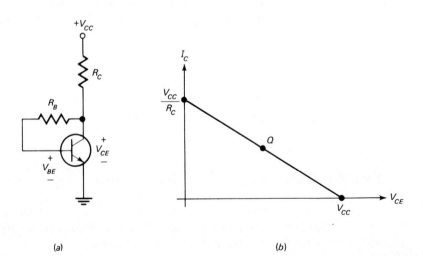

Fig. 5-27.
Collector-feedback bias.
(a) Circuit. (b) Dc load
line.

(a)

(b)

the collector current, partially offsetting the original increase in collector current.

By visualizing a short between the collector-emitter terminals, we can see that all of the supply voltage appears across the collector resistor. Therefore, the vertical intercept of the dc load line is given by

$$I_{C(\text{sat})} \cong \frac{V_{CC}}{R_C} \qquad (5\text{-}25)$$

Visualizing the collector-emitter terminals open, we get a cutoff voltage of

$$V_{CE(\text{cutoff})} \cong V_{CC} \qquad (5\text{-}26)$$

Figure 5-27*b* is the dc load line.

Starting at the base and summing voltages around the circuit gives

$$V_{BE} - V_{CC} + (I_C + I_B)R_C + I_B R_B = 0$$

Since I_C is much greater than I_B, the foregoing equation simplifies to

$$V_{BE} - V_{CC} + I_C R_C + I_B R_B \cong 0 \qquad (5\text{-}27)$$

Because $I_B = I_C/\beta_{\text{dc}}$, we can solve for I_C to get

$$I_C \cong \frac{V_{CC} - V_{BE}}{R_C + R_B/\beta_{\text{dc}}} \qquad (5\text{-}28)$$

Collector-feedback bias is more effective than emitter-feedback bias. Although the circuit is still somewhat sensitive to β_{dc}, it is used in practice. It has the advantage of simplicity (only two resistors) and improved frequency response (discussed later). In this book, we usually set the Q point near the middle of the dc load line. This requires a base resistance of

$$R_B = \beta_{\text{dc}} R_C \qquad (5\text{-}29)$$

This is easy to prove. When this expression for R_B is substituted into Eq. (5-28), the collector current is approximately half of the saturated value of collector current.

Emitter Bias

When positive and negative supplies are available, *emitter bias* (Fig. 5-28*a*) provides a Q point that is as stable as voltage-divider bias. The name "emitter bias" is used because a negative supply forward-biases the emitter diode through resistor R_E. Figure 5-28*b* shows the circuit as you will see it in industry.

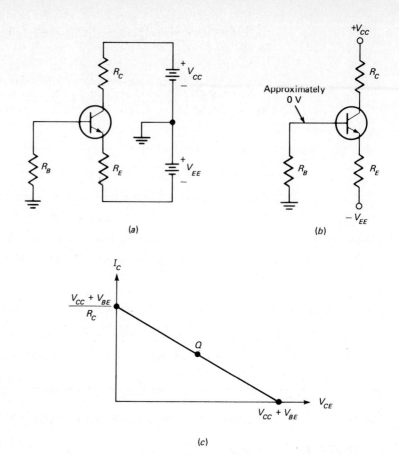

Fig. 5-28.
Emitter bias. (a) Circuit.
(b) Base voltage is
approximately zero. (c) Dc
load line.

Here is how to analyze the circuit. By deliberate design, the dc base voltage is approximately 0 V. Because of the V_{BE} drop, the voltage across the emitter resistor is $V_{EE} - V_{BE}$. Therefore, the emitter current is

$$I_E \cong \frac{V_{EE} - V_{BE}}{R_E} \qquad (5\text{-}30)$$

For example, if $V_{EE} = 15$ V and $R_E = 2.2$ kΩ, the emitter current is approximately

$$I_E \cong \frac{15 \text{ V} - 0.7 \text{ V}}{2.2 \text{ k}\Omega} = 6.5 \text{ mA}$$

By a derivation similar to that given for voltage-divider bias, an accurate formula for emitter current is

$$I_E \cong \frac{V_{EE} - V_{BE}}{R_E + R_B/\beta_{dc}} \qquad (5\text{-}31)$$

When R_E is at least 10 times greater than R_B/β_{dc}, I_E is almost insensitive to variations in β_{dc}.

We can derive the vertical and horizontal intercepts of the dc load line as follows: First, recall that the emitter voltage is approximately $-V_{BE}$. Second, visualize the collector-emitter terminals shorted, equivalent to $V_{CE} = 0$. Then, the voltage across R_C is

$$V_{CC} - (-V_{BE}) = V_{CC} + V_{BE}$$

and the collector saturation current is

$$I_{C(sat)} \cong \frac{V_{CC} + V_{BE}}{R_C} \qquad \textbf{(5-32)}$$

Next, visualize the collector-emitter terminals open. Then you can see that

$$V_{CE(cutoff)} \cong V_{CC} + V_{BE} \qquad \textbf{(5-33)}$$

To have a Q point near the middle of the load line in Fig. 5-28c, we need to set up a collector current that is approximately half the saturation value.

EXAMPLE 5-9 Calculate the values of I_C and V_{CE} in Fig. 5-29a. Assume $h_{FE} = 300$.

SOLUTION With Eq. (5-24),

$$I_C \cong \frac{V_{CC} - V_{BE}}{R_E + R_B/\beta_{dc}} = \frac{15\ \text{V} - 0.7\ \text{V}}{100\ \Omega + 430\ \text{k}\Omega/300} = 9.33\ \text{mA}$$

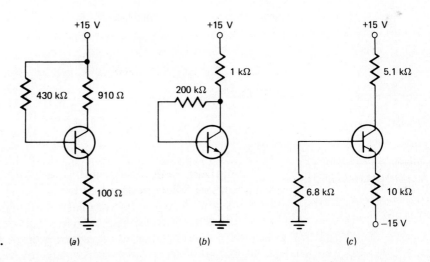

Fig. 5-29.   (a)   (b)   (c)

The voltage across the collector-emitter terminals equals the supply voltage minus the voltage drops across the R_C and R_E:

$$V_{CE} \cong V_{CC} - I_C(R_C + R_E) = 15 \text{ V} - (9.33 \text{ mA})(1010 \text{ }\Omega) = 5.58 \text{ V}$$

EXAMPLE 5-10 What do I_C and V_{CE} equal in Fig. 5-29b? Use a β_{dc} of 100.

SOLUTION With Eq. (5-28),

$$I_C \cong \frac{V_{CC} - V_{BE}}{R_C + R_B/\beta_{dc}} = \frac{15 \text{ V} - 0.7 \text{ V}}{1 \text{ k}\Omega + 200 \text{ k}\Omega/300} = 8.58 \text{ mA}$$

In Fig. 5-29b, V_{CE} equals the supply voltage minus the voltage drop across R_C. Ignoring the small base current through R_C,

$$V_{CE} \cong V_{CC} - I_C R_C = 15 \text{ V} - (8.58 \text{ mA})(1 \text{ k}\Omega) = 6.42 \text{ V}$$

EXAMPLE 5-11 Work out the values of I_C and V_{CE} in Fig. 5-29c.

SOLUTION With Eq. (5-30),

$$I_E \cong \frac{V_{EE} - V_{BE}}{R_E} \cong \frac{15 \text{ V} - 0.7 \text{ V}}{10 \text{ k}\Omega} = 1.43 \text{ mA}$$

The collector voltage to ground is

$$V_C = V_{CC} - I_C R_C \cong 15 \text{ V} - (1.43 \text{ mA})(5.1 \text{ k}\Omega) = 7.71 \text{ V}$$

Since the emitter is clamped at approximately -0.7 V,

$$V_{CE} = V_C - V_E \cong 7.71 \text{ V} - (-0.7 \text{ V}) = 8.41 \text{ V}$$

An alternative way to calculate V_{CE} is

$$V_{CE} \cong V_{CC} + V_{EE} - I_C(R_C + R_E)$$
$$\cong 15 \text{ V} + 15 \text{ V} - (1.43 \text{ mA})(15.1 \text{ k}) = 8.41 \text{ V}$$

5-9

pnp BIASING

Figure 5-30a is a *pnp* transistor. With the polarities shown for V_{BE} and V_{CE}, the emitter diode is forward-biased and the collector diode is reverse-biased. All currents flow in the opposite direction from an *npn* transistor. To use a *pnp* transistor in any of the biasing circuits discussed earlier, reverse the polarity of all power supplies. The calculation for currents and voltages is similar to that of *npn* transistors, except that all currents and voltages are reversed.

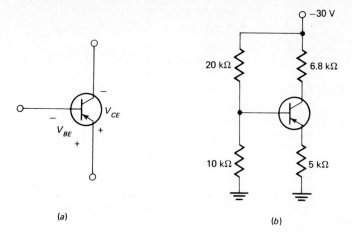

Fig. 5-30.
(a) *pnp* transistor.
(b) Voltage-divider biased
pnp circuit.

(a)   (b)

Voltage-Divider Bias

Figure 5-30*b* illustrates a *pnp* transistor in a voltage-divider biased circuit. Notice that the V_{CC} supply is -30 V instead of $+30$ V. In this circuit, all voltages and currents are opposite to what they were earlier. The calculation of currents and voltages is identical to earlier calculations, except for the reversals. For instance, the base voltage to ground is approximately

$$V_B = -10 \text{ V}$$

Allowing 0.7 V for V_{BE}, the emitter voltage to ground is

$$V_E = -10 \text{ V} - (-0.7 \text{ V}) = -9.3 \text{ V}$$

The emitter current is

$$I_E = \frac{9.3 \text{ V}}{5 \text{ k}\Omega} = 1.86 \text{ mA}$$

The collector voltage to ground is

$$V_C = -30 \text{ V} + (1.86 \text{ mA})(6.8 \text{ k}\Omega) = -17.4 \text{ V}$$

Upside-Down Convention

Ground is a reference point that we can move around. For instance, Fig. 5-31*a* shows voltage-divider bias with a *pnp* transistor. Figure 5-31*b* is the same circuit with ground removed. Even though the circuit is *floating* (no ground), all transistor currents have the same values as before, because the emitter diode is still forward-biased and the collector diode reverse-biased.

Nothing prevents us from grounding the negative terminal of the power supply to get the circuit of Fig. 5-31*c*. The transistor still operates in the active region with the same currents as before. It does not matter how a transistor is oriented in space; it works just as well

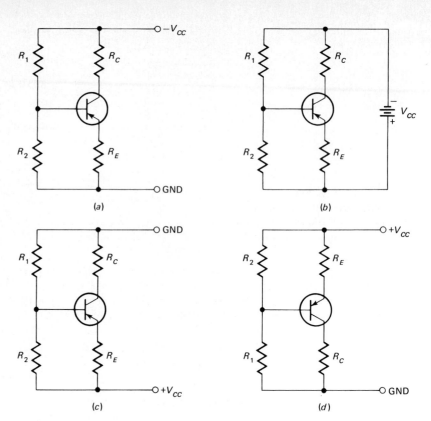

Fig. 5-31.
Evolution of upside-down
biased *pnp* circuit.

(a)   (b)

(c)   (d)

upside down as shown in Fig. 5-31*d*. In this upside-down biasing
circuit, all transistor currents have the same values as when we
started (Fig. 5-31*a*), although the voltages with respect to ground
have changed.

Drawing *pnp* transistors upside down with positive power supplies
is common industrial practice. Often, both *npn* and *pnp* transistors
are used in the same circuit. By drawing *pnp* transistors upside down,
we wind up with a simpler-looking schematic diagram.

EXAMPLE 5-12   What is the dc voltage from each collector to ground in Fig. 5-32?

SOLUTION   The capacitors are open to direct current. Therefore, they have nothing
to do with the problem. The first stage is voltage-divider biased with an
npn transistor. Because of the voltage divider, 6 V appears across the
2-kΩ base resistor. The emitter current in the first stage is

$$I_E = \frac{V_2 - V_{BE}}{R_E} = \frac{6\ V - 0.7\ V}{3\ k\Omega} = 1.77\ mA$$

and the collector voltage to ground is

$$V_C = V_{CC} - I_C R_C = 30\ V - (1.77\ mA)(4\ k\Omega) = 22.9\ V$$

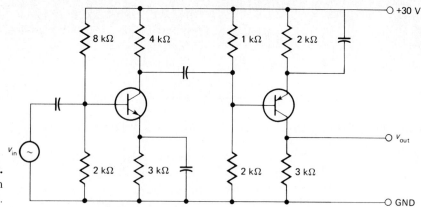

Fig. 5-32.
Two-stage amplifier with
npn and *pnp* transistors.

The second stage uses a *pnp* transistor with upside-down voltage-divider bias. Because of the voltage divider, 10 V appears across the 1-kΩ. Subtracting 0.7 V for the V_{BE} drop gives 9.3 V across the emitter resistor. Therefore, the emitter current is

$$I_E = \frac{9.3 \text{ V}}{2 \text{ k}\Omega} = 4.65 \text{ mA}$$

and the collector voltage to ground is

$$V_C = I_C R_C = (4.65 \text{ mA})(3 \text{ k}\Omega) = 14 \text{ V}$$

5-10
MANUFACTURERS'
SPECIFICATIONS

Small-signal transistors have a power rating of half a watt or less; *power* transistors have a power rating of more than half a watt. When you look at the data sheet for either type of transistor, start with the maximum ratings because these set the limits on the transistor currents, voltages, and other quantities.

As an example, the maximum ratings for a 2N3904 are as follows:

$$V_{CEO} = 40 \text{ V}$$
$$V_{CBO} = 60 \text{ V}$$
$$V_{EBO} = 6 \text{ V}$$
$$I_C = 200 \text{ mA}$$
$$P_D = 310 \text{ mW}$$

All voltage ratings are reverse breakdown voltages. The first rating is V_{CEO}, which stands for the voltage from collector to emitter with the base open. V_{CBO} is the voltage between the collector and base with the emitter open. V_{EBO} is the voltage from the emitter to the base with the collector open. I_C is the maximum dc collector current rating; this means that a 2N3904 can handle up to 200 mA of steady current.

The last rating is P_D, the maximum power rating of the device. You calculate the power dissipation of a transistor by

$$P_D = V_{CE}I_C \qquad (5\text{-}34)$$

For instance, if a 2N3904 has $V_{CE} = 20$ V and $I_C = 10$ mA, then

$$P_D = (20\ \text{V})(10\ \text{mA}) = 200\ \text{mW}$$

This is less than the power rating, 310 mW. As with breakdown voltage, a good design should include a safety factor to ensure a longer operating life for the transistor. Safety factors of 2 or more are common. For example, if 310 mW is the maximum power rating, a safety factor of 2 requires a power dissipation of less than 155 mW.

5-11 TROUBLE-SHOOTING

Many things can go wrong with a transistor. Since it contains two diodes, exceeding any of the breakdown ratings, maximum currents, or power ratings can damage either or both diodes. The troubles may include shorts, opens, high leakage currents, reduced β_{dc}, and other troubles.

One way to test a transistor is with an ohmmeter. You can begin by measuring the resistance between the collector and emitter. This should be very high in both directions because the collector and emitter diodes are back-to-back in series. One of the common troubles is a collector-emitter short, produced by exceeding the power rating. If you read from zero to a few thousand ohms in either direction, the transistor is shorted and should be thrown away.

Assuming the collector-emitter resistance is very high in both directions (megohms), you can read the reverse and forward resistances of the collector diode (collector-base terminals) and the emitter diode (base-emitter terminals). You should get a high reverse/forward ratio for both diodes, typically more than 1000 : 1 (silicon). If you do not, the transistor is defective.

Incidentally, some analog ohmmeters can produce enough current on lower ranges to damage a small-signal transistor. For this reason, you should test small-signal transistors on a higher range such as $R \times 100$. Then the internal resistance of the ohmmeter prevents excessive current through the transistor.

Even if the transistor passes the ohmmeter tests, it still may have some faults. After all, the ohmmeter only tests each transistor junction under dc conditions. You can use a curve tracer to look for more subtle faults, like too much leakage current, low β_{dc}, or insufficient breakdown voltage. Commercial testers are also available; these check the leakage current, β_{dc}, and other quantities.

The simplest in-circuit tests are to measure the transistor voltages to ground. For instance, measuring the collector voltage V_C and the emitter voltage V_E is a good start. The difference $V_C - V_E$ should be more than 1 V but less than V_{CC}. If the reading is less than 1 V, the transistor may be shorted. If the reading equals V_{CC}, the transistor may be open. Designers often set the Q point near the middle of the dc load line; so the typical value of V_C is near $V_{CC}/2$.

The foregoing test usually pins down a dc trouble if one exists. Many people include a test of V_{BE}, done as follows: Measure the base voltage V_B with respect to ground and the emitter voltage V_E with respect to ground. The difference of these readings is V_{BE}, which should be 0.6 to 0.7 V for small-signal transistors operating in the active region. If the V_{BE} reading is less than 0.6 V, the emitter diode is not being forward-biased. The trouble could be in the transistor or in the biasing components.

Some people include a cutoff test, performed as follows: Short the base-emitter terminals with a jumper wire. This removes the forward bias on the emitter diode and should force the transistor into cutoff. The collector-emitter voltage should then equal the collector supply voltage. If it does not, something is wrong with the transistor or the circuitry.

Summary

Normally, the emitter diode is forward-biased and the collector diode is reverse-biased. In this case, free electrons from the emitter are injected into the base. Because the base is thin and lightly doped, almost all of these free electrons diffuse into the collector. As a result, the collector current approximately equals the emitter current, while the base current is much smaller than the emitter current.

The α_{dc} is the ratio of collector current to emitter current. For most transistors, α_{dc} is greater than 0.95. For this reason, it is often approximated as equal to unity. Similarly, the β_{dc} is almost always greater than 20, with typical values between 50 and 300. On data sheets, β_{dc} is equivalent to h_{FE}.

A transistor has four operating regions: active, saturation, cutoff, and breakdown. For linear circuits, the transistor operates in the active region, which means a forward-biased emitter diode and a reverse-biased collector diode. The dc load line is the locus of all Q points between cutoff and saturation. In a base-biased circuit, the Q point is unpredictable and unstable because it is heavily dependent on the value of β_{dc}. On the other hand, voltage-divider bias produces a Q point that is almost independent of β_{dc}. Other forms of bias include emitter-feedback bias, collector-feedback bias, and emitter bias.

Small-signal transistors have a power rating of half a watt or less. Power transistors have a power rating of more than half a watt. The maximum ratings include V_{CEO}, V_{CBO}, V_{EBO}, I_C, and P_D. Exceeding any of these ratings can destroy the transistor. One way to test a transistor is with an ohmmeter. The resistance between the collector and emitter should be high in either direction. Furthermore, the emitter and collector diodes should each have a very high reverse/forward ratio.

Glossary

base bias The worst way to bias a transistor in linear circuits. The heavy dependence of the operating point on the β_{dc} means the location of the Q point on the dc load line is virtually impossible to predict because β_{dc} varies with transistor replacement and temperature change.

base-spreading resistance Because the depletion layers of the emitter and collector diodes penetrate the base region, the effective conducting area of the base is reduced to a thin channel. The resistance of this channel is called the base-spreading resistance, designated r_b'. It has a typical value between 50 and 150 Ω. This resistance has little effect at low frequencies.

breakdown Excessive collector voltage produces breakdown of the collector diode. Operation in the breakdown region usually destroys the transistor unless the current is limited to a safe value. Most designers never deliberately operate the transistor in the breakdown region.

common-base Abbreviated CB, this refers to the way the transistor is connected. With a CB connection, the base of a transistor is common to both the emitter and collector loops.

common-emitter Abbreviated CE, this describes a transistor connection where the emitter is common to the base and collector loops.

cutoff When the base current is zero, the collector current is approximately zero, except for the leakage current of the collector diode. With silicon transistors, cutoff is usually approximated by an open transistor.

dc alpha Abbreviated α_{dc}, this is the ratio of dc collector current to dc emitter current. It is usually greater than 0.95; so most people approximate it as it unity when troubleshooting and designing.

dc beta Abbreviated β_{dc}, this is the ratio of dc collector current to dc base current. Typically, β_{dc} is between 50 and 300.

dc load line Applying Kirchhoff's voltage law to the collector loop produces a linear equation in I_C and V_{CE}. When this equation is graphed, it results in a straight line that superimposes the collector curves. The Q point is the intersection of the dc load line and the active collector curve.

power transistor Any transistor with a power rating of more than half a watt.

recombination current Another name for base current. To get base current in an *npn* transistor, emitter-injected electrons in the base have to recombine with base holes. Since the base is thin and lightly doped, very little recombination occurs in the base, equivalent to saying base current is small.

saturation The almost vertical part of the collector curves near the origin. In this region, V_{CE} is typically a few tenths of a volt. Often, the transistor is approximated as a short circuit between the collector and emitter.

small-signal transistor Any transistor whose power rating is half a
watt or less.

voltage-divider bias The preferred form of bias for all single-supply
circuits. When properly designed, it produces a rock-solid Q point
virtually independent of β_{dc}.

Review Questions

1. Name the three doped regions of the transistor.
2. Describe the flow of electrons in an *npn* transistor with a forward-biased emitter diode and a reverse-biased collector diode.
3. Define the following: α_{dc}, r'_b, β_{dc}, and h_{FE}.
4. What are the four operating regions on the collector curves?
5. Why is base bias a poor way to bias a transistor in linear circuits?
6. How many points do you need to draw the dc load line? How can you find the values of these points?
7. How do you calculate the collector current in a voltage-divider biased circuit?
8. What is the difference between a small-signal transistor and a power transistor?
9. What is the meaning of V_{CEO}? V_{CBO}? V_{EBO}?

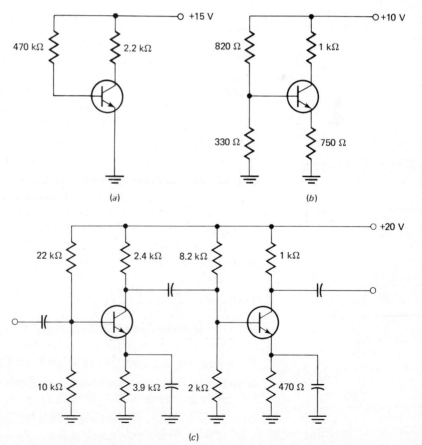

Fig. 5-33.

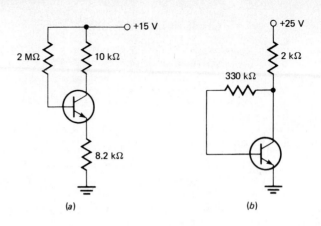

(a)

(b)

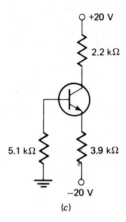

(c)

Fig. 5-34.

10. How do you use an ohmmeter to test a transistor?

11. How do you test a transistor in a built-up circuit?

Problems

5-1. When the collector current is 4.9 mA, the dc emitter current is 5 mA. What does α_{dc} equal? What does β_{dc} equal?

5-2. A transistor data sheet lists an h_{FE} of 175. What are the values of α_{dc} and β_{dc}?

5-3. A 2N3904 has a minimum h_{FE} of 100 and a maximum h_{FE} of 300. What are the minimum and maximum values of α_{dc} for this transistor?

5-4. The transistor of Fig. 5-33a has an h_{FE} of 200. Draw the dc load line. Calculate I_C and V_{CE}.

5-5. In Fig. 5-33a, what does I_C equal when h_{FE} is 100? When h_{FE} is 300?

5-6. What does I_C equal in Fig. 5-33b? V_C? V_E?

5-7. Figure 5-33c shows a two-stage amplifier. What do I_C and V_C equal for each stage?

5-8. If $h_{FE} = 100$, what are the values of I_C, V_C, and V_E in Fig. 5-34a?

5-9. If $\beta_{dc} = 150$ in Fig. 5-34b, what does I_C equal? V_C?

CHAPTER 5

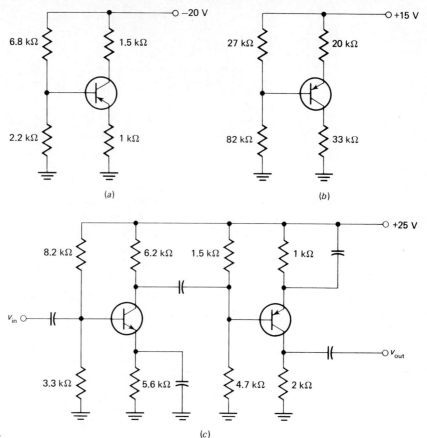

Fig. 5-35.

5-10. In Fig. 5-34c, what does I_E equal? V_C?

5-11. Draw the dc load line for each circuit of Fig. 5-34.

5-12. Calculate I_C and V_C in Fig. 5-35a.

5-13. Find I_C and V_C in Fig. 5-35b.

5-14. Work out the collector voltage to ground for each stage of Fig. 5-35c.

5-15. Calculate the transistor power dissipation for each transistor in Fig. 5-35c.

5-16. In Fig. 5-35c, the first collector has a voltage of 12 V to ground. Which of the following is a possible trouble:
   a. Collector-emitter open in first transistor
   b. Collector-emitter short in first transistor
   c. Collector-emitter open in second transistor
   d. Collector-emitter short in second transistor

5-17. The 3.3-kΩ resistor is shorted in Fig. 5-35c. Do the following voltages increase, decrease, or remain the same:
   a. Emitter voltage of first stage
   b. Collector voltage of first stage
   c. Emitter voltage of second stage
   d. Collector voltage of second stage

6 Common-Emitter Approximations

Descartes: This original thinker offered four rules for solving a problem: 1. Never accept anything as true unless it is clear and distinct enough to exclude all doubt from your mind. 2. Divide the problem into as many parts as necessary to reach a solution. 3. Start with the simplest things and proceed step by step toward the complex. 4. Review the solution so completely and generally that you are sure nothing was omitted.

After a transistor has been biased with a Q point near the middle of the dc load line, we can couple a small ac voltage into the base. If the input is a sine wave, the output will be an amplified or enlarged sine wave of the same frequency. The transistor circuit is called a *linear amplifier* if it does not change the shape of the input signal. This chapter discusses common-emitter (CE) amplifiers.

6-1 THE SUPERPOSITION THEOREM

The superposition theorem states the following: In a circuit with two or more sources, the current or voltage for any component is the sum of the currents or voltages produced by each source acting separately. To have only one source active at a time, all other sources must be reduced to zero; this means replacing voltage sources by short circuits and current sources by open circuits. When analyzing transistor amplifiers, we apply the superposition theorem in a special way to separate the dc effects from the ac effects. The following discussion explains this in detail.

DC and AC Equivalent Circuits

Figure 6-1*a* shows an ac source and a dc source, both driving a load resistor of 10 kΩ. The dc source has a voltage of 10 V, and the ac source has a peak of 10 mV. How can we find the total current through the 10 kΩ using the superposition theorem?

Here is how to get the dc component. Reduce the ac voltage source to zero, identical to replacing it by a short as shown in Fig. 6-1*b*. This

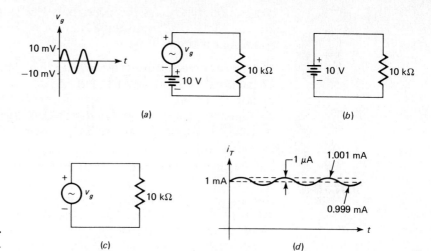

Fig. 6-1.
Superposition theorem.

(a)   *(b)*   *(c)*   *(d)*

simplified circuit is called the *dc equivalent circuit*; this is what remains after all ac sources have been reduced to zero. In this circuit, the dc current through the load resistor is

$$I = \frac{V_L}{R_L} = \frac{10 \text{ V}}{10 \text{ k}\Omega} = 1 \text{ mA}$$

Next, find the ac component. Reduce the dc voltage source to zero. What remains is the *ac equivalent circuit* of Fig. 6-1*c*. The ac source is sinusoidal with a peak of 10 mV. Therefore, the ac current through the resistor is sinusoidal with a peak of

$$I_P = \frac{10 \text{ mV}}{10 \text{ k}\Omega} = 1 \text{ }\mu\text{A}$$

The total current through the resistor is the sum of the dc and ac currents. Figure 6-1*d* shows this total current. It has an average or dc value of 1 mA. Superimposed on this average value is a sinusoidal component with a peak of 1 μA. On the positive peak, the total current is 1.001 mA. On the negative peak it is 0.999 mA.

Coupling Capacitors   Transistor amplifiers may use *coupling capacitors* to transmit ac components but block dc components. How is this possible? The reactance of a capacitor is given by

$$X_C = \frac{1}{2\pi f C}$$

For direct current the frequency is zero. Therefore, X_C is infinite for all dc signals. On the other hand, when the frequency is high enough, X_C approaches zero. Because of this, a designer can select a capacitor large enough to pass the ac signals but stop the dc signals. When drawing dc and ac equivalent circuits, remember these points about coupling capacitors:

1. Direct current cannot flow through a capacitor. Therefore, all capacitors look like *open circuits* in dc equivalent circuits.
2. For ac signals, however, capacitors usually couple the voltage and appear as *short circuits* in the ac equivalent circuit. Unless otherwise indicated, we automatically will assume all capacitors look like ac shorts.

Notation To keep dc and ac components distinct, we will use the following notation:

1. All dc and fixed quantities have capital letters. For instance, to represent dc voltage and current, we will use V and I.
2. All ac and varying quantities have lowercase letters. For example, to represent ac voltage and current, we will use v and i.

EXAMPLE 6-1 In Fig. 6-2a, what is the total voltage v_t produced by the dc and ac sources?

SOLUTION To get the dc equivalent circuit, reduce the ac source to zero and open the capacitor as shown in Fig. 6-2b. With no current in the circuit, the dc source voltage must appear across the output. So the dc component is

$$V = 10 \text{ V}$$

To get the ac equivalent circuit, reduce the dc source to zero and short the capacitor (Fig. 6-2c). Because of the 2 : 1 voltage divider, the ac output v is a sine wave with a peak of 5 mV.

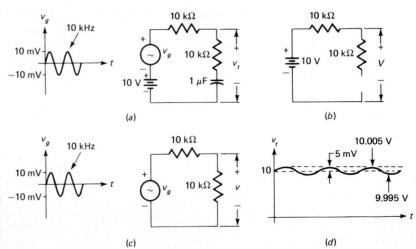

Fig. 6-2.
Equivalent circuits. (a) Original. (b) Dc equivalent. (c) Ac equivalent. (d) Total waveform.

The total voltage v_t is the sum of dc and ac voltages. Figure 6-2d shows how it looks. The total voltage has a dc or average value of 10 V. Superimposed on this average value is a sinusoidal component with a peak of 5 mV. On the positive peak, the total voltage is 10.005 V and on the negative peak it is 9.995 V.

EXAMPLE 6-2   Find the total voltage v_t across the 10 Ω of Fig. 6-3a.

SOLUTION   After reducing the ac source to zero and opening the capacitor, we get the dc equivalent circuit of Fig. 6-3b. With the voltage-divider theorem, the dc voltage across the 10 Ω is

$$V = \frac{10\ \Omega}{10{,}010\ \Omega}\ 10\ V = 9.99\ mV \cong 10\ mV$$

Next, reduce the dc source to zero and short the capacitor to get the ac equivalent circuit of Fig. 6-3c. The parallel resistance of 10Ω and 10 kΩ is approximately 10Ω; so the ac equivalent circuit simplifies to Fig. 6-3d. Because of the 10 : 1 voltage divider, the ac output voltage v is a sine wave with a peak of 1 mV.

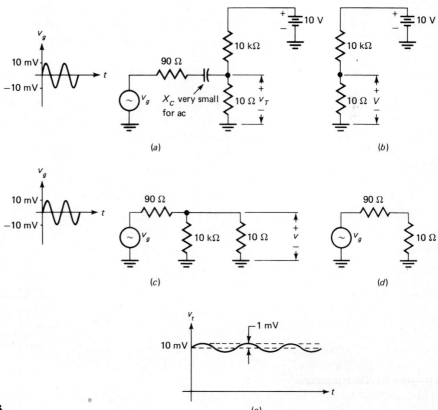

Fig. 6-3.

The total voltage v_t produced by both sources acting simultaneously is the sum of the dc and ac components. Figure 6-3e shows this total voltage. It has an average value of 10 mV. Superimposed on this average value is a sine wave with a peak of 1 mV.

6-2
AC RESISTANCE

To apply the superposition theorem to transistor amplifiers, we must first discuss how a diode acts as far as small ac signals are concerned.

Small-Signal Operation

In Fig. 6-4a, a dc source is in series with a small ac source. The dc source produces a dc current through the diode, and the ac source produces an ac current. Since the dc source is much larger than the ac source (10 V versus 1 mV), the dc current through the diode is much larger than the ac current.

Figure 6-4b shows the fluctuations in diode current and voltage. The dc operating point is at Q, with a current of 1 mA ideally or 0.93 mA to a second approximation. When the ac current is small, the diode curve between points A and B is approximately linear. This means ac voltage and current are directly proportional for small changes. For instance, if the ac voltage across the diode decreases to half its original value, the ac current decreases to half its original value.

As a guide in this book, we consider a signal small when the peak-to-peak ac current is less than *one-tenth of the quiescent current.* For example, to have small-signal operation with a dc current of 1 mA, the peak-to-peak ac current must be less than 0.1 mA. Transistor stages near the front end of systems typically operate as *small-signal* amplifiers. This is in contrast to *large-signal* operation (Chaps. 9 and 10) used by the later stages of a system.

AC Resistance

When the operation is small-signal, a diode acts like a resistance. In other words, we can replace a diode by a resistance in the ac equiva-

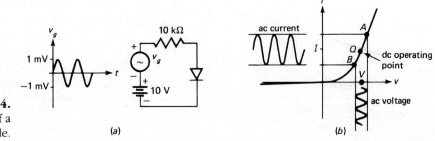

Fig. 6-4.
The ac resistance of a diode.

(a)

(b)

lent circuit. The Appendix derives this formula for the resistance of a diode to small ac voltages:

$$r_{ac} = \frac{25 \text{ mV}}{I} \qquad\qquad (6\text{-}1)$$

where I is the dc current through the diode. This is an unusual formula because it relates an ac quantity (the ac resistance of the diode) to a dc quantity (the dc current through the diode). As you will see later, this formula is the key to analyzing transistor amplifiers.

As an example, the dc source of Fig. 6-4a ideally sets up a dc current of approximately 1 mA through the diode. Therefore, the diode has an ac resistance of

$$r_{ac} = \frac{25 \text{ mV}}{1 \text{ mA}} = 25 \ \Omega$$

To a second approximation, the diode has a drop of about 0.7 V; so the dc current is

$$I = \frac{10 \text{ V} - 0.7 \text{ V}}{10 \text{ k}\Omega} = 0.93 \text{ mA}$$

and the diode has an ac resistance of

$$r_{ac} = \frac{25 \text{ mV}}{0.93 \text{ mA}} = 26.9 \ \Omega$$

Figure 6-5 shows how the ac resistance of a diode varies with the dc current through the diode. As you see, the ac resistance is 250 Ω for a dc current of 0.1 mA, 125 Ω for 0.2 mA, and so on. Remember that this resistance can be used only *in the ac equivalent circuit*.

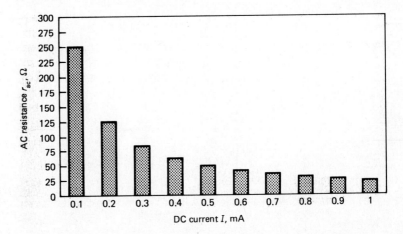

Fig. 6-5.
Variation of ac resistance
with dc current.

As preparation for transistor amplifiers, we need to analyze diode circuits with large dc sources and small ac sources. For such circuits, here is how we will apply the superposition theorem:

1. In the dc equivalent circuit, use the ideal or second approximation of the diode to analyze the dc operation of the circuit.
2. In the ac equivalent circuit, replace the diode by its ac resistance and analyze the ac operation of the circuit.
3. Add the dc and ac components to get the total current or voltage you are interested in.

EXAMPLE 6-3   Find the total current i_t through the diode of Fig. 6-6a.

SOLUTION   Start by drawing the dc equivalent circuit as shown in Fig. 6-6b. With the second approximation the dc current through the diode is

$$I = \frac{15 \text{ V} - 0.7 \text{ V}}{10 \text{ k}\Omega} = 1.43 \text{ mA}$$

With Eq. (6-1), the ac resistance is

$$r_{ac} = \frac{25 \text{ mV}}{1.43 \text{ mA}} = 17.5 \ \Omega$$

Next, consider the ac equivalent circuit. Notice that reducing the dc source to zero in Fig. 6-6a grounds the upper end of the 10 kΩ, equivalent to placing this resistor in parallel with the ac resistance of the diode. Furthermore, the coupling capacitor appears like a short. Therefore, we

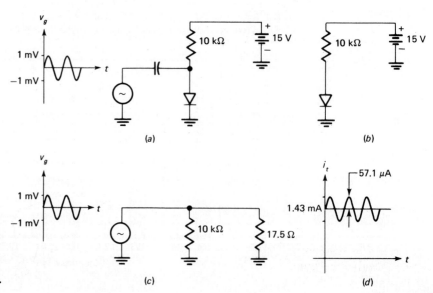

Fig. 6-6.

(a)   (b)   (c)   (d)

get the ac equivalent circuit of Fig. 6-6c. Since the ac source is in parallel with the diode, it produces a diode ac current with a peak value of

$$I_p = \frac{V_p}{r_{ac}} = \frac{1\ mV}{17.5\ \Omega} = 57.1\ \mu A$$

The total current through the diode is the sum of the dc and ac currents. Figure 6-6d shows how this total current looks. It has a dc value of 1.43 mA. Superimposed on this average value is a sine wave with a peak value of 57.1 μA.

<table>
<tr><td>

6-3

**ANALYZING A
CE AMPLIFIER**

</td><td>

In the typical CE amplifier the Q point is located near the middle of the load line as shown in Fig. 6-7. An ac source drives the base and produces a sinusoidal variation in base and collector currents. This forces the instantaneous operating point to swing above and below the Q point, producing an ac output voltage. This ac output voltage is normally much larger than the ac input voltage. The increase in size of the signal going from input to output is called *amplification*.

In Fig. 6-7, the operation remains in the active region throughout the ac cycle. In other words, the transistor never enters the saturation or cutoff regions. If it did, the peaks of the signal would be clipped off, resulting in *distortion*.

</td></tr>
</table>

AC Beta The ac beta of a transistor is symbolized β and is defined as the ratio of ac collector current i_c to ac base current i_b. In symbols,

$$\beta = \frac{i_c}{i_b} \qquad\qquad (6\text{-}2)$$

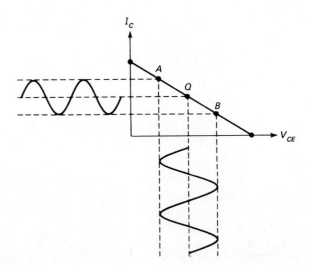

Fig. 6-7.
Ac signal swing on
load line.

On data sheets, β is listed as h_{fe}. (Recall that $\beta_{dc} = h_{FE}$.) As an example, when the Q point is at 1 mA, a 2N3904 has a minimum h_{fe} of 100 and a maximum h_{fe} of 400. This means that the ac collector current is 100 to 400 times greater than the ac base current. Sometimes, β or h_{fe} is referred to as the *ac current gain* of the transistor.

AC Equivalent Circuit of Transistor

The ac resistance of the emitter diode is symbolized r'_e and its approximate value is given by

$$r'_e = \frac{25 \text{ mV}}{I_E} \qquad (6\text{-}3)$$

Notice that this is similar to Eq. (6-1), the formula for the ac resistance of any diode. For instance, if you calculate a dc emitter current of 1 mA, then the emitter diode has an ac resistance of 25 Ω. This is the resistance that an ac signal would see when driving the emitter terminal.

As you know, the base current is much smaller than the emitter current. Because of the ac current gain between the base and the emitter, any resistance in the emitter circuit appears β times larger at the base. This is why the ac input resistance looking into the base is

$$r_{ac} = \beta r'_e \qquad (6\text{-}4)$$

This is the resistance seen by an ac signal driving the base of a transistor when the emitter is at ac ground.

The r_{ac} of the base is called the *input impedance of the base*, written as

$$z_{in(base)} = \beta r'_e \qquad (6\text{-}5)$$

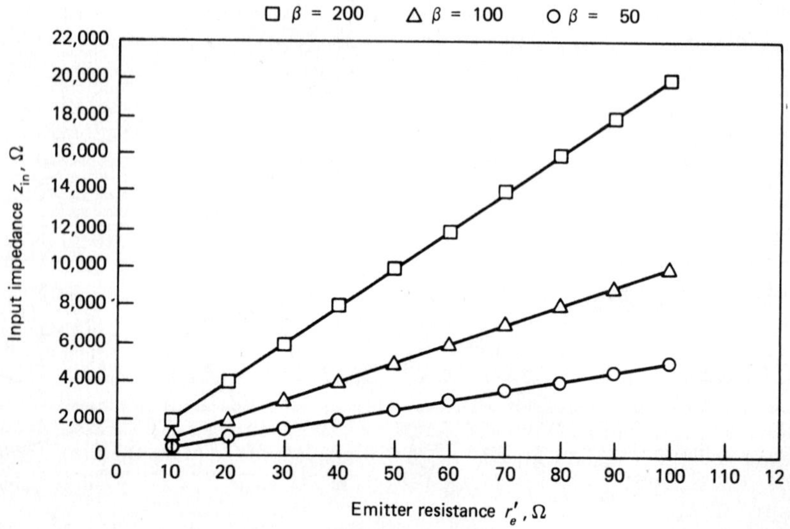

□ β = 200   △ β = 100   ○ β = 50

Fig. 6-8.
Variation of input impedance.

158

As an example, if $I_E = 1$ mA and $\beta = 200$, then r'_e is 25 Ω and

$$z_{\text{in(base)}} = 200(25 \ \Omega) = 5 \ \text{k}\Omega$$

Figure 6-8 shows how $z_{\text{in(base)}}$ varies for other values of β and r'_e. Because of the current gain, the input impedance of the base is much larger than the ac resistance of the emitter diode.

Figure 6-9a is the ac equivalent circuit of a transistor used in a CE amplifier. Notice how the base-emitter part of the transistor is replaced by an input impedance of $\beta r'_e$. Since the base current is approximately β times smaller than the emitter current, any impedance in the emitter circuit is increased by a factor of β when seen from the base. Remember this impedance-transforming property of the transistor because it has applications to be discussed later.

Sources of Error   Figure 6-9b is a more accurate equivalent circuit that includes the base-spreading resistance r'_b and the internal resistance r'_c of the collector current source. Usually, we can ignore r'_b because it is small compared with $\beta r'_e$. Likewise, we can ignore r'_c in most circuits because it is typically much larger than the load resistance connected to the collector. Incidentally, r'_c is related to the slope of the collector curves. If the collector curves were perfectly horizontal, r'_c would be infinite. Since the collector curves actually have a slight upward slope, r'_c is high but less than infinity.

Some data sheets specify a quantity called the *output admittance* h_{oe}. You can get a rough estimate of r'_c by using

$$r'_c \cong \frac{1}{h_{oe}}$$

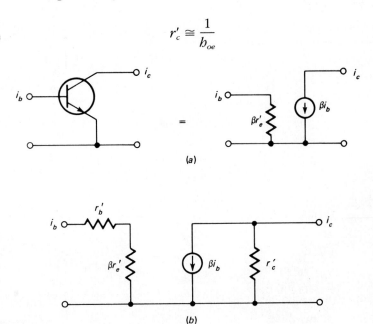

(a)

(b)

Fig. 6-9.
(a) Ideal ac equivalent circuit. (b) Model with r'_b and r'_c.

For example, a 2N3904 data sheet gives $h_{oe} = 8.5\ \mu S$. As an approximation,

$$r_c' \cong \frac{1}{8.5\ \mu S} = 118\ k\Omega$$

We will not use r_b' or r_c' in our ac analysis. Simply be aware that these quantities exist and that they produce some error. When you really want accuracy, it is best to use h-parameter analysis, which is discussed in Chap. 11. For most troubleshooting and preliminary design, the simplified transistor model of Fig. 6-9a is adequate.

Voltage Gain Defined
Figure 6-10a shows a CE amplifier. Capacitor C_1 couples the ac signal into the base, and C_2 couples the amplified signal to a load resistor R_L. C_3 is called a *bypass capacitor* because it provides a low-impedance path between the emitter and ground. Stated another way, the bypass capacitor appears shorted to an ac signal. This effectively grounds the emitter in the ac equivalent circuit.

Voltage gain A is defined as the ratio of ac output voltage v_{out} to ac input voltage v_{in}. As a formula,

$$A = \frac{v_{out}}{v_{in}} \tag{6-6}$$

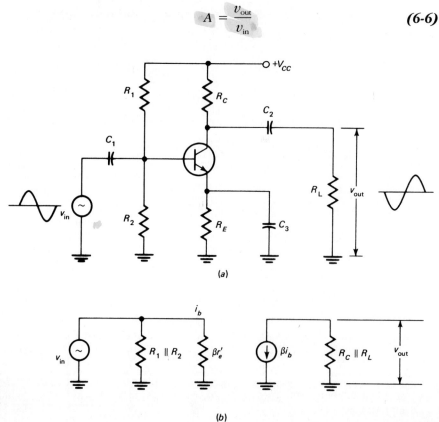

Fig. 6-10.
CE amplifier. (a) Original circuit. (b) Ac equivalent circuit.

For instance, if v_{out} = 150 mV and v_{in} = 2 mV, then the voltage gain is

$$A = \frac{150 \text{ mV}}{2 \text{ mV}} = 75$$

Equation (6-6) is useful when you are troubleshooting a circuit. With an oscilloscope you can measure v_{out} and v_{in}, then calculate the voltage gain.

Another Formula for Voltage Gain

Figure 6-10b shows the ac equivalent circuit. Notice how the biasing resistors R_1 and R_2 appear in parallel because the dc supply voltage has been reduced to zero. Furthermore, since the upper end of R_C appears grounded to ac signals, R_C is in parallel with R_L. This parallel combination is called the *ac load resistance*, symbolized r_L and given by

$$r_L = R_C \parallel R_L \qquad\qquad \textbf{(6-7)}$$

Since the ac output voltage appears across r_L, it equals

$$v_{out} = i_c r_L = \beta i_b r_L \qquad\qquad \textbf{(6-8)}$$

On the input side, the ac source voltage is across $\beta r'_e$, allowing us to write

$$v_{in} = i_b \beta r'_e \qquad\qquad \textbf{(6-9)}$$

Taking the ratio of Eq. (6-8) to (6-9) gives an alternative formula for voltage gain:

$$A = \frac{v_{out}}{v_{in}} = \frac{\beta i_b r_L}{i_b \beta r'_e} \qquad\qquad \textbf{(6-10)}$$

or

$$A = \frac{r_L}{r'_e} \qquad\qquad \textbf{(6-11)}$$

When you are troubleshooting, you can use Eq. (6-11) to calculate the theoretical voltage gain. Remember that it applies only when the emitter is at ac ground. (Later, you will see circuits where the emitter is no longer at ac ground.) After measuring the ac input and output voltages of a circuit, you can use Eq. (6-6) to check the actual voltage gain of the amplifier. If the amplifier is working correctly, the actual voltage gain should be approximately equal to the theoretical voltage gain. Figure 6-11 shows how the voltage gain varies with changes in r_L and r'_e. Notice that voltage gain is directly proportional to r_L and inversely proportional to r'_e.

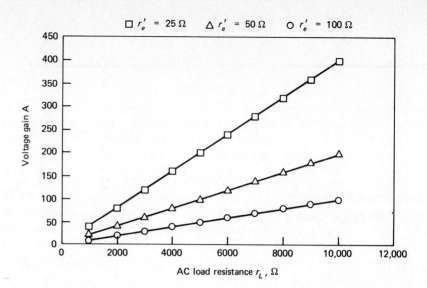

Fig. 6-11.
Voltage gain of CE amplifier.

Input Impedance   The input impedance of an amplifier is the impedance that the ac source has to drive. In Fig. 6-10*b*, the ac source has to supply current to the biasing resistors in parallel with the input impedance of the base. This means the input impedance of the amplifier includes the biasing resistors as well as the input impedance of the base:

$$z_{in} = R_1 \parallel R_2 \parallel \beta r'_e \qquad\qquad \textbf{\textit{(6-12)}}$$

As an example, if $R_1 = 10$ kΩ, $R_2 = 2.2$ kΩ, $b_{fe} = 200$, and $r'_e = 25$ Ω, then the ac source sees an input impedance of

$$z_{in} = 10 \text{ k}\Omega \parallel 2.2 \text{ k}\Omega \parallel 200(25 \ \Omega) = 1.33 \text{ k}\Omega$$

Phase Inversion   A final point. During the positive half cycle of ac input voltage, the total base current increases, causing the total collector current to increase. This produces a greater voltage drop across the collector resistor, which reduces the collector voltage. On the negative half cycle of ac input voltage, collector voltage increases. For this reason, the ac output voltage in any CE amplifier is always 180° out of phase with the ac input voltage, as shown in Fig. 6-10*a*. This reversal of phase is called *phase inversion*.

EXAMPLE 6-4   The 2N3904 of Fig. 6-12 has an h_{fe} of 200. What is the ac output voltage? The input impedance of the amplifier?

SOLUTION   The dc voltage across the 15 kΩ is

$$V_2 = \frac{R_2}{R_1 + R_2} V_{CC} = \frac{15 \text{ k}\Omega}{62 \text{ k}\Omega} \, 30 \text{ V} = 7.26 \text{ V}$$

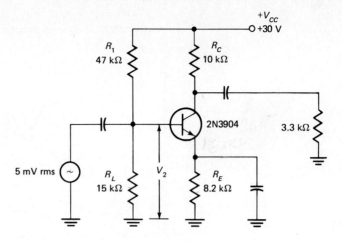

Fig. 6-12.

The dc voltage across the emitter resistor is

$$V_E = V_2 - V_{BE} = 7.26 \text{ V} - 0.7 \text{ V} = 6.56 \text{ V}$$

Therefore, the dc emitter current is

$$I_E = \frac{V_E}{R_E} = \frac{6.56 \text{ V}}{8.2 \text{ k}\Omega} = 0.8 \text{ mA}$$

and the ac resistance of the emitter diode is

$$r'_e = \frac{25 \text{ mV}}{I_E} = \frac{25 \text{ mV}}{0.8 \text{ mA}} = 31.3 \text{ }\Omega$$

Now, we can calculate the voltage gain:

$$A = \frac{r_L}{r'_e} = \frac{10 \text{ k}\Omega \parallel 3.3 \text{ k}\Omega}{31.3 \text{ }\Omega} = 79.3$$

The ac output voltage is

$$V_{out} = AV_{in} = 79.3(5 \text{ mV}) = 397 \text{ mV}$$

The input impedance of the base is

$$z_{in(base)} = \beta r'_e = 200(31.3 \text{ }\Omega) = 6.26 \text{ k}\Omega$$

and the input impedance of the amplifier is

$$z_{in} = R_1 \parallel R_2 \parallel z_{in(base)} = 47 \text{ k}\Omega \parallel 15 \text{ k}\Omega \parallel 6.26 \text{ k}\Omega = 4.04 \text{ k}\Omega$$

EXAMPLE 6-5 A dc-coupled oscilloscope displays the total signal with its dc and ac components. Figure 6-13 shows the total voltages at the base, emitter, and collector of a CE amplifier. Explain what these voltages represent.

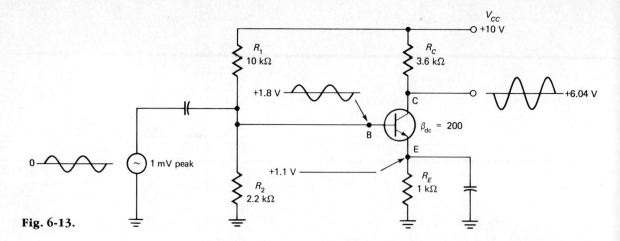

Fig. 6-13.

SOLUTION To begin with, a β_{dc} of 200 implies that the dc base current is much smaller than the dc current through the voltage divider. Therefore, the voltage divider is lightly loaded and produces a dc voltage of approximately

$$V_B = \frac{R_2}{R_1 + R_2} V_{CC} = \frac{2.2 \text{ k}\Omega}{12.2 \text{ k}\Omega} \, 10 \text{ V} = 1.8 \text{ V}$$

Notice that the ac input signal is a small sinusoidal voltage. This passes through the input coupling capacitor and appears at the base. So, the total input voltage to the base is 1.8 V dc, plus the small ac source signal.

Subtracting the V_{BE} drop gives a dc voltage of 1.1 V at the emitter. No ac signal appears here because the emitter is bypassed to ground through the emitter bypass capacitor. Stated another way, the bypass capacitor ac grounds the emitter.

An amplified and inverted ac voltage appears at the collector. As shown, it has a quiescent level of 6.04 V, calculated as follows:

$$V_C = V_{CC} - I_C R_C = 10 \text{ V} - (1.1 \text{ mA})(3.6 \text{ k}\Omega) = 6.04 \text{ V}$$

EXAMPLE 6-6 What does the ac output voltage equal in Fig. 6-13?

SOLUTION The dc emitter current is

$$I_E = \frac{V_2}{R_2} = \frac{1.1 \text{ V}}{1 \text{ k}\Omega} = 1.1 \text{ mA}$$

and the ac resistance of the emitter diode is

$$r'_e = \frac{25 \text{ mV}}{I_E} = \frac{25 \text{ mV}}{1.1 \text{ mA}} = 22.7 \ \Omega$$

Since R_L is infinite, $r_L = R_C$ and the voltage gain is

$$A = \frac{R_C}{r'_e} = \frac{3.6 \text{ k}\Omega}{22.7 \ \Omega} = 159$$

This means the ac output voltage is

$$V_{out} = A V_{in} = 159(1 \text{ mV}) = 159 \text{ mV}$$

STOP

6-4
SWAMPING THE EMITTER DIODE

As an approximation, we have been using the following formula for the ac resistance of the emitter diode:

$$r'_e = \frac{25 \text{ mV}}{I_E}$$

This is an ideal formula for an *abrupt junction* at 25°C. (An abrupt junction is one where the doping immediately changes from p-type to n-type.) When the junction is not abrupt or when the temperature is not 25°C, the value of r'_e lies between 25 mV/I_E and 50 mV/I_E. Any change in r'_e changes the voltage gain of a CE amplifier because $A = r_L/r'_e$.

Swamping Resistor

To desensitize an amplifier from the changes in r'_e, we can *swamp* the emitter diode by adding a resistor between the emitter and ac ground, as shown in Fig. 6-14a. Recall that any impedance in the emitter circuit is increased by a factor of β when viewed from the base. Since r_E is in series with r'_e in the emitter circuit, the input impedance looking into the base is

$$z_{in(base)} = \beta(r_E + r'_e) \qquad (6\text{-}13)$$

The input impedance of the amplifier includes the biasing resistors in parallel; so

$$z_{in} = R_1 \parallel R_2 \parallel \beta(r_E + r'_e) \qquad (6\text{-}14)$$

In the ac equivalent circuit of Fig. 6-14b, the input voltage is

$$v_{in} = i_b \beta(r_E + r'_e)$$

and the output voltage is

$$v_{out} = \beta i_b r_L$$

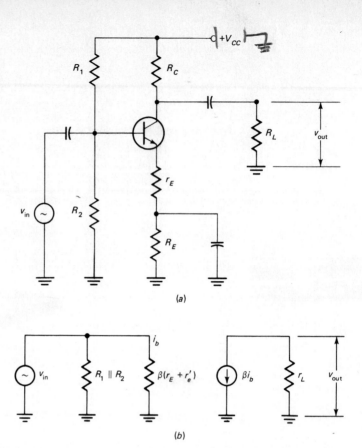

Fig. 6-14.
Swamped CE amplifier. (a)
Circuit. (b) Ac equivalent.

Taking the ratio of v_{out} to v_{in} gives

$$A = \frac{v_{out}}{v_{in}} = \frac{\beta i_b r_L}{i_b \beta (r_E + r_e')}$$

which simplifies to

$$A = \frac{r_L}{r_E + r_e'} \qquad \qquad \textbf{(6-15)}$$

The use of a swamping resistor produces *feedback*, which reduces the effect of r_e' on voltage gain.

Heavy Swamping Look at the denominator of Eq. (6-15). When r_E is much greater than r_e', changes in r_e' have little effect on the value of the denominator. For instance, if r_E is 10 times greater than r_e', changes in r_e' have only one-tenth their original effect. The large value of r_E swamps out (overpowers) the changes in r_e'. With heavy swamping, Eqs. (6-14) and (6-15) simplify to

$$z_{in} \cong R_1 \,\|\, R_2 \,\|\, \beta r_E \qquad \qquad \textbf{(6-16)}$$

$$A \cong \frac{r_L}{r_E} \qquad\qquad \textbf{(6-17)}$$

Because r_E is a fixed resistor, the voltage gain is fixed. This means we can change transistors and still have the same value of voltage gain. Likewise, changes in temperature no longer affect the voltage gain because r'_e has been swamped out.

Figure 6-15 shows the effect of a swamping resistor on voltage gain for two different values of r'_e. When r_E is zero, an r'_e of 25 Ω produces a voltage gain of 400, while an r'_e of 50 Ω produces a voltage gain of 200. Although the voltage gain is high, it has a 2 : 1 variation with r'_e. As the swamping resistance r_E increases, the voltage gain decreases but the gain becomes more stable. When $r_E = 200$ Ω, the two voltage gains are almost the same value. Although you sacrifice voltage gain by using a swamping resistor, you get a gain that is less dependent on the value of r'_e. This is the whole point of swamping.

EXAMPLE 6-7   In Fig. 6-16, r'_e varies from 50 to 100 Ω. What is the minimum voltage gain? The maximum?

SOLUTION   The ac load resistance is

$$r_L = R_C \| R_L = 10 \text{ k}\Omega \| 82 \text{ k}\Omega = 8.91 \text{ k}\Omega$$

The minimum voltage gain is

$$A = \frac{8.91 \text{ k}\Omega}{1.1 \text{ k}\Omega} = 8.1$$

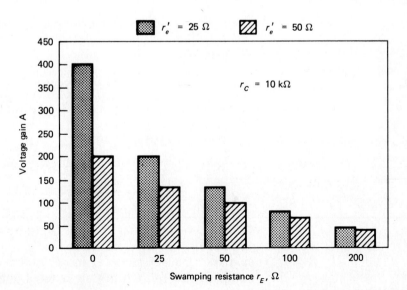

Fig. 6-15.
Voltage gain of swamped CE amplifier.

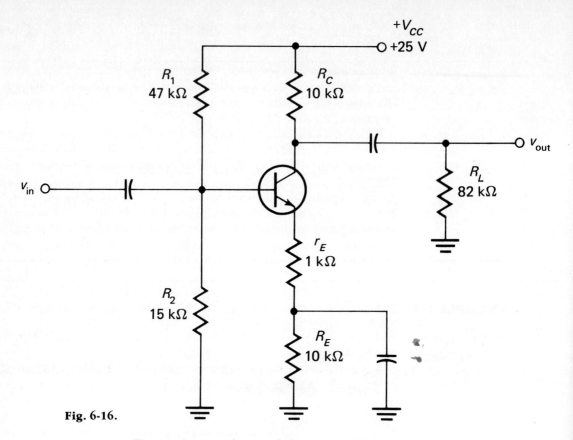

Fig. 6-16.

The maximum voltage gain is

$$A = \frac{r_L}{r_E + r'_e} = \frac{8.91 \text{ k}\Omega}{1.05 \text{ k}\Omega} = 8.49$$

As you see, the voltage gain is almost constant.

Without the swamping resistor, the changes in r'_e would produce a minimum voltage gain of

$$A = \frac{8.91 \text{ k}\Omega}{100 \text{ }\Omega} = 89.1$$

and a maximum voltage gain of

$$A = \frac{8.91 \text{ k}\Omega}{50 \text{ }\Omega} = 178$$

The use of a swamping resistor is up to the designer. In some applications, you want all the voltage gain you can get even though it may have a 2:1 variation or more. In this case, you would not use a swamping resistor. On the other hand, other applications require a relatively constant voltage gain; so a swamping resistor would be useful.

EXAMPLE 6-8 In Fig. 6-16, h_{fe} = 200 and r'_e = 50 Ω. What is the input impedance of the amplifier?

SOLUTION The input impedance of the base is

$$Z_{in(base)} = \beta(r_E + r'_e) = 200(1.05 \text{ k}\Omega) = 210 \text{ k}\Omega$$

The input impedance of the amplifier includes the biasing resistors:

$$Z_{in} = R_1 \parallel R_2 \parallel Z_{in(base)} = 47 \text{ k}\Omega \parallel 15 \text{ k}\Omega \parallel 210 \text{ k}\Omega = 10.8 \text{ k}\Omega$$

6-5
CASCADED
STAGES

To increase the voltage gain, we can use the output of one transistor stage as the input to another transistor stage. For instance, Fig. 6-17a shows a three-stage amplifier. The coupling capacitors transmit ac voltages but block dc voltages. Because of this, the stages are isolated as far as dc biasing is concerned. This dc isolation is necessary to

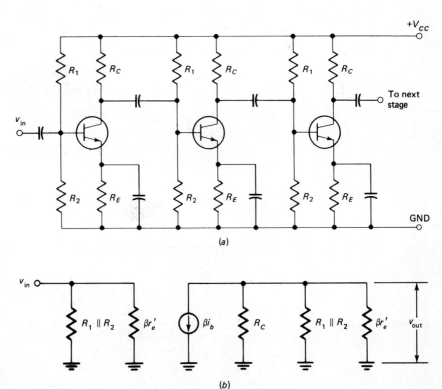

Fig. 6-17.
(a) Cascaded stages. (b) Ac equivalent circuit.

prevent dc interference between stages and shifting of Q points. The bypass capacitors are used to bypass the emitters to ground, effectively ac grounding the emitters.

Voltage Gains Multiply The ac voltage across the collector resistor of each stage is coupled into the base of the next stage. In this way, the cascaded (one after another) stages amplify the signal, and the overall voltage gain equals the product of the individual gains:

$$A = A_1 A_2 A_3 \qquad\qquad\qquad (6\text{-}18)$$

As an example, if each stage has a voltage gain of 50, the total voltage gain is

$$A = (50)(50)(50) = 125,000$$

Input Impedance Loads Preceding Stage How do you calculate the voltage gain of each stage? Figure 6-17b shows the ac equivalent circuit for an individual stage. The input impedance of each stage is

$$z_{in} = R_1 \parallel R_2 \parallel \beta r'_e \qquad\qquad (6\text{-}19)$$

This input impedance loads the preceding stage. You can see this by looking at the collector circuit of Fig. 6-17b. The collector current source has to drive R_C in parallel with the z_{in} of the next stage. In other words, the ac load resistance now is

$$r_L = R_C \parallel z_{in} \qquad\qquad\qquad (6\text{-}20)$$

Since $A = r_L/r'_e$, the voltage gain is reduced because of the input impedance of the next stage.

For instance, suppose $R_C = 10$ kΩ and $r'_e = 25\ \Omega$. If $z_{in} = 1$ MΩ, the voltage gain is

$$A = \frac{r_L}{r'_e} = \frac{10\ \text{k}\Omega \parallel 1\ \text{M}\Omega}{25\ \Omega} = 400$$

If $z_{in} = 100$ kΩ, the voltage gain drops slightly to

$$A = \frac{10\ \text{k}\Omega \parallel 100\ \text{k}\Omega}{25\ \Omega} = 364$$

If $z_{in} = 10$ kΩ, the voltage gain decreases to

$$A = \frac{10\ \text{k}\Omega \parallel 10\ \text{k}\Omega}{25\ \Omega} = 200$$

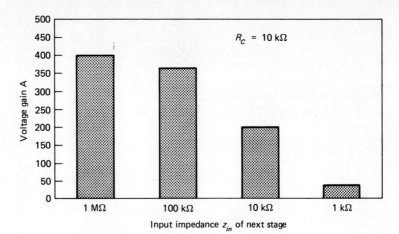

Fig. 6-18.
Input impedance of next
stage reduces voltage gain.

If $z_{in} = 1\text{ k}\Omega$, the voltage gain is

$$A = \frac{10\text{ k}\Omega \parallel 1\text{ k}\Omega}{25\ \Omega} = 36.4$$

Figure 6-18 summarizes the foregoing results. As indicated, the input impedance of the next stage loads down the preceding stage. The smaller the input impedance, the lower the voltage gain.

EXAMPLE 6-9   Calculate the input impedance of each stage in Fig. 6-19a.

SOLUTION   Each stage has biasing resistors of 10 kΩ and 2.2 kΩ. The dc voltage at each base is

$$V_B = \frac{R_2}{R_1 + R_2}\, V_{CC} = \frac{2.2\text{ k}\Omega}{12.2\text{ k}\Omega}\, 10\text{ V} = 1.8\text{ V}$$

Subtracting the V_{BE} drop gives a dc emitter voltage of

$$V_E = V_B - V_{BE} = 1.8\text{ V} - 0.7\text{ V} = 1.1\text{ V}$$

The dc emitter current is

$$I_E = \frac{V_E}{R_E} = \frac{1.1\text{ V}}{1\text{ k}\Omega} = 1.1\text{ mA}$$

So, the ac resistance of the emitter diode is

$$r'_e = \frac{25\text{ mV}}{I_E} = \frac{25\text{ mV}}{1.1\text{ mA}} = 22.7\ \Omega$$

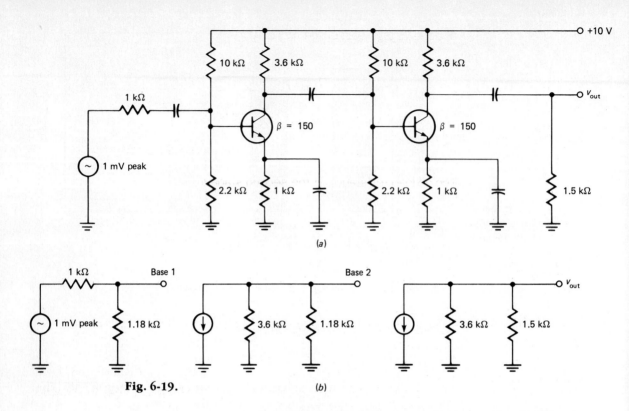

Fig. 6-19.

Since each transistor has a β of 150, the input impedance of the base is

$$Z_{in(base)} = \beta r'_e = 150(22.7\ \Omega) = 3.41\ k\Omega$$

Because of the biasing resistors, the input impedance of each stage is

$$Z_{in} = R_1 \| R_2 \| Z_{in} = 10\ k\Omega \| 2.2\ k\Omega \| 3.41\ k\Omega = 1.18\ k\Omega$$

EXAMPLE 6-10   The ac source voltage has a peak of 1 mV in Fig. 6-19a. Calculate the ac output voltage across the final load resistor.

SOLUTION   Figure 6-19b shows the ac equivalent circuit for the two-stage amplifier. As you see, the ac source voltage drives the first base through a voltage divider. Therefore, the ac voltage reaching the base of the first transistor is

$$V_{b1} = \frac{Z_{in}}{R_S + Z_{in}} V_{in} = \frac{1.18\ k\Omega}{2.18\ k\Omega}\ 1\ mV = 0.541\ mV$$

In Fig. 6-19b, the ac load resistance of the first stage is

$$r_L = R_C \| Z_{in} = 3.6\ k\Omega \| 1.18\ k\Omega = 889\ \Omega$$

and the voltage gain of the first stage is

$$A_1 = \frac{r_L}{r'_e} = \frac{889\ \Omega}{22.7\ \Omega} = 39.2$$

Therefore, the ac voltage at the second base is

$$V_{b2} = A_1 V_{b1} = 39.2(0.541\ \text{mV}) = 21.2\ \text{mV}$$

The ac load resistance of the second stage is

$$r_L = R_C \parallel R_L = 3.6\ \text{k}\Omega \parallel 1.5\ \text{k}\Omega = 1.06\ \text{k}\Omega$$

and the voltage gain of the second stage is

$$A_2 = \frac{r_L}{r'_e} = \frac{1.06\ \text{k}\Omega}{22.7\ \Omega} = 46.7$$

So, the ac voltage at the final output is

$$V_{\text{out}} = A_2 V_{b2} = 46.7(21.2\ \text{mV}) = 0.99\ \text{V}$$

An alternative way to calculate the output voltage is to get the overall voltage gain:

$$A = A_1 A_2 = (39.2)(46.7) = 1830$$

This is the voltage gain from the first base to the second collector. Since the ac voltage at the first base is 0.541 mV, the final output voltage is

$$V_{\text{out}} = A V_{b1} = 1820(0.541\ \text{mV}) = 0.99\ \text{V}$$

EXAMPLE 6-11 In Fig. 6-19a, each 1 kΩ is changed to 820 Ω. If a swamping resistor of 180 Ω is inserted in each stage, what is the ac voltage at the final output?

SOLUTION In the dc equivalent circuit, the total emitter resistance is still 1 kΩ. Therefore, r'_e is still 22.7 Ω. But the input impedance of each base increases to

$$Z_{\text{in(base)}} = \beta(r_E + r'_e) = 150(180\ \Omega + 22.7\ \Omega) = 30.4\ \text{k}\Omega$$

and the input impedance of each stage increases to

$$Z_{\text{in}} = R_1 \parallel R_2 \parallel Z_{\text{in(base)}} = 10\ \text{k}\Omega \parallel 2.2\ \text{k}\Omega \parallel 30.4\ \text{k}\Omega = 1.7\ \text{k}\Omega$$

Figure 6-20 shows the ac equivalent circuit with this value of input impedance.

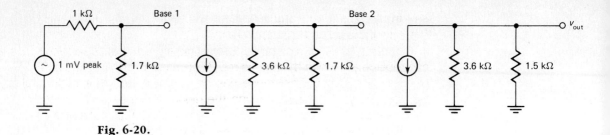

Fig. 6-20.

The ac load resistance of the first stage is

$$r_L = R_C \| z_{in} = 3.6 \text{ k}\Omega \| 1.7 \text{ k}\Omega = 1.15 \text{ k}\Omega$$

and the voltage gain is

$$A_1 = \frac{r_L}{r_E + r'_e} = \frac{1.15 \text{ k}\Omega}{180 \text{ }\Omega + 22.7 \text{ }\Omega} = 5.67$$

Similarly, the second stage has an ac load resistance of

$$r_L = R_C \| R_L = 3.6 \text{ k}\Omega \| 1.5 \text{ k}\Omega = 1.06 \text{ k}\Omega$$

and a voltage gain of

$$A_2 = \frac{r_L}{r'_e} = \frac{1.06 \text{ k}\Omega}{180 \text{ }\Omega + 22.7 \text{ }\Omega} = 5.23$$

The overall voltage gain from the first base to the second collector is

$$A = A_1 A_2 = (5.67)(5.23) = 29.7$$

The ac voltage at the first base is

$$V_{b1} = \frac{z_{in}}{R_S + z_{in}} V_{in} = \frac{1.7 \text{ k}\Omega}{1 \text{ k}\Omega + 1.7 \text{ k}\Omega} 1 \text{ mV} = 0.63 \text{ mV}$$

Therefore, the final output voltage is

$$V_{out} = A V_{b1} = 29.7(0.63 \text{ mV}) = 18.7 \text{ mV}$$

The superposition theorem enables us to separate the dc and ac effects in transistor circuits. First, we analyze the dc equivalent circuit to get any dc currents and voltages of interest. Second, we analyze the ac equivalent for the ac currents and voltages. The total currents and voltages are the sum of the dc and ac components.

A diode acts like a resistance to small ac signals. Because of this, we can replace all diodes by their ac resistances when drawing the ac equivalent circuit. This ac resistance has an approximate value of 25 mV/I, where I is the dc current through the diode. Since the emitter diode is forward-biased, it has an r'_e of 25 mV/I_E. Because of ac current gain of a transistor, the input impedance seen at the base is $\beta r'_e$. On the other hand, the collector acts like an ac current source of βi_b.

A CE amplifier has a voltage gain of r_L/r'_e. The input impedance of a CE amplifier includes the effect of the biasing resistors in parallel with the input impedance of the base. Because r'_e may vary with temperature and transistor replacement, a swamping resistor is sometimes added to the emitter circuit to stabilize the voltage gain. When CE stages are cascaded, the overall voltage gain equals the product of the individual stage gains. Especially important, the input impedance of each stage loads down the collector of the preceding stage because z_{in} appears in parallel with R_C in the ac equivalent circuit.

Glossary

ac current gain In a CE amplifier, this is the ratio of ac collector to ac base current. Symbolized by β or h_{fe}.

ac equivalent circuit The circuit that remains after all dc sources have been reduced to zero. This means replacing all dc voltage sources by short circuits and all dc current sources by open circuits.

ac load resistance The total resistance seen by the collector in the ac equivalent circuit.

ac resistance of a diode As far as a small ac signal is concerned, a forward-biased diode acts like a small resistance rather than a switch.

amplification The increase in voltage (or power) of a signal between the input and output of a circuit.

bypass capacitor A capacitor that ac grounds some point in a circuit.

coupling capacitor A capacitor that transmits an ac signal between one ungrounded point and another.

dc equivalent circuit The circuit that remains after all ac sources have been reduced to zero. This means replacing all ac voltage sources by short circuits and all ac current sources by open circuits.

distortion Any change in the shape of the signal as it passes through an amplifier.

input impedance of the base This is the equivalent impedance looking into the base.

input impedance of a stage This is the total impedance of a circuit. It includes the biasing resistors and the input impedance of the base. This impedance loads the source that drives it. With cascaded stages, the input impedance appears in parallel with the collector resistor of the preceding stage.

phase inversion This refers to the 180° shift in phase that takes place in a CE amplifier.

small signal Ideally this refers to an infinitesimally small signal. As an approximation, however, we consider a signal small if its peak-to-peak variation in collector current is less than 10 percent of the quiescent collector current.

swamping resistor A resistor inserted in the emitter circuit to stabilize the voltage gain against changes in r'_e.

Review Questions

1. How do you get the dc equivalent circuit? The ac equivalent circuit?
2. What has to be true about the signal when using the ac resistance of a diode?
3. In the ac equivalent circuit, how does a forward-biased diode appear to a small ac signal?
4. What is the β of a transistor? How is it related to the h_{fe} appearing on transistor data sheets?
5. Define the voltage gain of an amplifier.
6. How would you calculate the input impedance of an unswamped CE amplifier?
7. Explain why a CE amplifier inverts the phase of the ac voltage between the base and collector.
8. What is a swamping resistor and why is it used?
9. When stages are cascaded, what effect does the input impedance of a stage have on the voltage gain of the preceding stage?

Problems

6-1. How much dc current is there through the 50 Ω of Fig. 6-21?

6-2. In Fig. 6-21, the ac source is sinusoidal with a value of 3 mV rms. What is the ac current through the 50 Ω?

6-3. What is the total current through the 50 Ω of Fig. 6-21?

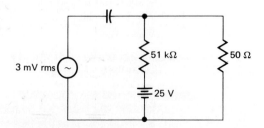

Fig. 6-21.

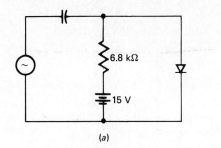

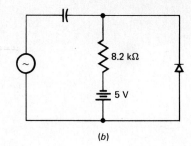

Fig. 6-22.
(a)   (b)

6-4. Calculate the ac resistance of a diode for each of these dc currents:
  a. 0.05 mA
  b. 0.275 mA
  c. 1.2 mA
  d. 6.65 mA

6-5. In Fig. 6-22a, what is the dc current through the diode using the second approximation? The ac resistance of the diode?

6-6. What is the ac resistance of the diode in Fig. 6-22b?

6-7. If the ac source of Fig. 6-22a has an rms value of 2 mV, what is the rms value of ac current through the diode?

6-8. In Fig. 6-22b, the ac source has a peak-to-peak value of 5 mV. What is the rms value of the ac current through the diode?

6-9. In Fig. 6-22a, the ac source has a peak value of 1 mV. What is the total current through the diode?

6-10. The ac source of Fig. 6-22b has an rms value of 10 mV. What is the dc current through the diode? The rms current?

6-11. What is the voltage gain in Fig. 6-23? The ac output voltage?

6-12. If h_{fe} = 250 in Fig. 6-23, what is the input impedance of the base? The input impedance of the stage?

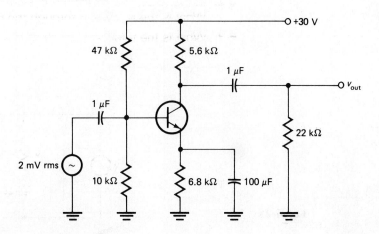

Fig. 6-23.

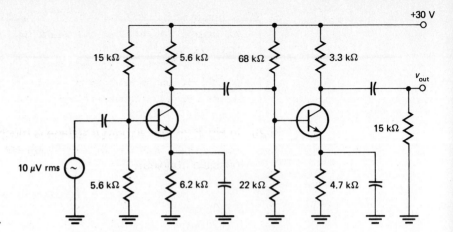

Fig. 6-24.

6-13. The ac source of Fig. 6-23 has a frequency of 1 kHz. Calculate the capacitive reactance for the input coupling capacitor, the output coupling capacitor, and the bypass capacitor.

6-14. Each stage of Fig. 6-24 has an h_{fe} of 200. What is the input impedance of the first stage? Of the second stage?

6-15. Assuming a β of 200 for each stage, what is the voltage gain of the first stage in Fig. 6-24? Of the second stage?

6-16. If $h_{fe} = 200$ for each stage in Fig. 6-24, what does the ac output voltage equal?

6-17. What does the ac load resistance of the first stage equal if β is 125 in Fig. 6-25?

6-18. Each transistor of Fig. 6-25 has an h_{fe} of 300. What does v_{out} equal if v_{in} is 1 mV rms?

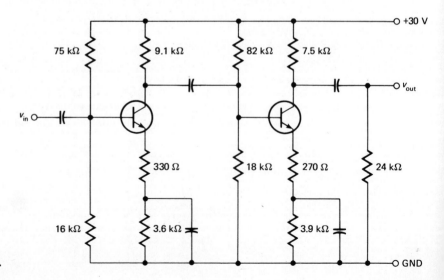

Fig. 6-25.

6-19. Suppose v_{in} is 1 mV in Fig. 6-25. If the first emitter bypass capacitor opens, do the following quantities increase, decrease, or remain the same:

a. Dc collector voltage of first stage
b. Voltage gain of first stage
c. Voltage gain of second stage
d. Ac output voltage

6-20. In Fig. 6-25, the ac output voltage is much lower than it should be, but it is not zero. If all dc voltages are normal, what are three possible troubles?

7 Common-Collector Approximations

Steinmetz: A hunchback from birth, he took the middle name "Proteus," a Greek god who could change his appearance at will. Steinmetz was the one who introduced complex numbers to solve ac problems. On one occasion he was hired to troubleshoot a large system no one else could fix. After a brief study of the schematics, he climbed a few ladders and chalked an X on a metal plate. "There, look there," he said and left. The trouble was there. A few days later, the company was jolted by his $1000 bill and demanded an itemized statement. Steinmetz sent this reply: $1 for marking plate; $999 for knowing where to make mark.

If you connect a high-impedance source to a low-impedance load, most of the ac voltage is dropped across the internal impedance of the source. One way to get around this problem is to use an *emitter follower* between the high-impedance source and the low-impedance load. The emitter follower steps up the impedance level and reduces the signal loss. In addition to emitter followers, this chapter discusses Darlington amplifiers and improved voltage regulation.

7-1 THE CC CONNECTION

Figure 7-1a shows a *common-collector* (CC) stage. Because the collector resistor has a value of zero, the collector is ac grounded through the power supply. This is why the circuit is sometimes called a grounded-collector amplifier.

Basic Idea

An ac input voltage v_{in} is coupled into the base. This produces sinusoidal variations in emitter current, and a sinusoidal voltage appears across R_E. The ac signal is then coupled to the load resistor R_L. In effect, a CC amplifer is a heavily swamped amplifier with the collector resistor shorted and the output taken from the emitter. On the positive half cycle of ac input voltage the emitter current increases, pro-

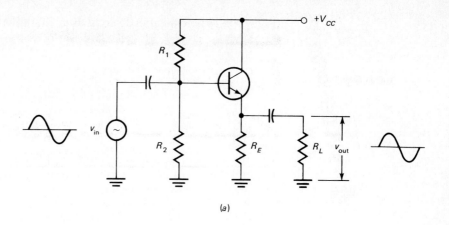

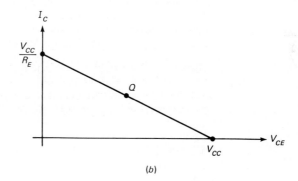

Fig. 7-1.
CC amplifier. (a) Circuit.
(b) Dc load line.

ducing the positive half cycle of ac output voltage. Similarly, on the negative half cycle of input voltage, the emitter current decreases and we get the negative half cycle of output voltage. For this reason, the ac output voltage of a CC amplifier is in phase with the ac input voltage.

DC Load Line   By an analysis similar to that given earlier for CE amplifers, the graph of I_C versus V_{CE} results in a dc load line. To get the upper end of the dc load line, visualize the collector-emitter terminals shorted. Then all of the supply voltage appears across R_E and the collector saturation current is

$$I_{C(\text{sat})} = \frac{V_{CC}}{R_E} \qquad (7\text{-}1)$$

Similarly, to get the lower end of the dc load line, visualize the collector-emitter terminals open. Then all of the supply voltage appears across the collector-emitter terminals and

$$V_{CE(\text{cutoff})} = V_{CC} \qquad (7\text{-}2)$$

Figure 7-1*b* shows the dc load line. Usually, the designer selects R_1 equal to R_2. This produces a Q point near the middle of this load line.

Voltage Gain Figure 7-2 is the ac equivalent circuit. Ac load resistance r_L is the equivalent parallel resistance of R_E and R_L:

$$r_L = R_E \parallel R_L$$

In Fig. 7-2, the ac input voltage v_{in} produces an ac base current i_b through $\beta r'_e$ and an ac emitter current i_e through r_L. The ac output voltage is across r_L and equals

$$v_{out} = i_e r_L$$

The ac input voltage appears across $\beta r'_e$ and r_L. It equals

$$v_{in} = i_b \beta r'_e + i_e r_L$$

Since $i_e \cong \beta i_b$, the voltage gain is

$$A = \frac{v_{out}}{v_{in}} = \frac{i_e r_L}{i_b \beta r'_e + i_e r_L} \cong \frac{i_e r_L}{i_e r'_e + i_e r_L}$$

which reduces to

$$A \cong \frac{r_L}{r_L + r'_e} \qquad (7\text{-}3)$$

By dividing numerator and denominator by r_L, we get an alternative formula that is sometimes useful:

$$A \cong \frac{1}{1 + r'_e/r_L} \qquad (7\text{-}4)$$

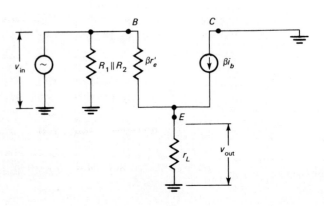

Fig. 7-2.
Ac equivalent circuit of CC amplifier.

Figure 7-3 shows how the voltage gain varies. Usually, r'_e is much smaller than r_L, equivalent to having a small ratio r'_e/r_L. In this case, the voltage gain is close to unity. This means the output signal has the same phase and approximately the same amplitude as the input signal. This is why a CC stage is called an *emitter follower*; the ac voltage on the emitter follows the ac voltage on the base. Typically, r'_e/r_L is less than 0.1 and A is greater than 0.9. For heavy loading, such as r'_e equal to r_L, the ratio r'_e/r_L is unity and A equals 0.5.

EXAMPLE 7-1   Draw the dc load line and the Q point for Fig. 7-4*a*.

SOLUTION   When the collector-emitter terminals are shorted, all of the supply voltage appears across the emitter resistor; so the collector saturation current is

$$I_{C(sat)} \cong \frac{V_{CC}}{R_E} = \frac{10 \text{ V}}{430 \text{ }\Omega} = 23.3 \text{ mA}$$

When the collector-emitter terminals are open, all of the supply voltage appears across the collector-emitter terminals and

$$V_{CE(cutoff)} = V_{CC} = 10 \text{ V}$$

Because of the voltage divider, the dc voltage at the base is half the supply voltage, or 5 V. After subtracting the V_{BE} drop, we get 4.3 V across the emitter resistor. This gives a quiescent current of

$$I_C = \frac{V_B - V_{BE}}{R_E} \cong \frac{4.3 \text{ V}}{430 \text{ }\Omega} = 10 \text{ mA}$$

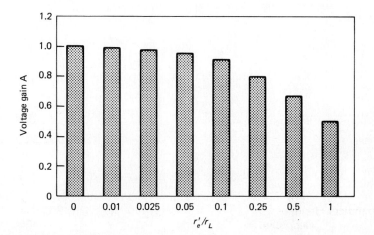

Fig. 7-3.
Voltage gain of emitter follower.

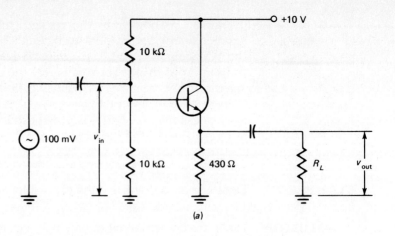

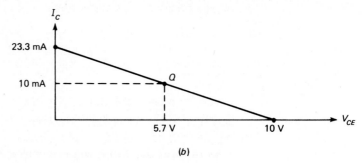

Fig. 7-4.

(b)

The quiescent voltage across the collector-emitter terminals is

$$V_{CE} = V_{CC} - V_E = 10 \text{ V} - 4.3 \text{ V} = 5.7 \text{ V}$$

Figure 7-4b summarizes these results.

EXAMPLE 7-2 What is the ac output voltage in 7-4a when R_L is 1 kΩ?

SOLUTION The total ac load resistance is

$$r_L = R_E \,\|\, R_L = 430 \text{ Ω} \,\|\, 1 \text{ kΩ} = 301 \text{ Ω}$$

Since $I_E = 10$ mA, r_e' is 2.5 Ω. The voltage gain is

$$A = \frac{r_L}{r_L + r_e'} = \frac{301 \text{ Ω}}{301 \text{ Ω} + 2.5 \text{ Ω}} = 0.992$$

So, the ac output voltage is

$$v_{out} = Av_{in} = 0.992(100 \text{ mV}) = 99.2 \text{ mV}$$

7-2

INPUT IMPEDANCE

An emitter follower can step up the impedance, similar to a transformer. As discussed in Chap. 6, any impedance in the emitter circuit is increased by a factor of β when seen from the base. The mathematical proof follows.

Figure 7-5 is the ac equivalent circuit of an emitter follower, shown earlier. The ac input voltage is

$$v_{in} = i_b \beta r'_e + i_e r_L$$

Since $i_e \cong \beta i_b$, the foregoing may be written as

$$v_{in} \cong i_b \beta r'_e + \beta i_b r_L$$

Dividing by i_b gives the input impedance looking into the base:

$$z_{in(base)} \cong \frac{v_{in}}{i_b} = \frac{i_b \beta r'_e + \beta i_b r_L}{i_b}$$

or
$$z_{in(base)} \cong \beta(r_L + r'_e) \qquad \textbf{(7-5)}$$

As you see, this impedance is similar to that of a swamped CE amplifier. The total ac emitter resistance is multiplied by β to get the impedance seen at the base terminal.

As before, the input impedance of the stage includes the biasing resistors:

$$z_{in} = R_1 \parallel R_2 \parallel z_{in(base)} \qquad \textbf{(7-6)}$$

Small Distortion

The emitter follower is inherently a low-distortion amplifier. Since the emitter resistor is unbypassed, the swamping is extremely heavy and the nonlinearity of the emitter diode is almost eliminated. Stated another way, the voltage gain is approximately equal to unity throughout the ac cycle; therefore, the ac output voltage is a replica of the ac

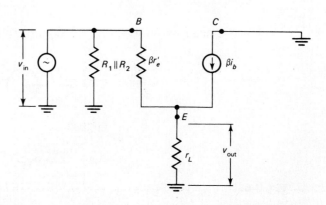

Fig. 7-5.
Ac equivalent circuit of
emitter follower.

input voltage. Given a perfect input sine wave, we get an almost perfect output sine wave.

EXAMPLE 7-3   In Fig. 7-6, what is the voltage gain of the emitter follower? The input impedance?

SOLUTION   In Examples 7-1 and 7-2, we analyzed this circuit and found that $r'_e = 2.5\ \Omega$. Since the ac load resistance is

$$r_L = R_E \parallel R_L = 430\ \Omega \parallel 100\ \Omega = 81.1\ \Omega$$

The voltage gain is

$$A = \frac{r_L}{r_L + r'_e} = \frac{81.1\ \Omega}{81.1\ \Omega + 2.5\ \Omega} = 0.97$$

The input impedance of the base is

$$z_{in(base)} = \beta(r_L + r'_e) = 200(81.1\ \Omega + 2.5\ \Omega) = 16.7\ k\Omega$$

The input impedance includes the biasing resistors:

$$z_{in} = R_1 \parallel R_2 \parallel z_{in(base)} = 10\ k\Omega \parallel 10\ k\Omega \parallel 16.7\ k\Omega = 3.85\ k\Omega$$

7-3

POWER GAIN

We do not get voltage gain with an emitter follower, but we do get *power gain*, symbolized G. The ac output power of an emitter follower is

$$p_{out} = i_e^2 r_L$$

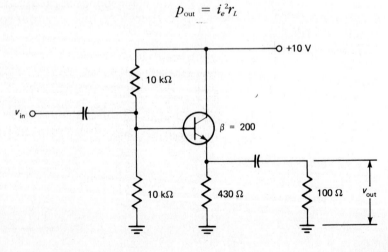

Fig. 7-6.

and the ac input power to the base is

$$p_{in} = i_b^2 z_{in(base)} = i_b^2 \beta(r_L + r_e')$$

The power gain G is the ratio of p_{out} to p_{in}:

$$G = \frac{p_{out}}{p_{in}} = \frac{i_e^2 r_L}{i_b^2 \beta(r_L + r_e')}$$

Since $i_c \cong i_e$, this reduces to

$$G \cong \beta \frac{r_L}{r_L + r_e'} \qquad\qquad \textit{(7-7)}$$

In this formula, the first factor β is the current gain, and the second factor is the voltage gain. For the typical case of r_L much greater than r_e', the voltage gain approaches unity and the power gain is approximately equal to β.

EXAMPLE 7-4   In Fig. 7-6, what is the power gain?

SOLUTION   In Example 7-3, we calculated a voltage gain of 0.97. Therefore, the power gain is

$$G \cong \beta \frac{r_L}{r_L + r_e'} = 200(0.97) = 194$$

7-4
SOURCE
IMPEDANCE

Every ac source has an internal impedance known as the *source impedance* (also called the output impedance or the Thevenin impedance). Figure 7-7 shows an ac source with a voltage v_s and a source impedance R_s driving a load resistance R_L. Because of the voltage drop across R_s, the load voltage v_L is less than the source voltage v_s. If R_s is much smaller than R_L, most of the source voltage appears across the load terminals. On the other hand, if R_s is much larger than R_L, most of the ac source voltage is dropped across the source impedance.

Mathematical Analysis   With the voltage-divider theorem, the ac load voltage is given by

$$v_L = \frac{R_L}{R_s + R_L} v_s \qquad\qquad \textit{(7-8)}$$

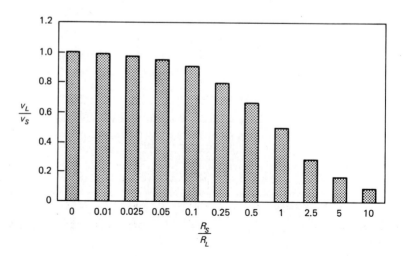

Fig. 7-7.
Voltage source with internal
impedance drives load
resistance.

We can rearrange this equation to get the following alternative formula:

$$\frac{v_L}{v_S} = \frac{1}{1 + R_S/R_L} \tag{7-9}$$

When a source is lightly loaded, R_S is much smaller than R_L and the ratio R_S/R_L approaches zero; in this case, the ratio v_L/v_S approaches unity, meaning that the load voltage almost equals the source voltage. On the other hand, when a source is heavily loaded, R_S is much larger than R_L and the ratio R_S/R_L is large; for this case, the ratio v_L/v_S is small, implying that v_L is small compared with v_S.

Figure 7-8 summarizes the idea of loading a source. As long as R_S/R_L is small, the loading is light and the load voltage approximately equals the source voltage. As the loading increases, R_S/R_L becomes larger and less voltage appears across the load resistor. For the special case of matched impedances ($R_S = R_L$), the load voltage equals half the source voltage. Beyond this point, the loading becomes extremely heavy and most of the source voltage is dropped across the source impedance, which means the load voltage is approaching zero.

With voltage amplifiers, you try to keep R_S/R_L less than 0.1. This ensures that more than 90 percent of the source voltage reaches the load resistance.

Fig. 7-8.
Load voltage decreases
when load resistance
decreases.

7-5

**OUTPUT
IMPEDANCE**

Every amplifier has an *output impedance*, equivalent to the source impedance discussed in the preceding section. This output impedance is important because some of the ac voltage is dropped across it. If too much signal is lost across the output impedance, the load voltage will be too small.

**Standard
Analysis**

The CE amplifier of Fig. 7-9*a* drives a load resistance R_L. Figure 7-9*b* shows the ac equivalent circuit. The standard method of analysis discussed in Chap. 6 is to calculate the voltage gain as follows:

$$A \cong \frac{r_L}{r'_e} \qquad \textit{Loaded}$$

where $r_L = R_C \parallel R_L$. Because of the loading effect of R_L in Fig. 7-9*b*, the output voltage decreases when R_L decreases.

**Thevenin
Analysis**

Thevenin's theorem can give us more insight into the loading effect of R_L. To apply Thevenin's theorem, begin by disconnecting the load resistor as shown in Fig. 7-10*a*. The Thevenin voltage v_{TH} is the unloaded or open-circuit voltage that appears across R_C. Now, the only ac load resistance is R_C, and the unloaded voltage gain becomes

$$A \cong \frac{R_C}{r'_e} \qquad \textit{unloaded}$$

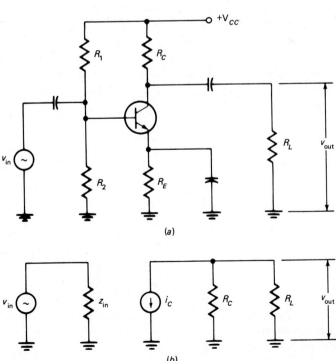

(a)

(b)

Fig. 7-9.
CE amplifier. (a) Circuit.
(b) Ac equivalent circuit.

Therefore, the Thevenin voltage is

$$v_{TH} \cong \frac{R_C}{r'_e} v_{in}$$

Next, find the Thevenin resistance as follows: In Fig. 7-10a, this is the resistance looking back into the output terminal when all sources are reduced to zero. Ideally, the collector-current source of Fig. 7-10a has an internal impedance that is high enough to ignore in typical circuits. Therefore, when this current source is reduced to zero, it appears as an open circuit and we get a Thevenin resistance of approximately

$$R_{TH} \cong R_C$$

Figure 7-10b shows the Thevenin equivalent circuit for the CE amplifier. On the output side, an ac voltage source is in series with a Thevenin resistance. With amplifiers, this Thevenin resistance is usually called the *output impedance*.

When we reconnect the load resistance as shown in Fig. 7-10c, the output impedance plays the same role as the source impedance dis-

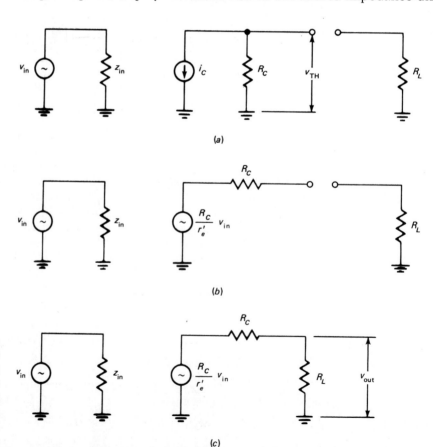

Fig. 7-10. Applying Thevenin's theorem. (a) Disconnect load resistor. (b) Replace current source by voltage source. (c) Output impedance is in series with load resistance.

190

CHAPTER 7

cussed in Sec. 7-4. If the output impedance is small compared with the load resistance, most of the ac voltage will appear across R_L. On other hand, if the output impedance is large compared with the load resistance, most of the ac voltage will be dropped across R_C.

EXAMPLE 7-5   Calculate the ac output voltage in Fig. 7-11a for $R_L = 1$ kΩ.

SOLUTION   If you analyze the voltage-divider bias, you will get the following dc quantities: $V_B = 1.8$ V, $V_E = 1.1$ V, and $I_E = 1$ mA. Therefore, r'_e is 25 Ω and the unloaded voltage gain is

$$A = \frac{R_C}{r'_e} = \frac{5 \text{ k}\Omega}{25 \text{ }\Omega} = 200$$

The Thevenin voltage is

$$v_{TH} = Av_{in} = 200(1 \text{ mV}) = 200 \text{ mV}$$

Figure 7-11b shows the Thevenin equivalent circuit. When $R_L = 1$ kΩ, the voltage-divider theorem gives

$$V_{out} = \frac{R_L}{R_C + R_L} v_{TH} = \frac{1 \text{ k}\Omega}{5 \text{ k}\Omega + 1 \text{ k}\Omega} 200 \text{ mV} = 33.3 \text{ mV}$$

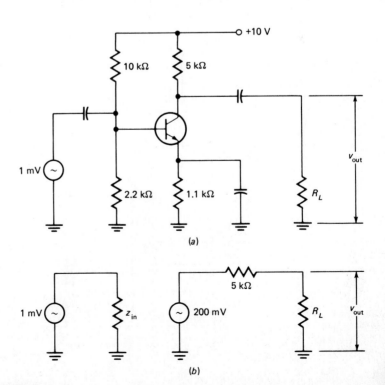

(a)

(b)

Fig. 7-11.

COMMON-COLLECTOR APPROXIMATIONS

EXAMPLE 7-6   Graph the ac output voltages in Fig. 7-11*b* for the following load resistances: 1 MΩ, 100 kΩ, 10 kΩ, 1 kΩ, and 100 Ω.

SOLUTION   For all values of R_L, the Thevenin voltage v_{TH} is the same as calculated in the preceding example, 200 mV. In Fig. 7-11*b*, the ac output voltage is given by

$$V_{out} = \frac{R_L}{5 \text{ k}\Omega + R_L} \, 200 \text{ mV}$$

By substituting the each value of R_L, we can calculate the corresponding output voltage v_{out}. Figure 7-12 shows each of the output voltages. As you see, a load resistance of 1 MΩ is very light loading because the output voltage is approximately equal to 200 mV, the Thevenin voltage. As the load resistance decreases, the output voltage decreases. For small load resistances like 1 kΩ and 100 Ω, the loading is heavy and the output voltage is much smaller than the Thevenin voltage.

7-6

USING AN EMITTER FOLLOWER

With an emitter follower, we can avoid the loss of ac voltage across the output impedance of a CE amplifier. When the output impedance is dropping too much ac voltage, we can insert an emitter follower between the CE amplifier and the load resistance as shown in Fig. 7-13. The ac load resistance r_L of the emitter follower is increased by a factor of β. As a result, the $z_{in(base)}$ of the emitter follower is high and there is less loading of the CE amplifier. Since the emitter follower usually has a voltage gain near unity, the final output voltage is higher than it would be without the emitter follower.

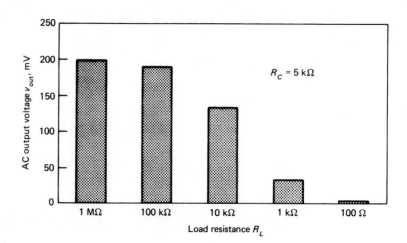

Fig. 7-12.
Load voltage decreases when load resistance decreases.

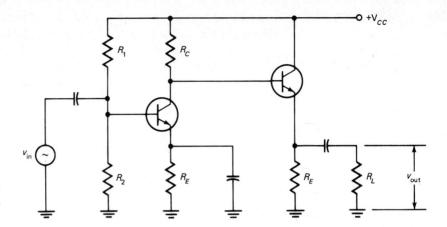

Fig. 7-13.
Emitter follower between
CE stage and load resistor.

Omit Biasing Resistors and Coupling Capacitor

Notice that we can leave out the biasing resistors of the emitter follower and couple directly from the collector of the CE amplifier into the base of the emitter follower. The dc collector voltage of the CE amplifier supplies the bias voltage for the emitter follower. This not only eliminates two biasing resistors and a coupling capacitor, it also increases the input impedance of the emitter follower, which now is

$$z_{in} = z_{in(base)}$$

Calculating the Output Voltage

As before, the Thevenin voltage of the CE amplifier is

$$v_{TH} = \frac{R_C}{r'_e} v_{in}$$

where v_{in} = ac input voltage to the CE amplifier
r'_e = ac emitter resistance in the CE amplifier

Next, you can calculate the input to the emitter follower using the voltage-divider theorem:

$$v_{in} = \frac{z_{in}}{R_C + z_{in}} v_{TH}$$

where v_{in} = ac input voltage to the emitter follower
$z_{in} = z_{in(base)}$ of the emitter follower

Then you can get the final output voltage by multiplying the voltage gain of the emitter follower and the input voltage:

$$v_{out} = \frac{r_L}{r_L + r'_e} v_{TH}$$

where $r_L = R_E \parallel R_L$ in the emitter follower
r'_e = ac emitter resistance in the emitter follower

EXAMPLE 7-7 For the emitter follower of Fig. 7-14a, both h_{FE} and h_{fe} equal 200. Calculate the ac output voltage in Fig. 7-14a for an R_L of 1 kΩ.

SOLUTION We analyzed the CE stage in Example 7-5 and found that

$$V_B = 1.8 \text{ V}$$
$$V_E = 1.1 \text{ V}$$
$$I_E = 1 \text{ mA}$$
$$r'_e = 25 \text{ }\Omega$$
$$V_{TH} = 200 \text{ mV}$$

The dc collector voltage of the CE stage is

$$V_C = V_{CC} - I_C R_C = 10 \text{ V} - (1 \text{ mA})(5 \text{ k}\Omega) = 5 \text{ V}$$

This is an approximation because we are assuming the dc base current of the emitter follower is much smaller than 1 mA, the dc collector current in the first stage.

Since the dc base voltage of the emitter follower is approximately 5 V, its dc emitter voltage is 4.3 V. Therefore, the dc emitter current is

$$I_E = \frac{V_E}{R_E} = \frac{4.3 \text{ V}}{430 \text{ }\Omega} = 10 \text{ mA}$$

Now, calculate the dc base current:

$$I_B = \frac{I_E}{\beta_{dc}} = \frac{10 \text{ mA}}{200} = 0.05 \text{ mA}$$

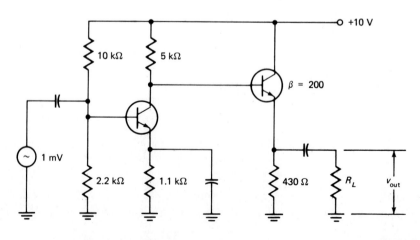

Fig. 7-14.

This proves the assumption about the base current in the emitter follower. As you see, it is much smaller than 1 mA, the dc collector current in the first stage. Because of this, 5 V is an accurate value for the dc base voltage in the emitter follower. In other words, there is almost no dc loading of the CE amplifier by the emitter follower.

With 10 mA of emitter current in the second stage, the ac resistance of the emitter diode is

$$r'_e = \frac{25 \text{ mV}}{10 \text{ mA}} = 2.5 \text{ } \Omega$$

When $R_L = 1$ kΩ, the ac load resistance of the emitter follower is

$$r_L = R_E \| R_L = 430 \text{ } \Omega \| 1 \text{ k}\Omega = 301 \text{ } \Omega$$

The input impedance of the emitter follower is

$$z_{in} = z_{in(base)} = \beta(r_L + r'_e) = 200(301 \text{ } \Omega + 2.5 \text{ } \Omega) = 60.7 \text{ k}\Omega$$

Now, calculate the ac input voltage to the emitter follower:

$$v_{in} = \frac{z_{in}}{R_C + z_{in}} v_{TH} = \frac{60.7 \text{ k}\Omega}{5 \text{ k}\Omega + 60.7 \text{ k}\Omega} 200 \text{ mV} = 185 \text{ mV}$$

Finally, calculate the ac output voltage:

$$v_{out} = \frac{r_L}{r_L + r'_e} = \frac{301 \text{ } \Omega}{301 \text{ } \Omega + 2.5 \text{ } \Omega} 185 \text{ mV} = 183 \text{ mV}$$

This final output voltage is almost as large as the Thevenin voltage of the CE amplifier, which means the emitter follower has reduced the loading effect. Without the emitter follower, the CE amplifier can produce only 33.3 mV across a load resistance of 1 kΩ (see Example 7-5).

EXAMPLE 7-8 Compare the ac output voltages of Fig. 7-14 and Fig. 7-11a for the following load resistances: 1 MΩ, 100 kΩ, 10 kΩ, 1 kΩ, and 100 Ω.

SOLUTION In Fig. 7-14, the ac input impedance of the emitter follower is

$$z_{in} = \beta(430 \text{ } \Omega \| R_L + 2.5 \text{ } \Omega)$$

The ac Thevenin voltage of the first stage is still 200 mV. Therefore, the ac input voltage to the emitter follower is

$$v_{in} = \frac{z_{in}}{5 \text{ k}\Omega + z_{in}} 200 \text{ mV}$$

The voltage gain of the emitter follower is

$$A = \frac{430\ \Omega\ \|\ R_L}{430\ \Omega\ \|\ R_L\ +\ 2.5\ \Omega}$$

and the final output voltage is

$$v_{out} = Av_{in}$$

With the foregoing equations, we can calculate and graph the corresponding ac load voltages. In Example 7-6 we analyzed the same CE amplifier without an emitter follower. Figure 7-15 compares this CE amplifier used with and without an emitter follower. The left bar in each group is with the emitter follower, and the right bar is without. Now you can see how effective an emitter follower is. With an emitter follower, the output voltage remains high even for a load resistance as small as 100 Ω. Without the emitter follower, the ac output voltage has almost reached zero when the load resistance is 100 Ω.

Incidentally, it is possible to derive the output impedance of an emitter follower by applying Thevenin's theorem. If this is done, the following formula results:

$$z_{out} \cong r_e' + \frac{R_C}{\beta}$$

Any impedance in the base circuit appears β times smaller when viewed from the emitter. Since r_e' is already small, the output imped-

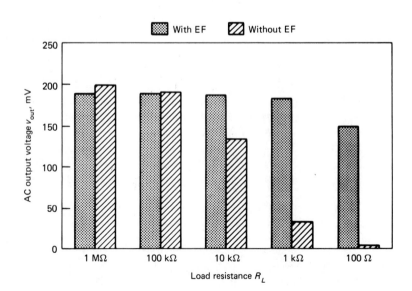

Fig. 7-15.
Comparison of output voltage with and without emitter follower.

ance of an emitter follower tends to be low compared with a CE amplifier.

In Fig. 7-14, the output impedance of the emitter follower is

$$z_{out} = 2.5 \ \Omega + \frac{5 \ k\Omega}{200} = 27.5 \ \Omega$$

Because its output impedance is small in Fig. 7-14, the emitter follower is able to drive smaller load resistances than the CE amplifier whose output impedance is 5 kΩ.

7-7 THE DARLINGTON AMPLIFIER

The *Darlington amplifier* is a popular transistor circuit. It consists of cascaded emitter followers, typically a pair like Fig. 7-16. The overall voltage gain is close to unity. The main advantage of this connection is a very large increase in input impedance.

DC Analysis To begin with, the dc voltage at the first base is

$$V_2 = \frac{R_2}{R_1 + R_2} V_{CC}$$

Because each transistor has a V_{BE} drop, the dc voltage across the emitter resistor is

$$V_E = V_2 - 2V_{BE}$$

and the dc emitter current is

$$I_E = \frac{V_2 - 2V_{BE}}{R_E} \qquad (7\text{-}10)$$

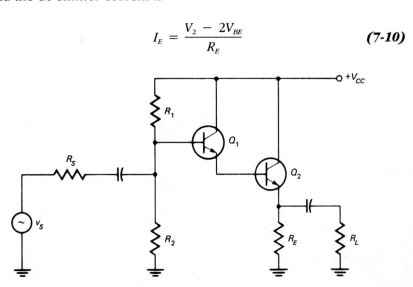

Fig. 7-16.
Darlington amplifier.

AC Analysis   The ac load seen by the second emitter is

$$r_L = R_E \parallel R_L$$

The input impedance of the second stage is

$$z_{in(2)} = \beta_2(r_L + r'_{e2}) \qquad (7\text{-}11)$$

where β_2 is the ac beta of the second transistor and r'_{e2} is the ac resistance of the second emitter diode. In a typical design, r_L is much greater than r'_{e2}, so the equation simplifies to

$$z_{in(2)} \cong \beta_2 r_L \qquad (7\text{-}12)$$

The quantity $z_{in(2)}$ is the ac load seen by the first emitter. Therefore, the input impedance looking into the first base is

$$z_{in(1)} = \beta_1(z_{in(2)} + r'_{e1}) \qquad (7\text{-}13)$$

In most circuits, $z_{in(2)}$ is much greater than r'_{e1} and the equation simplifies to

$$z_{in(1)} \cong \beta_1\beta_2 r_L \qquad (7\text{-}14)$$

The input impedance of the stage includes the biasing resistors in parallel:

$$z_{in} \cong R_1 \parallel R_2 \parallel \beta_1\beta_2 r_L \qquad (7\text{-}15)$$

When direct-coupling from a CE stage to a Darlington amplifier, the biasing resistors are omitted as discussed in Sec. 7-6. In this case, the input impedance becomes

$$z_{in} \cong \beta_1\beta_2 r_L \qquad (7\text{-}16)$$

Darlington Pair   The two transistors of Fig. 7-16 are called a *Darlington pair*. This connection has an overall beta that is equal to the product of the individual betas:

$$\beta = \beta_1\beta_2 \qquad (7\text{-}17)$$

As an example, if the first transistor has $\beta_1 = 100$ and the second transistor has $\beta_2 = 50$, the Darlington pair has an effective beta of

$$\beta = (100)(50) = 5000$$

Transistor manufacturers can put a Darlington pair inside a single transistor package. The three-terminal device then acts like a single transistor with an extremely high beta. For instance, a 2N2785 is an *npn* Darlington transistor with a minimum β of 2000 and a maximum β of 20,000.

EXAMPLE 7-9 Each transistor of Fig. 7-17 has $h_{FE} = 100$. Calculate the r'_{e1} and r'_{e2} of the transistors.

SOLUTION The dc voltage at the first base is

$$V_2 = \frac{100 \text{ k}\Omega}{100 \text{ k}\Omega} \, 10 \text{ V} = 5 \text{ V}$$

With Eq. (7-10), the dc emitter current in the second transistor is

$$I_{E2} = \frac{5 \text{ V} - 1.4 \text{ V}}{360 \text{ }\Omega} = 10 \text{ mA}$$

Therefore, the ac resistance of the second emitter diode is

$$r'_{e2} = \frac{25 \text{ mV}}{10 \text{ mA}} = 2.5 \text{ }\Omega$$

The dc emitter current in the first transistor is

$$I_{E1} = I_{B2} = \frac{10 \text{ mA}}{100} = 0.1 \text{ mA}$$

So, the ac resistance of the first emitter diode is

$$r'_{e1} = \frac{25 \text{ mV}}{0.1 \text{ mA}} = 250 \text{ }\Omega$$

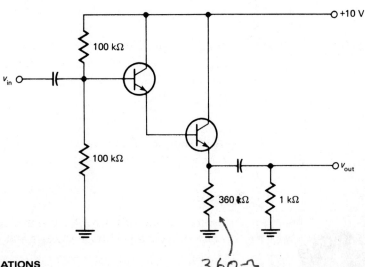

Fig. 7-17.

EXAMPLE 7-10 Each transistor of Fig. 7-17 has $h_{FE} = 100$ and $h_{fe} = 100$. What is the approximate input impedance of the first base? Of the Darlington amplifier?

SOLUTION The ac load resistance is

$$r_L = R_E \| R_L = 360 \ \Omega \| 1 \ k\Omega = 265 \ \Omega$$

If we ignore r'_{e1} and r'_{e2}, Eq. (7-14) gives

$$z_{in(1)} = 100(100)(265 \ \Omega) = 2.65 \ M\Omega$$

With Eq. (7-15),

$$z_{in} = 100 \ k\Omega \| 100 \ k\Omega \| 2.65 \ M\Omega = 49.1 \ k\Omega$$

If we include r'_{e1} and r'_{e2}, then Eq. (7-11) gives

$$z_{in(2)} = 100(265 \ \Omega + 2.5 \ \Omega) = 26,750 \ \Omega$$

With Eq. (7-13),

$$z_{in(1)} = 100(26,750 \ \Omega + 250 \ \Omega) = 2.7 \ M\Omega$$

With Eq. (7-15),

$$z_{in} = 100 \ k\Omega \| 100 \ k\Omega \| 2.7 \ M\Omega = 49.1 \ k\Omega$$

7-8

THE ZENER FOLLOWER

An emitter follower can improve the performance of a zener regulator. Figure 7-18 shows a *zener follower,* a circuit that combines a zener regulator and an emitter follower. Here is how it works. The zener voltage is the dc input voltage to the base. Therefore, the dc output voltage is

$$V_{out} = V_Z - V_{BE} \qquad\qquad (7\text{-}18)$$

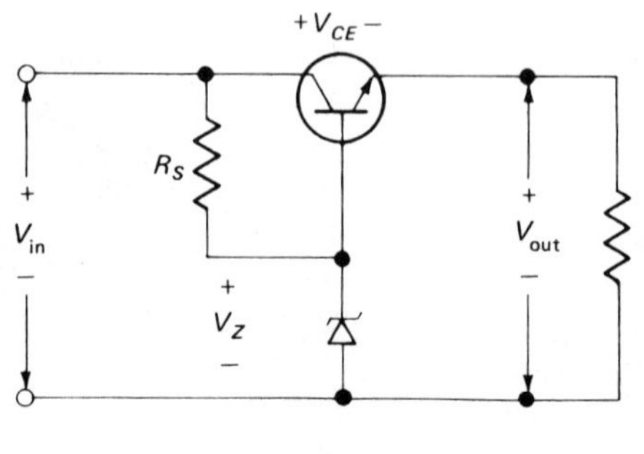

Fig. 7-18.
Zener follower.

This output voltage is fixed, equal to the zener voltage minus the V_{BE} drop of the transistor. If the supply voltage V_{in} changes, the zener voltage remains approximately constant and so too does the output voltage. Figure 7-19 emphasizes the relation of output voltage to zener voltage. This graph assumes a V_{BE} drop of 0.7 V. As indicated, the output voltage is always 0.7 V less than the zener voltage.

Advantage   In an ordinary zener regulator, the zener current is the difference of the current through the series resistor and load current:

$$I_Z = I_S - I_L$$

To handle large load currents, the designer has to use a large zener current.

The zener follower has a major advantage over an ordinary zener regulator. In Fig. 7-18, the zener current is the difference of the current through R_S and the base current:

$$I_Z = I_S - I_B$$

Since the base current equals

$$I_B = \frac{I_L}{\beta_{dc}}$$

less zener current is required. Therefore, the designer can select a smaller zener diode, one with a lower power rating. For instance, if you are trying to supply amperes to a load resistor, an ordinary zener regulator requires a zener diode capable of handling amperes. On the other hand, with the zener follower of Fig. 7-18, the zener diode needs to handle only tens of milliamperes because the current has been reduced by a factor of β_{dc}.

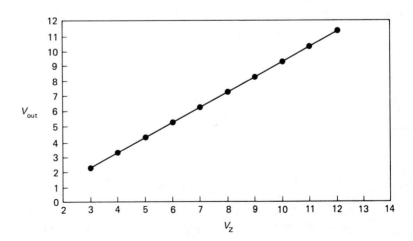

Fig. 7-19.
Output of zener follower.

Temperature Effects

The value of V_{BE} is important because the output voltage of a zener follower is one V_{BE} drop less than the zener voltage. When the emitter temperature increases, V_{BE} decreases. The change in V_{BE} depends on the emitter current, the particular transistor, and other factors. A useful approximation for the change is this: V_{BE} decreases 2 mV for each degree Celsius rise. For example, suppose $V_{BE} = 0.7$ V at 25°C. If the temperature rises to 75°C, V_{BE} decreases by

$$V_{BE} = (75 - 25)(2 \text{ mV}) = 100 \text{ mV}$$

Therefore, $V_{BE} = 0.6$ V at 75°C. As an aid, Figure 7-20 shows how V_{BE} varies with temperature.

Series Regulators

The zener follower is an example of a *series* voltage regulator. The collector-emitter terminals are in series with the load resistor (see Fig. 7-18). Because of this, the load current must pass through the transistor, and this is why the transistor is called a *pass transistor*. The voltage across the pass transistor is

$$V_{CE} = V_{in} - V_{out}$$

and its power dissipation is

$$P_D = (V_{in} - V_{out})I_L \qquad \textbf{(7-19)}$$

The main disadvantage of a series voltage regulator is the power dissipation of the pass transistor. As long as load current is not too high, the pass transistor does not get too hot. But when the load current is heavy, the pass transistor has to dissipate a lot of power, and this raises the internal temperature of the equipment.

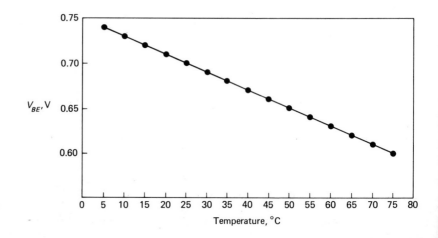

Fig. 7-20.
Effect of temperature on V_{BE}.

EXAMPLE 7-11   In Fig. 7-21, the transistor has an h_{FE} of 80. What is the current through the zener diode? The transistor power dissipation?

SOLUTION   The current through the series-limiting resistor is

$$I_S = \frac{V_{in} - V_Z}{R_S} = \frac{20\ V - 10\ V}{680\ \Omega} = 14.7\ mA$$

The load voltage is

$$V_{out} = V_Z - V_{BE} = 10\ V - 0.7\ V = 9.3\ V$$

The load current is

$$I_E = \frac{V_{out}}{R_L} = \frac{9.3\ V}{15\ \Omega} = 620\ mA$$

The base current is

$$I_B = \frac{I_E}{\beta_{dc}} = \frac{620\ mA}{80} = 7.75\ mA$$

The zener current is

$$I_Z = I_S - I_B = 14.7\ mA - 7.75\ mA = 6.95\ mA$$

Notice how much smaller this zener current is than the load current. This is the whole point of the circuit. A simple zener regulator capable of handling milliamperes is cascaded with an emitter follower to get a load current in hundreds of milliamperes.

The voltage across the collector-emitter terminals is

$$V_{CE} = V_{in} - V_{out} = 20\ V - 9.3\ V = 10.7\ V$$

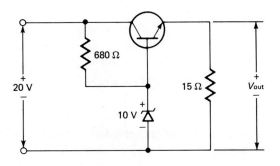

Fig. 7-21.

and the transistor power dissipation is

$$P_D = (V_{in} - V_{out})I_L = (20\ \text{V} - 9.3\ \text{V})(0.62\ \text{A}) = 6.63\ \text{W}$$

Summary

The CC connection, better known as the emitter follower, has the collector ac grounded. The input signal drives the base, and the output signal is taken from the emitter. The voltage gain of an emitter follower is equal to or less than unity, and the output signal has the same phase as the input signal. Because of the ac current gain, the input impedance looking into the base is very high. Another advantage is the low distortion of the ac signal.

The output impedance of a CE amplifier is approximately equal to the collector resistance R_C. When this output impedance is large compared with the load resistance R_L, most of the ac voltage is dropped across the output impedance. To avoid this, we can insert an emitter follower between the CE amplifier and the load resistance. The emitter follower steps up the impedance; so there is less loading of the CE amplifier.

A Darlington amplifier is a circuit with cascaded emitter followers. The overall voltage gain approaches unity and the input impedance of the first base approaches infinity. When calculating the dc emitter current in the output transistor, you have to subtract two V_{BE} drops. The overall ac beta of a Darlington pair equals the product of the individual ac betas of the transistors. For this reason, the effective ac beta is very high.

An emitter follower may be cascaded with a zener regulator to increase the load current of the circuit. Because of the current gain of the transistor, the zener current is much smaller in a zener follower than in an ordinary zener regulator. One temperature effect worth remembering is the change in V_{BE} because this changes the final output voltage. A guideline is to assume a decrease of 2 mV for each degree Celsius rise.

A zener follower is an example of a series regulator because all the load current must pass through the transistor. The disadvantage of series regulators is the high power dissipation in the pass transistor.

Glossary

common collector One of the three basic ways to connect a transistor. The input signal is applied to the base, and the output signal is taken from the emitter.

Darlington amplifier Two transistors connected as cascaded emitter followers. The effective current gain of the circuit equals the product of the individual current gains.

emitter follower Equivalent to a CC amplifier, this circuit usually has a voltage gain near unity and a very high input impedance. The output signal is in phase with the input signal.

output impedance The Thevenin resistance at the output of an amplifier.

power gain The ratio of ac output power to ac input power.

source impedance The same as the Thevenin impedance of the source. For audio systems (20 to 20,000 Hz), source impedances are often 600 Ω. For microwave and other high-frequency systems, source impedances are typically 50 Ω.

zener follower A zener regulator driving an emitter follower. This circuit is capable of regulating the output voltage over a large current range.

Review Questions

1. How does a common-collector stage differ from a common-emitter stage?
2. What can you say about the voltage gain of an emitter follower? About the phase of the input and output signals?
3. In an emitter follower, why is the input impedance of the base much higher than the resistance in the emitter circuit?
4. How much distortion does an emitter follower produce compared with a CE amplifier? (Include r'_e in your explanation.)
5. Why is the emitter follower a useful circuit?
6. What is a Darlington amplifier? What is its main advantage?
7. What is a zener follower? What advantage does it have over an ordinary zener regulator?

Problems

7-1. The ac load resistance of an emitter follower is $r_L = 1$ kΩ. If $r'_e = 35$ Ω, what is the voltage gain?

7-2. In an emitter follower, $I_E = 0.5$ mA, $R_E = 2.2$ kΩ, and $R_L = 8.2$ kΩ. What does the voltage gain equal?

7-3. What is the voltage gain in Fig. 7-22a? The input impedance of the base? The input impedance of the stage?

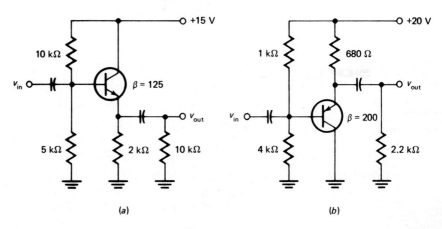

Fig. 7-22. (a) (b)

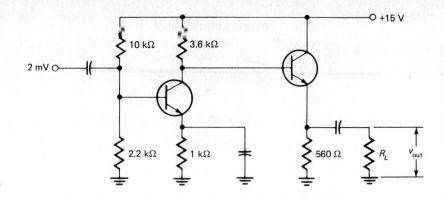

Fig. 7-23.

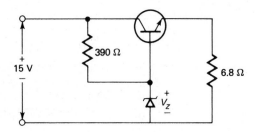

Fig. 7-24.

7-4. Figure 7-22b shows a *pnp* emitter follower. What are the voltage gain and input impedance?

7-5. What is the power gain in Fig. 7-22a? Fig. 7-22b?

7-6. In Fig. 7-23, all transistors have a β of 150. What does v_{out} equal when R_L is 1 kΩ?

7-7. If all transistors of Fig. 7-23 have a β of 300, what is the output voltage when R_L is 470 Ω?

7-8. Suppose the second transistor of Fig. 7-23 is replaced by a Darlington pair where each h_{fe} is 200. Ignore the r'_e of the Darlington transistors and calculate v_{out} for an R_L of 100 Ω.

7-9. In Fig. 7-24, $V_Z = 7.5$ V, $V_{BE} = 0.7$ V, and $\beta_{dc} = 100$. What is the load voltage? The base current? The zener current?

7-10. If $h_{FE} = 200$ and $V_Z = 6.2$ V in Fig. 7-24, what does the zener current equal?

7-11. R_L is 1 kΩ in Fig. 7-23. You measure no output voltage. Name three possible troubles.

8 Common-Base Approximations

Gauss: Known as the prince of mathematics, his great genius showed early in life. In Gauss' first arithmetic class, a somewhat sadistic teacher challenged the students to find the sum of all numbers from 1 to 100. The teacher had barely finished stating the problem when Gauss raised his hand and said "5050." The dumbfounded teacher asked Gauss how he had done it. He answered as follows: The numbers (1, 2, 3, . . . , 100) can be paired off as 1 and 100, 2 and 99, 3 and 98, and so on. Each pair of numbers has a sum of 101. Since there are 50 pairs, the total is 5050.

The common-base (CB) connection is not as popular as the CE or CC connections but nevertheless does find applications at high frequencies and in a special circuit called a *differential amplifier.* Because the differential amplifier is the backbone of a widely used integrated circuit known as an *operational amplifier* (op amp), you will need to know how to analyze a CB amplifier.

8-1
COMMON-BASE COLLECTOR CURVES

Figure 8-1*a* shows adjustable dc sources driving an *npn* transistor. The voltage across the emitter diode is V_{BE}, and the voltage across the collector diode is V_{CB}. If we adjust the emitter supply to get 1 mA of emitter current, we will find the collector current is approximately 1 mA. Even though we vary the collector supply, the collector current

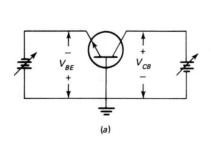

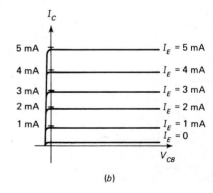

Fig. 8-1.
CB connection. (a) Circuit.
(b) Collector curves.

remains at approximately 1 mA. This ties in with earlier discussions of transistor action. Since α_{dc} is close to unity, the dc collector current is almost equal to the dc emitter current. The collector captures almost all of the free electrons diffusing into the base whether V_{CB} is small or large.

Figure 8-1b summarizes the idea. When the emitter current is 1 mA, the collector current is approximately 1 mA for all collector voltages greater than zero. If we adjust the emitter supply to get 2 mA of emitter current, the collector current automatically increases to approximately 2 mA. Once the emitter current is set by the emitter supply voltage, changing the collector supply voltage has almost no effect.

The curves of Fig. 8-1b imply that the collector of a CB transistor acts like a controlled current source, similar to the CE connection. Since the collector curves are almost horizontal, the current source is almost perfect. Stated another way, the internal resistance of the current source approaches infinity (typical value is over 1 MΩ).

In a CB connection, the current source is controlled by the emitter current rather than the base current. When we change the emitter current, the collector current changes to approximately the same value. Again, collector voltage has little effect on the collector current. No matter what the value of V_{CB}, the collector current remains approximately constant for a fixed emitter current.

Incidentally, notice that the collector curves of a CB transistor extend slightly to the left of the origin. In other words, V_{CB} can be a few tenths of a volt negative before the collector current decreases to zero. The reason is that the emitter voltage is approximately -0.7 V to ground. As long as V_{CB} is more positive than the emitter, it can attract free electrons from the emitter.

8-2
THE IDEAL CB TRANSISTOR

In most transistors, α_{dc} is greater than 0.98. Almost always, you can approximate it as unity. This is why the CB model of an *npn* transistor consists of a diode and a current source as shown in Fig. 8-2a. The emitter diode acts like any forward-biased diode. Because of transistor action, the collector diode acts like a current source. The current source pumps free electrons out of the collector (conventional flow is in the direction of the arrow).

Figure 8-2b is the equivalent circuit of a *pnp* transistor in the CB connection. Again, the emitter diode acts like an ordinary diode, while the collector diode is a controlled current source.

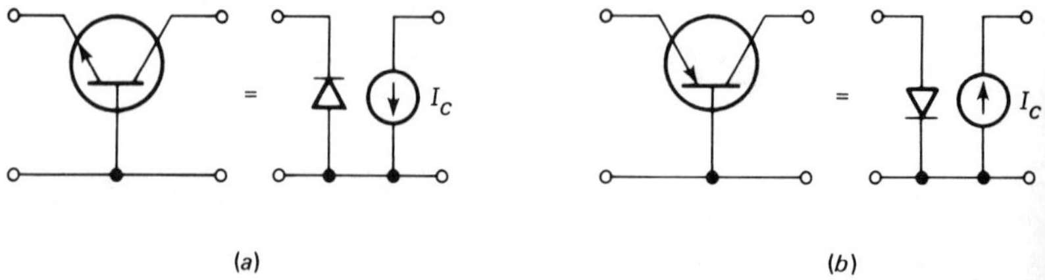

Fig. 8-2.
Ideal CB equivalent circuits.
(a) *npn*. (b) *pnp*.

(a)

(b)

EXAMPLE 8-1   In Fig. 8-3a, what does V_{CB} equal?

SOLUTION   Allow 0.7 V for the V_{BE} drop. Then, the emitter current is

$$I_E = \frac{V_{EE} - V_{BE}}{R_E} = \frac{10\ V - 0.7\ V}{20\ k\Omega} = 0.465\ mA$$

Since the collector current approximately equals 0.465 mA, we can calculate the collector-base voltage as follows:

$$V_{CB} = V_{CC} - I_C R_C = 25\ V - (0.465\ mA)(10\ k\Omega) = 20.35\ V$$

EXAMPLE 8-2   Figure 8-3b is a simplified way to draw a CB circuit. What does V_{CB} equal?

SOLUTION   The emitter current is

$$I_E = \frac{V_{EE} - V_{BE}}{R_E} = \frac{12\ V - 0.7\ V}{5.6\ k\Omega} = 2.02\ mA$$

and the collector voltage is

$$V_{CB} = V_{CC} - I_C R_C = 15\ V - (2.02\ mA)(6.8\ k\Omega) = 1.26\ V$$

8-3

THE CB AMPLIFIER

Figure 8-4a is a *common-base* (CB) amplifier. Since the base is grounded, this circuit is also called a grounded-base amplifier. The V_{EE} source forward-biases the emitter diode, and the V_{CC} source reverse-biases the collector diode. The ac source produces small fluctuations in transistor currents and voltages. As will be shown, the ac output voltage is an amplified version of the ac input voltage.

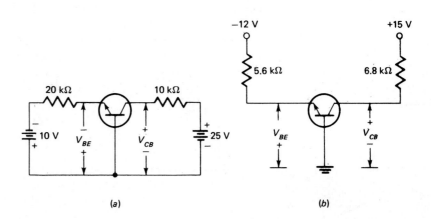

Fig. 8-3.   (a)   (b)

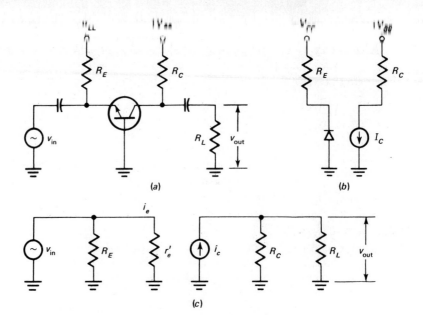

Fig. 8-4.
CB amplifier. (a) Circuit.
(b) Dc equivalent circuit.
(c) Ac equivalent circuit.

DC Equivalent Circuit

You get the dc equivalent circuit by reducing all ac sources to zero and opening all capacitors. In the dc equivalent of Fig. 8-4b, we can calculate the dc emitter current by using

$$I_E = \frac{V_{EE} - V_{BE}}{R_E} \qquad (8\text{-}1)$$

The dc collector voltage to ground is

$$V_C = V_{CC} - I_C R_C \qquad (8\text{-}2)$$

AC Equivalent Circuit

To get the ac equivalent circuit, reduce all dc sources to zero, short all coupling capacitors, and replace the emitter diode by its ac resistance. In the ac equivalent circuit of Fig. 8-4c, the ac input voltage is

$$v_{\text{in}} = i_e r'_e$$

The ac load resistance is

$$r_L = R_C \parallel R_L$$

Therefore, the ac collector voltage is

$$v_{\text{out}} = i_c r_L$$

and the voltage gain is

$$A = \frac{v_{\text{out}}}{v_{\text{in}}} = \frac{i_c r_L}{i_e r'_e}$$

Since $i_c \cong i_e$, the equation simplifies to

$$A \cong \frac{r_L}{r_e'} \qquad (8\text{-}3)$$

In Fig. 8-4a, the positive half cycle of input voltage opposes the $-V_{EE}$ supply, causing the emitter current to decrease. This produces less collector current, which means the collector voltage is increasing. In other words, the positive half cycle of input voltage produces the positive half cycle of output voltage. Therefore, the output signal of a CB amplifier is in phase with the input signal.

Input Impedance The main disadvantage of a CB amplifier is its very low input impedance. In Fig. 8-4c, the input impedance of the stage is

$$z_{in} = R_E \parallel r_e'$$

Since R_E is much larger than r_e' in any practical circuit,

$$z_{in} \cong r_e' \qquad (8\text{-}4)$$

This means the ac source driving a CB amplifier sees an input impedance of only r_e', which can be quite small. For instance, if $I_E = 1$ mA, then $r_e' = 25\ \Omega$.

Output Impedance In Fig. 8-4c, the output impedance of a CB amplifier is the Thevenin resistance facing the load resistor. With a CB connection, the collector curves are almost perfectly horizontal, implying that the collector current source has a very high internal impedance (typically in megohms). For this reason, the output impedance is approximately equal to the value of the collector resistor:

$$z_{out} \cong R_C \qquad (8\text{-}5)$$

EXAMPLE 8-3 Calculate the output voltage in Fig. 8-5.

SOLUTION First, get the dc emitter current. Allowing 0.7 V for the voltage drop across the emitter diode,

$$I_E = \frac{V_{EE} - V_{BE}}{R_E} = \frac{15\ V - 0.7\ V}{22\ k\Omega} = 0.65\ mA$$

Next, get the ac resistance of the emitter diode:

$$r_e' = \frac{25\ mV}{I_E} = \frac{25\ mV}{0.65\ mA} = 38.5\ \Omega$$

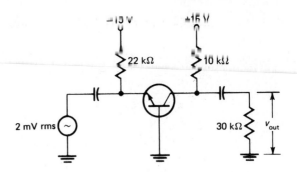

Fig. 8-5.

The voltage gain is

$$A = \frac{r_L}{r_e'} = \frac{10 \text{ k}\Omega \parallel 30 \text{ k}\Omega}{38.5 \text{ }\Omega} = 195$$

$$v_{\text{out}} = Av_{\text{in}} = 195(2 \text{ mV}) = 390 \text{ mV}$$

EXAMPLE 8-4 In Fig. 8-6, an ac source voltage v_S and a source resistance R_S drive a CB amplifier. What does v_{out} equal if v_S is 1 mV and R_S is 100 Ω?

SOLUTION The dc emitter current is

$$I_E = \frac{V_{EE} - V_{BE}}{R_E} \cong \frac{10 \text{ V} - 0.7 \text{ V}}{6.8 \text{ k}\Omega} = 1.37 \text{ mA}$$

and the ac resistance of the emitter diode is

$$r_e' \cong \frac{25 \text{ mV}}{I_E} = \frac{25 \text{ mV}}{1.37 \text{ mA}} = 18.2 \text{ }\Omega$$

The input impedance of the stage is

$$Z_{\text{in}} \cong r_e' = 18.2 \text{ }\Omega$$

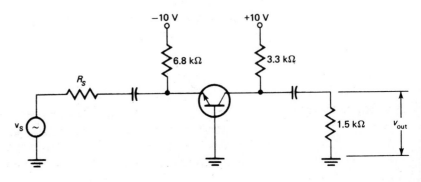

Fig. 8-6.

Wit: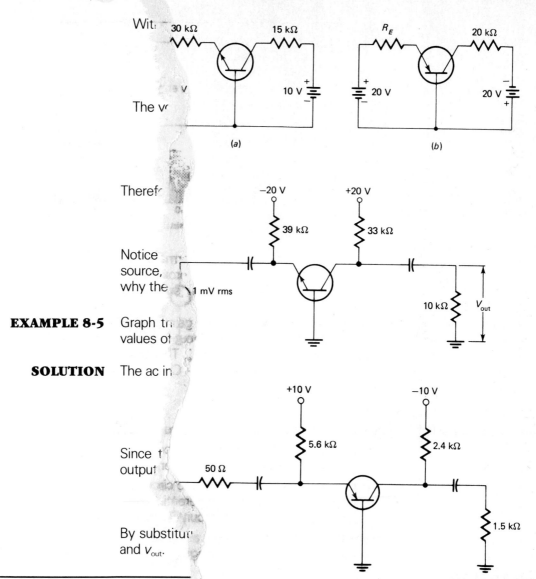

The v

The v

Theref

Notice
source,
why the

EXAMPLE 8-5 Graph tr
values of

SOLUTION The ac in

Since t
output

By substitu
and V_{out}.

Figure 8-7
maximum w
decreases w
than half of
you an ide
that is dr   ie value of R_E that produces a V_{CB} of 10 V in Fig. 8-8b.
         .t is the dc emitter current in Fig. 8-9? The ac resistance of the
         itter diode?
         n Fig. 8-9, what does the voltage gain equal? The ac output voltage?

Summary
Th   Figure 8-10 shows a CB amplifier using a *pnp* transistor. What does
C   the dc emitter current equal? The dc collector voltage to ground?
     . Calculate the ac output voltage in Fig. 8-10.

8-8. Suppose the 50 Ω of Fig. 8-10 opens. Do each of the following increase, decrease, or remain the same:

    a. Dc collector voltage

    b. Ac output voltage

    c. Dc emitter voltage

    d. Ac input voltage

8-9. There is no ac output voltage in Fig. 8-10. Name as many troubles as you can think of.

Class A Power Amplifiers

Euler: The most prolific mathematician of all time. His known writings would fill 60 to 80 books. Once in the court of Catherine the Great, the atheistic French philosopher Diderot was rambling on and boring the court. Suddenly, Euler stepped forward and said, "Sir, $x = (a + b^n)/n$; therefore, God exists. Your reply?" Diderot knew nothing about mathematics; Euler, everything. Because it sounded like sense, Diderot remained silent. The court then realized he was ignorant of even simple mathematics and laughed him all the way back to France.

The earlier stages of most systems amplify small ac voltages, while the later stages amplify large ac voltages. In these later stages, the collector currents are much larger because load resistances are smaller. In a typical AM radio, for instance, the final load resistance is 3.2 Ω, the impedance of the loudspeaker. Therefore, the final stage of most systems is a *power amplifier,* a circuit that produces large ac currents to drive small load resistances.

9-1
THE AC LOAD LINE

Every amplifier sees two loads: a dc load and an ac load. This means there are two load lines: a *dc load line* and an *ac load line.* You can derive the dc load line by analyzing the dc equivalent circuit. This is what we did in preceding chapters. To find the ac load line, you have to analyze the ac equivalent circuit.

The DC Load Line

Figure 9-1*a* shows a CE amplifier. To get the dc load line, visualize the dc equivalent circuit of Fig. 9-1*b*. When the collector-emitter terminals are shorted, the transistor is saturated and the dc collector current is maximum, given by

$$I_{C(\text{sat})} \cong \frac{V_{CC}}{R_C + R_E}$$

(9-1)

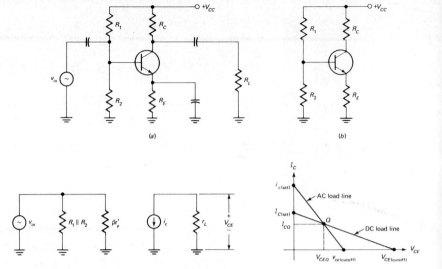

Fig. 9-1.
CE amplifier. (a) Circuit.
(b) Dc equivalent circuit.
(c) Ac equivalent circuit.
(d) Dc and ac load lines.

When the collector-emitter terminals are open, the transistor is cut off and the dc collector-emitter voltage is maximum, equal to

$$V_{CE(\text{cutoff})} \cong V_{CC} \tag{9-2}$$

The vertical and horizontal intercepts of the dc load line are shown in Fig. 9-1d.

Next, we can find the dc current and voltage at the Q point. From now on, we will use I_{CQ} and V_{CEQ} to represent the values of I_C and V_{CE} at the Q point, as indicated in Fig. 9-1d. In Fig. 9-1b, the dc collector current is approximately

$$I_{CQ} \cong \frac{V_B - V_{BE}}{R_E} \tag{9-3}$$

where

$$V_B \cong \frac{R_2}{R_1 + R_2} V_{CC}$$

The dc collector-emitter voltage is

$$V_{CEQ} \cong V_{CC} - I_C(R_C + R_E) \tag{9-4}$$

The AC Load Line   Figure 9-1c shows the ac equivalent circuit. The emitter is at ac ground because of the bypass capacitor, and the collector drives a load resistance of

$$r_L = R_C \parallel R_L \tag{9-5}$$

When the ac input voltage of Fig. 9-1c is reduced to zero, the transistor operates at the Q point shown in Fig. 9-1d. When the ac input signal is increased from zero, changes occur in the total collector current I_C and total collector voltage V_{CE}. Because the ac load resistance is different from the dc load resistance, the operation is along the ac load line rather than the dc load line. Notice that the ac load line has a saturation point labeled $i_{c(sat)}$ and a cutoff point labeled $v_{ce(cutoff)}$.

How can we find the value of $i_{c(sat)}$ and $v_{ce(cutoff)}$? In Fig. 9-1c, the ac collector voltage is given by

$$v_{ce} = -i_c r_L$$

where the minus sign indicates phase inversion. Since ac voltage and ac current are equivalent to the changes in total voltage and current, the foregoing equation may be written as

$$\Delta V_{CE} = -\Delta I_C r_L \qquad (9\text{-}6)$$

In Fig. 9-1d, the change in total current between the Q point and the saturation point of the ac load line is

$$\Delta I_C = i_{c(sat)} - I_{CQ}$$

This represents the increase in total current when moving from the Q point to the saturation point on the ac load line.

Similarly, the change in total voltage between the Q point and the saturation point is

$$\Delta V_{CE} = 0 - V_{CEQ} = -V_{CEQ}$$

This is negative because the total voltage decreases when moving from the Q point to the saturation point on the ac load line.

We can substitute the foregoing changes into Eq. (9-6) to get

$$-V_{CEQ} = -[i_{c(sat)} - I_{CQ}]r_L$$

Solving for $i_{c(sat)}$ gives

$$i_{c(sat)} = I_{CQ} + \frac{V_{CEQ}}{r_L} \qquad (9\text{-}7)$$

This is the current at the upper end of the ac load line.

By a similar derivation, the voltage at the lower end of the ac load line is

$$v_{ce(cutoff)} = V_{CEQ} + I_{CQ}r_L \qquad (9\text{-}8)$$

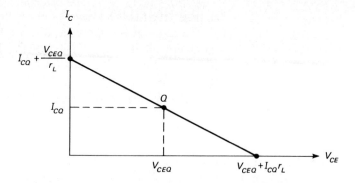

Fig. 9-2.
Ac load line for any
amplifier.

Figure 9-2 summarizes the ac load line. Although we analyzed a CE amplifier, this ac load line applies to any amplifier because the derivations for the end points are similar to those given for Eqs. (9-7) and (9-8). Throughout this book, you may refer to Fig. 9-2 for the ac load line of any amplifier.

EXAMPLE 9-1 Draw the dc and ac load lines of Fig. 9-3a.

SOLUTION The dc base voltage is approximately

$$V_B = \frac{2.2 \text{ k}\Omega}{10 \text{ k}\Omega + 2.2 \text{ k}\Omega} 10 \text{ V} = 1.8 \text{ V}$$

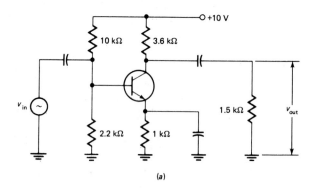

(a)

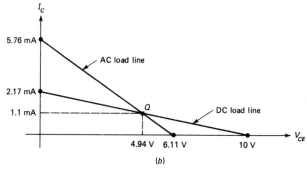

Fig. 9-3.
Drawing the load lines for a
CE amplifier.

(b)

The dc emitter current is

$$I_E = \frac{1.8\ V - 0.7\ V}{1\ k\Omega} = 1.1\ mA$$

Since the dc collector current approximately equals the dc emitter current,

$$I_{CQ} = 1.1\ mA$$

This quiescent current produces a dc collector-emitter voltage of

$$V_{CE} = 10\ V - (1.1\ mA)(3.6\ k\Omega + 1\ k\Omega) = 4.94\ V$$

or

$$V_{CEQ} = 4.94\ V$$

The dc saturation current is

$$I_{C(sat)} = \frac{10\ V}{4.6\ k\Omega} = 2.17\ mA$$

and the dc cutoff voltage is

$$V_{CE(cutoff)} = 10\ V$$

Figure 9-3b shows the dc load line with its intercepts and Q point.
   Next, visualize the ac equivalent circuit. You can see that the ac load resistance is

$$r_L = 3.6\ k\Omega \parallel 1.5\ k\Omega = 1.06\ k\Omega$$

Now you can calculate the ends of the ac load line as follows:

$$i_{c(sat)} = I_{CQ} + \frac{V_{CEQ}}{r_L} = 1.1\ mA + \frac{4.94\ V}{1.06\ k\Omega} = 5.76\ mA$$

and $v_{ce(cutoff)} = V_{CEQ} + I_{CQ}r_L$

$$= 4.94\ V + (1.1\ mA)(1.06\ k\Omega) = 6.11\ V$$

Figure 9-3b shows the ac load line.

EXAMPLE 9-2 Figure 9-4a shows the outline of a D42C. The metal tab can be fastened to the chassis to increase the maximum power rating. The data sheet of a D42C gives the following maximum ratings: $I_C = 5\ A$ and $V_{CEO} = 30\ V$. Show that none of these ratings is exceeded in the circuit of Fig. 9-4b.

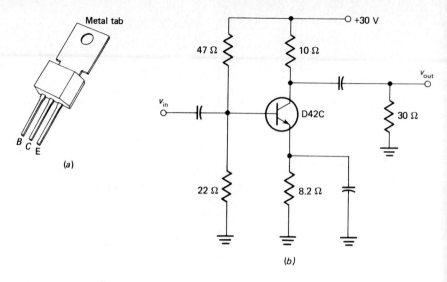

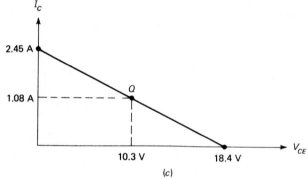

Fig. 9-4.
(a) Power-tab transistor.
(b) Circuit. (c) Ac load
line.

SOLUTION Visualize the dc equivalent circuit. The dc voltage from base to ground is approximately

$$V_B = \frac{22\ \Omega}{47\ \Omega + 22\ \Omega}\ 30\ V = 9.57\ V$$

The dc emitter current is

$$I_E = \frac{9.57\ V - 0.7\ V}{8.2\ \Omega} = 1.08\ A$$

or $$I_{CQ} = 1.08\ A$$

and $$V_{CEQ} = 30\ V - (1.08\ A)(10\ \Omega + 8.2\ \Omega) = 10.3\ V$$

The ac load resistance is

$$r_L = 10\ \Omega\ \|\ 30\ \Omega = 7.5\ \Omega$$

At the upper end of the ac load line,

$$i_{c(sat)} = 1.08 \text{ A} + \frac{10.3 \text{ V}}{7.5 \text{ }\Omega} = 2.45 \text{ A}$$

At the lower end of the ac load line,

$$V_{ce(cutoff)} = 10.3 \text{ V} + (1.08 \text{ A})(7.5 \text{ }\Omega) = 18.4 \text{ V}$$

Figure 9-4c shows the ac load line. Since the D42C has a maximum I_c rating of 5 A, there is no danger of exceeding this rating, because the maximum collector current is 2.45 A (the upper end of the ac load line). On the other hand, the V_{CEO} rating is 30 V, more than enough because the maximum collector-emitter voltage is 18.4 V (the lower end of the ac load line).

EXAMPLE 9-3 Figure 9-5 shows an emitter follower. The transistor has an I_c rating of 1 A and a V_{CEO} of 20 V. Show that neither of these ratings is exceeded during the ac cycle.

SOLUTION The analysis of an emitter follower is the same as before. First, get I_{CQ} and V_{CEQ}. The dc base voltage is approximately 10 V and the dc emitter voltage is approximately 9.3 V. So, the dc collector current is approximately

$$I_{CQ} = \frac{9.3 \text{ V}}{50 \text{ }\Omega} = 0.186 \text{ A}$$

and the dc collector-emitter voltage is

$$V_{CEQ} = 20 \text{ V} - (0.186 \text{ A})(50 \text{ }\Omega) = 10.7 \text{ V}$$

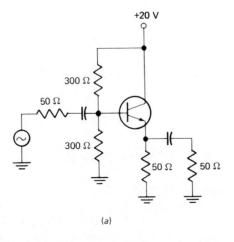

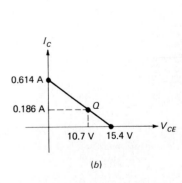

Fig. 9-5.
Drawing the ac load line of
an emitter follower.

(a)

(b)

Figure 9-5b shows the Q point.
In the ac equivalent circuit,

$$r_L = 50 \ \Omega \ \| \ 50 \ \Omega = 25 \ \Omega$$

The ac saturation current is

$$i_{c(sat)} = I_{CQ} + \frac{V_{CEQ}}{r_L} = 0.186 \ A + \frac{10.7 \ V}{25 \ \Omega} = 0.614 \ A$$

The ac cutoff voltage is

$$V_{ce(cutoff)} = V_{CEQ} + I_{CQ}r_L = 10.7 \ V + (0.186 \ A)(25 \ \Omega) = 15.4 \ V$$

Figure 9-5b shows the ac load line with the saturation and cutoff values. The maximum collector current is 0.614 A, which is less than the I_C rating of 1 A. Similarly, the maximum collector-emitter voltage is 15.4 V, which is less than the V_{CEO} rating of 20 V.

9-2
COMPLIANCE

In a linear amplifier, a transistor acts like a current source. As long as the ac signal is small, the transistor continues to act like a current source throughout the ac cycle. If the signal is large, however, the transistor may be driven into saturation or cutoff where it no longer acts like a current source.

DC Compliance

The compliance of a current source is the voltage range over which it can operate. For instance, if a current source can operate between a minimum voltage of 5 V and a maximum voltage of 25 V, then it has a compliance of 20 V.

The *dc compliance* of a transistor amplifier is the dc collector voltage range over which it can operate. For example, Fig. 9-6 shows the dc load line of a voltage-divider biased circuit. Since the Q point can be located anywhere along the dc load line, the dc compliance equals V_{CC}. If V_{CC} is 15 V, the transistor amplifier has a dc compliance of 15 V.

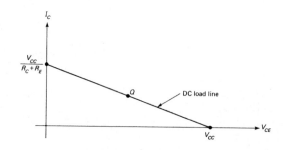

Fig. 9-6.
Dc compliance equals all of
the dc load line.

Maximum Unclipped Signal

When driven by an ac voltage, an amplifier operates along its ac load line. If the signal is too large, we will get clipping at either end of the ac load line. If the Q point is below the center of the ac load line, we get *cutoff clipping* as shown in Fig. 9-7a. If the Q point is above the center of the ac load line, we get *saturation clipping*, shown in Fig. 9-7b. If we locate the Q point at the center of the ac load line, we get the largest possible unclipped signal (Fig. 9-7c).

AC Compliance

The *ac compliance PP* is the maximum peak-to-peak ac output voltage of an amplifier (without clipping). For instance, Fig. 9-8a shows the ac load line with the Q point closer to cutoff than to saturation. Since cutoff clipping occurs first, the ac compliance is

$$PP = 2I_{CQ}r_L \qquad\qquad (9\text{-}9)$$

If $I_{CQ} = 1$ mA and $r_L = 3$ kΩ, the amplifier has an ac compliance of

$$PP = 2(1\text{ mA})(3\text{ kΩ}) = 6\text{ V}$$

This value of PP represents the maximum unclipped peak-to-peak output voltage of any amplifier with an I_{CQ} of 1 mA and an r_L of 3 kΩ.

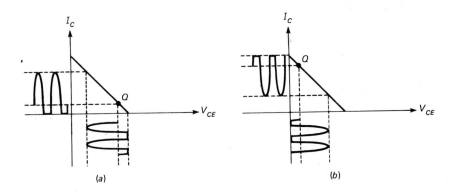

Fig. 9-7.
(a) Cutoff clipping.
(b) Saturation clipping.
(c) Maximum signal swing.

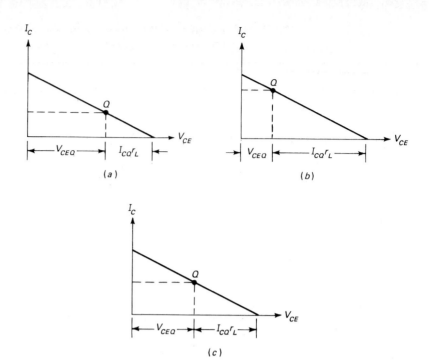

Fig. 9-8.
Location of Q point on ac
load line. (a) Below center.
(b) Above center. (c) At
center.

In Fig. 9-8*b,* the Q point is closer to saturation. This means the largest unclipped output signal has a peak-to-peak voltage of

$$PP = 2V_{CEQ} \tag{9-10}$$

For example, if $V_{CEQ} = 5$ V, the amplifier has an ac compliance of 10 V.

When the Q point is centered on the ac load line as shown in Fig. 9-8*c,* the amplifier has maximum ac compliance, meaning that it can produce the largest possible unclipped signal. Furthermore,

$$V_{CEQ} = I_{CQ}r_L \tag{9-11}$$

Therefore, you can use either $2V_{CEQ}$ or $2I_{CQ}r_L$ to calculate the ac compliance.

In conclusion, when analyzing transistor amplifiers, you often want to know what the ac compliance is. For any amplifier, it equals $2V_{CEQ}$ or $2I_{CQ}r_L$, whichever is smaller.

**Definition of
Class A**

When a transistor has a Q point near the middle of the dc load line, a small ac signal implies that the transistor operates in the active region throughout the ac cycle. As the signal increases, the transistor continues to operate in the active region as long as the peak excursions do not hit the saturation or cutoff points on the ac load line. To distinguish this type of operation from other kinds, we call it *class A operation.* Class A operation means that no clipping occurs at either peak of the ac signal. If we did get clipping, the operation would no longer be called class A operation.

EXAMPLE 9-4   If V_E = 1 V and R_L = 10 kΩ in Fig. 9-9, what does the ac compliance equal?

SOLUTION   The dc collector current is

$$I_{CQ} \cong \frac{V_E}{R_E} = \frac{1\ V}{1\ k\Omega} = 1\ mA$$

The dc collector-emitter voltage is

$$V_{CEQ} = V_{CC} - I_C(R_C + R_E) \cong 10\ V - (1\ mA)(5\ k\Omega) = 5\ V$$

The ac compliance is the smaller of $2V_{CEQ}$ or $2I_{CQ}r_L$:

$$2V_{CEQ} = 2(5\ V) = 10\ V$$

and $$2I_{CQ}r_L = 2(1\ mA)(4\ k\Omega\ ||\ 10\ k\Omega) = 5.71\ V$$

Therefore,

$$PP = 5.71\ V$$

EXAMPLE 9-5   As before, V_E = 1 V in Fig. 9-9. Graph the ac compliance for each of the following values of R_L: 1 MΩ, 100 kΩ, 10 kΩ, 1 kΩ, and 100 Ω.

SOLUTION   The Q point does not change when R_L changes; therefore, I_{CQ} = 1 mA and V_{CEQ} = 5 V. Also, $2V_{CEQ}$ = 10 V. So, the problem comes down to calculating the values of $2I_{CQ}r_L$ for each value of R_L. When R_L = 1 MΩ,

$$2I_{CQ}r_L = 2(1\ mA)(4\ k\Omega\ ||\ 1\ M\Omega) = 7.97\ V$$

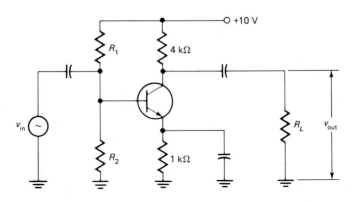

Fig. 9-9.
Calculating ac compliance
of CE amplifier.

When $R_L = 100 \text{ k}\Omega$,

$$2I_{CQ}r_L = 2(1 \text{ mA})(4 \text{ k}\Omega \parallel 100 \text{ k}\Omega) = 7.69 \text{ V}$$

When $R_L = 10 \text{ k}\Omega$,

$$2I_{CQ}r_L = 2(1 \text{ mA})(4 \text{ k}\Omega \parallel 10 \text{ k}\Omega) = 5.71 \text{ V}$$

When $R_L = 1 \text{ k}\Omega$,

$$2I_{CQ}r_L = 2(1 \text{ mA})(4 \text{ k}\Omega \parallel 1 \text{ k}\Omega) = 1.6 \text{ V}$$

When $R_L = 100 \text{ }\Omega$,

$$2I_{CQ}r_L = 2(1 \text{ mA})(4 \text{ k}\Omega \parallel 100 \text{ }\Omega) = 0.195 \text{ V}$$

In all cases, $2I_{CQ}r_L$ is less than $2V_{CEQ}$.

Figure 9-10 shows how the ac compliance varies with the load resistance. Notice how the ac compliance decreases when the load resistance decreases. In other words, as you load down the amplifier with smaller load resistors, the ac compliance becomes smaller.

9-3
CENTERING THE Q POINT ON THE DC LOAD LINE

For convenience, we have been using a Q point at the center of the dc load line. Most of the time, this is all right because we don't need much ac compliance with small-signal amplifiers.

CE Amplifier

Here are some ideas on how to center the Q point on the dc load line of a CE amplifier.

With voltage-divider bias,

$$I_{CQ} \cong \frac{V_E}{R_E} = \frac{V_B - V_{BE}}{R_E}$$

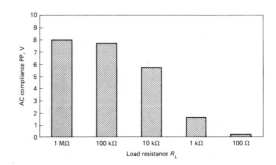

Fig. 9-10.
Ac compliance decreases when load resistance decreases.

Since V_{BE} varies slightly with temperature changes and transistor replacement, larger values of V_B mean I_{CQ} is less sensitive to changes in V_{BE}. On the other hand, if you make V_B too large, then V_E becomes very large. In this case, the ac compliance becomes too small because the transistor saturates too fast.

As a compromise, many designers use a dc emitter voltage of approximately one-tenth of V_{CC}. In symbols,

$$V_E = 0.1 V_{CC} \qquad \qquad \textbf{(9-12)}$$

This usually gives I_{CQ} stability and enough ac compliance for most situations. For instance, if V_{CC} is 15 V, then select R_1 and R_2 to produce a V_E of

$$V_E = 0.1 V_{CC} = 0.1(15 \text{ V}) = 1.5 \text{ V}$$

A Q point centered on the dc load line means that V_{CEQ} is $0.5 V_{CC}$ as shown in Fig. 9-11a. When V_E is $0.1 V_{CC}$, $0.4 V_{CC}$ is across the collector resistor. Since dc collector current approximately equals dc emitter current, the collector resistance should be about four times as large as the emitter resistance (see Fig. 9-11b).

Here is a summary of how to center the Q point on the dc load line of a voltage-divider biased CE stage:

1. Make V_E approximately one-tenth of V_{CC}.
2. Calculate R_E.
3. Make R_C equal to $4R_E$.
4. Add 0.7 V to V_E to get V_B.
5. Select R_1 and R_2 to set up the required V_B.

Emitter Follower To center the Q point on the dc load line of an emitter follower, all you have to do is use a dc emitter voltage of half the supply voltage:

$$V_E = 0.5 V_{CC}$$

Then calculate R_E, R_1, and R_2.

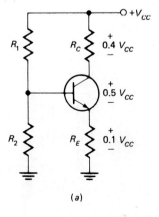

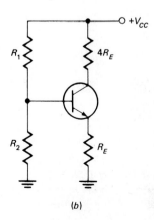

Fig. 9-11.
Q point centered on dc load line. (a) $V_E = 0.1 V_{CC}$.
(b) $R_C = 4R_E$.

(a)

(b)

AC Compliance

When the Q point is centered on the dc load line as shown in Fig. 9-12, cutoff clipping always occurs before saturation clipping because the ac load line always has a steeper slope than the dc load line. For this reason, you can always calculate the ac compliance of an amplifier with

$$PP = 2I_{CQ}r_L \qquad (9\text{-}13)$$

This applies to a CE amplifier, emitter follower, etc.

EXAMPLE 9-6   Design a voltage-divider biased stage with a Q point centered on the dc load line. Use $V_{CC} = 12$ V, $I_{CQ} = 5$ mA, and $\beta_{dc} = 200$.

SOLUTION   Start by calculating the dc emitter voltage:

$$V_E = 0.1V_{CC} = 0.1(12 \text{ V}) = 1.2 \text{ V}$$

The emitter resistance is

$$R_E = \frac{V_E}{I_E} = \frac{1.2 \text{ V}}{5 \text{ mA}} = 240 \text{ }\Omega$$

The collector resistance is

$$R_C = 4R_E = 4(240 \text{ }\Omega) = 960 \text{ }\Omega$$

The required base voltage is

$$V_B = V_E + V_{BE} \cong 1.2 \text{ V} + 0.7 \text{ V} = 1.9 \text{ V}$$

Since β_{dc} is 200,

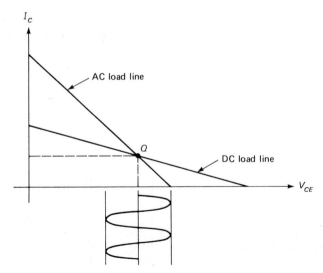

Fig. 9-12.
Ac compliance with Q point
at center of dc load line.

$$I_B = \frac{I_C}{\beta_{dc}} = \frac{5\text{ mA}}{200} = 0.025\text{ mA}$$

The current through the voltage divider should be at least 10 times greater than the base current:

$$10I_B = 10(0.025\text{ mA}) = 0.25\text{ mA}$$

Therefore, the total resistance of the voltage divider is

$$R = \frac{V_{CC}}{10I_B} = \frac{12\text{ V}}{0.25\text{ mA}} = 48\text{ k}\Omega$$

Since $V_B = 1.9$ V, we can calculate R_1 and R_2 as follows:

$$R_2 = \frac{V_2}{I_2} = \frac{1.9\text{ V}}{0.25\text{ mA}} = 7.6\text{ k}\Omega$$

and

$$R_1 = R - R_2 = 48\text{ k}\Omega - 7.6\text{ k}\Omega = 40.4\text{ k}\Omega$$

Using the nearest standard resistors with a tolerance of 5 percent, we get

$$R_1 = 39\text{ k}\Omega$$
$$R_2 = 7.5\text{ k}\Omega$$
$$R_E = 240\text{ }\Omega$$
$$R_C = 1\text{ k}\Omega$$

EXAMPLE 9-7 Calculate the ac compliance of Fig. 9-13.

SOLUTION The circuit contains the design values of the preceding example; so we know the Q point is approximately centered on the dc load line with $I_{CQ} \cong 5$ mA and $V_{CEQ} \cong 6$ V. The ac compliance is

$$PP = 2I_{CQ}r_L = 2(5\text{ mA})(1\text{ k}\Omega \parallel 1\text{ k}\Omega) = 5\text{ V}$$

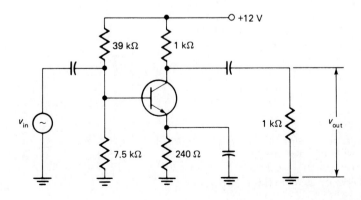

Fig. 9-13.
Calculating ac compliance
of CE amplifier with
matched load and Q point
at center of dc load line.

CLASS A POWER AMPLIFIERS

231

If you analyze Fig. 9-13 with the usual methods, you will get these slightly more accurate values: I_{CQ} = 5.15 mA and V_{CEQ} = 5.62 V. This results in

$$PP = 2I_{CQ}r_L = 2(5.15 \text{ mA})(1 \text{ k}\Omega \parallel 1 \text{ k}\Omega) = 5.15 \text{ V}$$

EXAMPLE 9-8 Graph the ac compliance of any CE amplifier with a Q point centered on the dc load line. Assume V_E is $0.1V_{CC}$.

SOLUTION Since $V_E = 0.1V_{CC}$, $R_C = 4R_E$. When the Q point is centered,

$$I_{CQ} = \frac{V_{CC}}{2(R_C + R_E)} = \frac{V_{CC}}{2(4R_E + R_E)} = \frac{V_{CC}}{10R_E}$$

Therefore,

$$PP = 2I_{CQ}r_L = 2\frac{V_{CC}}{10R_E}\frac{R_C R_L}{R_C + R_L}$$

Since $R_C = 4R_E$, the foregoing reduces to

$$PP = 0.8V_{CC}\frac{R_L}{R_C + R_L}$$

We can rewrite the foregoing equation as

$$\frac{PP}{V_{CC}} = \frac{0.8}{1 + R_C/R_L}$$

Figure 9-14 shows how the ac compliance varies with the load resistance. When R_L is much greater than R_C, the amplifier is lightly loaded and R_C/R_L approaches zero. In this case, the ac compliance is approximately $0.8V_{CC}$. For the matched case of $R_L = R_C$, the ratio R_C/R_L is unity and PP is $0.4V_{CC}$. When R_L is much smaller than R_C, the ratio R_C/R_L is greater than unity and PP becomes very small.

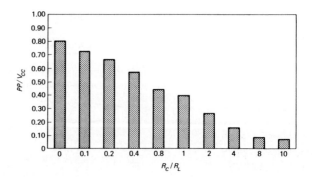

Fig. 9-14.
Ac compliance of CE amplifier with Q point at center of dc load line.

9-4
CENTERING THE Q POINT ON THE AC LOAD LINE

Setting the Q point at the center of the dc load line is convenient, and we will use it most of the time. Occasionally, we may want to set the Q point at the center of the ac load line, because this produces maximum ac compliance. To do this, we have to move the Q point higher, as shown in Fig. 9-15. When the Q point is at the center of the ac load line, we get equal positive and negative swings before clipping occurs. As a result,

$$V_{CEQ} = I_{CQ} r_L$$

For a CE amplifier using voltage-divider bias, this means

$$V_{CC} - I_{CQ}(R_C + R_E) = I_{CQ} r_L$$

Solving for I_{CQ}, we get

$$I_{CQ} = \frac{V_{CC}}{R_C + R_E + r_L} \qquad \textbf{(9-14)}$$

A designer can use this formula to calculate the I_{CQ} required to set the Q point at the center of the ac load line. You can use this formula for an emitter follower by setting R_C equal to zero.

For those CE circuits where $R_C = 4R_E$, Eq. (9-14) gives

$$I_{CQ} = \frac{V_{CC}}{5R_E + r_L} \qquad \textbf{(9-15)}$$

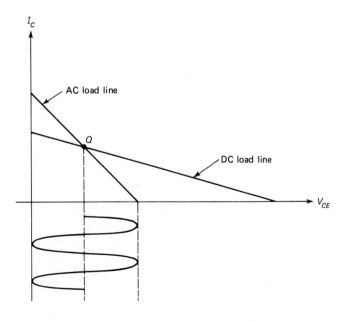

Fig. 9-15.
Ac compliance with Q point at center of ac load line.

CLASS A POWER AMPLIFIERS

To upgrade a CE amplifier with its Q point centered on the dc load line, all you have to do is increase V_B as needed to produce the I_{CQ} given by this equation.

EXAMPLE 9-9   Redesign the circuit of Fig. 9-13 to get the Q point centered on the ac load line. Calculate the new value of ac compliance.

SOLUTION   We need a dc collector current of

$$I_{CQ} = \frac{V_{CC}}{R_C + R_E + r_L} = \frac{12 \text{ V}}{1 \text{ k}\Omega + 240 \ \Omega + 500 \ \Omega} = 6.9 \text{ mA}$$

This gives

$$V_E = I_{CQ}R_E = (6.9 \text{ mA})(240 \ \Omega) = 1.66 \text{ V}$$

and

$$V_B = V_E + V_{BE} = 1.66 \text{ V} + 0.7 \text{ V} = 2.36 \text{ V}$$

Since β_{dc} is 200,

$$I_B = \frac{I_C}{\beta_{dc}} = \frac{6.9 \text{ mA}}{200} = 0.0345 \text{ mA}$$

The current through the voltage divider should be at least 10 times greater than the base current:

$$10I_B = 10(0.0345 \text{ mA}) = 0.345 \text{ mA}$$

Therefore, the total resistance of the voltage divider is

$$R = \frac{V_{CC}}{10I_B} = \frac{12 \text{ V}}{0.345 \text{ mA}} = 34.8 \text{ k}\Omega$$

Since $V_B = 2.36$ V, we can calculate R_1 and R_2 as follows:

$$R_2 = \frac{V_2}{I_2} = \frac{2.36 \text{ V}}{0.345 \text{ mA}} = 6.84 \text{ k}\Omega$$

and

$$R_1 = R - R_2 = 34.8 \text{ k}\Omega - 6.84 \text{ k}\Omega = 28 \text{ k}\Omega$$

Using the nearest standard resistors with a tolerance of 5 percent, we have

$$R_1 = 27 \text{ k}\Omega$$
$$R_2 = 6.8 \text{ k}\Omega$$
$$R_E = 240 \ \Omega$$
$$R_C = 1 \text{ k}\Omega$$

With these values, the circuit of Fig. 9-13 has a Q point that is approximately at the center of the ac load line. The ac compliance for the redesigned circuit is

$$PP = 2I_{CQ}r_L = 2(6.9 \text{ mA})(1 \text{ k}\Omega \ || \ 1 \text{ k}\Omega) = 6.9 \text{ V}$$

9-5
POWER
FORMULAS

What is the maximum power you can get out of a class A amplifier? How much power does a transistor dissipate? These are some of the questions answered in this section.

Quiescent Power Dissipation

With no ac input signal, the transistor has a power dissipation of

$$P_{DQ} = V_{CEQ}I_{CQ} \qquad (9\text{-}16)$$

For instance, if $V_{CEQ} = 10.3$ V and $I_{CQ} = 1.08$ A, then

$$P_{DQ} = (10.3 \text{ V})(1.08 \text{ A}) = 11.1 \text{ W}$$

To avoid damage to a transistor, P_{DQ} must be less than the maximum power rating specified on the data sheet.

Load Power

The load on an amplifier may be a loudspeaker, a motor, or some other device. The power delivered to the load resistance is

$$P_L = \frac{V_L^2}{R_L} \qquad (9\text{-}17)$$

where V_L is the rms load voltage. Recall the relation between the rms voltage and the peak voltage:

$$V_L = 0.707V_P \qquad (9\text{-}18)$$

So, when you know the peak voltage, you can calculate the rms voltage and then the load power.

Often, you look at the ac load voltage with an oscilloscope. In this case, it is convenient to have a formula for load power that uses peak-to-peak voltage instead of rms voltage. By rewriting Eq. (9-17),

$$P_L = \frac{V_{PP}{}^2}{8R_L} \qquad\qquad \textbf{(9-19)}$$

You will find this useful when you measure the peak-to-peak load voltage with an oscilloscope.

Maximum AC
Load Power   What is the maximum ac load power you can get from an amplifier operated class A? The ac compliance PP equals the maximum unclipped load output voltage. Since $PP = V_{PP}$, we can rewrite Eq. (9-19) as

$$P_{L(max)} = \frac{PP^2}{8R_L} \qquad\qquad \textbf{(9-20)}$$

This is the maximum ac load power that a class A amplifier can produce without clipping.

EXAMPLE 9-10   In Fig. 9-16, what is the quiescent power dissipation of the transistor? The maximum ac load power?

SOLUTION   We analyzed this circuit in Example 9-1 and calculated quiescent values of $I_{CQ} = 1.1$ mA and $V_{CEQ} = 4.94$ V. Therefore, the quiescent power dissipation of the transistor is

$$P_{DQ} = V_{CEQ}I_{CQ} = (4.94 \text{ V})(1.1 \text{ mA}) = 5.43 \text{ mW}$$

The ac compliance is

$$PP = 2I_{CQ}r_L = 2(1.1 \text{ mA})(3.6 \text{ k}\Omega \parallel 1.5 \text{ k}\Omega) = 2.33 \text{ V}$$

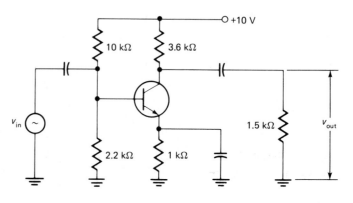

Fig. 9-16.
Calculating P_{DQ} and $P_{L(max)}$
for CE amplifier.

The maximum ac load power is

$$P_{L(max)} = \frac{PP^2}{8R_L} = \frac{(2.33 \text{ V})^2}{8(1.5 \text{ k}\Omega)} = 0.452 \text{ mW}$$

9-6
EFFICIENCY

In most power supplies, ac line voltage is rectified and filtered to get the dc voltage needed for the different stages of a system. But in some applications, batteries are used for the power supply rather than rectified line voltage. In this case, the efficiency of the design becomes important because of the current drain on the batteries.

Current Drain

In the CE amplifier of Fig. 9-17, the V_{CC} source has to supply dc current to the circuit. Part of the dc current flows through the voltage divider in the base circuit, and the rest flows through the collector resistor. The total *current drain* from the supply is

$$I_{CC} = I_1 + I_{CQ} \tag{9-21}$$

where I_1 is the dc current to the voltage divider and I_{CQ} is the dc current to the collector resistor.

Definition of Efficiency

The total *dc power* supplied to the CE stage of Fig. 9-17 is the product of supply voltage and current drain:

$$P_{CC} = V_{CC}I_{CC} \tag{9-22}$$

For instance, if $V_{CC} = 10$ V and $I_{CC} = 2$ mA,

$$P_{CC} = (10 \text{ V})(2 \text{ mA}) = 20 \text{ mW}$$

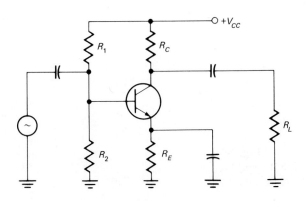

Fig. 9-17.
RC-coupled class A amplifier.

The efficiency of a stage is defined as the maximum unclipped ac load power divided by the dc power from the supply:

$$\eta = \frac{P_{L(max)}}{P_{CC}} \times 100 \text{ percent} \qquad (9\text{-}23)$$

The higher the efficiency, the better the amplifier is in converting dc power from the supply into useful ac load power. Efficiency is always less than 100 percent because some of the dc power from the supply is wasted in the biasing resistors and the transistor.

RC-Coupled Amplifier

The circuit of Fig. 9-17 is called a *resistance-capacitance* (*RC*) coupled amplifier because the ac voltage developed across the collector resistor is coupled through a capacitor into the load resistor. One way to improve the efficiency of an *RC*-coupled amplifier is to make

$$R_L = R_C \qquad (9\text{-}24)$$

This is called the *matched-load condition*. It is based on the maximum power transfer theorem discussed in basic courses. A designer uses the matched-load condition only when he or she wants to get maximum load power.

Another way to improve the efficiency of an *RC*-coupled amplifier is to reduce the power losses in R_1, R_2, and R_E. By making I_1 at least 10 times smaller than I_{CQ}, we can minimize the power losses in R_1 and R_2. Also, we can make R_E as small as possible compared with R_C.

What is the maximum efficiency of an *RC*-coupled amplifier? Assume we have a design where R_1, R_2, and R_E can be ignored, equivalent to having very large R_1 and R_2, and very small R_E. With a matched load, $r_L = R_C \| R_L = R_C \| R_C = 0.4R_C$. For maximum efficiency, we have to center the Q point on the ac load line, which means

$$I_{CQ} = \frac{V_{CC}}{R_C + R_E + r_L} = \frac{V_{CC}}{R_C + 0.5R_C} = \frac{2V_{CC}}{3R_C}$$

In this case, the dc power from the source is

$$P_{CC} = V_{CC}I_{CQ} = V_{CC}\frac{2V_{CC}}{3R_C} = \frac{2V_{CC}^2}{3R_C}$$

The ac compliance is

$$PP = 2I_{CQ}r_L = 2\frac{2V_{CC}}{3R_C}(0.5R_C) = \frac{2V_{CC}}{3}$$

The maximum ac load power is

$$P_{L(max)} = \frac{PP_2}{8R_L} = \frac{(2V_{CC}/3)^2}{8R_C} = \frac{V_{CC}^2}{18R_C}$$

The efficiency is

$$\eta = \frac{P_{L(max)}}{P_{CC}} \times 100 \text{ percent} = \frac{V_{CC}^2/18R_C}{2V_{CC}^2/3R_{CC}} \times 100 \text{ percent}$$

or

$$\eta = 8.33 \text{ percent} \qquad (9\text{-}25)$$

Here is what we have discovered. The maximum possible efficiency is 8.33 percent. This efficiency ignores the power losses in the biasing resistors including R_E. It assumes a load resistor R_L matched to the collector resistor R_C. Therefore, any practical design you encounter will always have an efficiency of less than 8.33 percent. A similar derivation for emitter followers and CB amplifiers produces the same limiting efficiency. In other words, the maximum possible efficiency for any RC-coupled class A amplifier is 8.33 percent.

Transformer-Coupled Amplifier

Figure 9-18 shows a transformer-coupled amplifier. This has better efficiency than an RC-coupled amplifier because no power is wasted in a collector resistor R_C. Assuming a lossless transformer and ignoring R_1, R_2, and R_E, we can derive an efficiency of

$$\eta = 50 \text{ percent} \qquad (9\text{-}26)$$

Transformers are not practical at audio frequencies (20 Hz to 20 kHz) because they are bulky and expensive. But you will see transformers used at RF frequencies (greater than 20 kHz) because they are

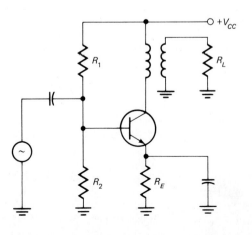

Fig. 9-18.
Transformation-coupled
class A amplifier.

smaller and less expensive. In this case, a class A amplifier can have efficiencies approaching 50 percent.

EXAMPLE 9-11   Calculate the efficiency in Fig. 9-19.

SOLUTION   The voltage divider produces a dc base voltage of 2.41 V. Subtracting the V_{BE} drop gives a dc emitter voltage of approximately 1.71 V. The approximate value of I_{CQ} is

$$I_{CQ} = \frac{V_E}{R_E} = \frac{1.71 \text{ V}}{240 \text{ } \Omega} = 7.13 \text{ mA}$$

The dc collector-emitter voltage is

$$V_{CEQ} = V_{CC} - I_C(R_C + R_E) = 12 \text{ V} - (7.13 \text{ mA})(1.24 \text{ k}\Omega) = 3.16 \text{ V}$$

Now, find the ac compliance:

$$2V_{CEQ} = 2(3.16 \text{ V}) = 6.32 \text{ V}$$

$$2I_{CQ}r_L = 2(7.13 \text{ mA})(500 \text{ } \Omega) = 7.13 \text{ V}$$

The ac compliance is the smaller of the two:

$$PP = 6.32 \text{ V}$$

The maximum ac load power is

$$P_{L(\text{max})} = \frac{PP^2}{8R_L} = \frac{(6.32 \text{ V})^2}{8(1 \text{ k}\Omega)} = 4.99 \text{ mW}$$

The dc power from the supply is

$$P_{CC} = V_{CC}I_{CC} = (12 \text{ V})(0.355 \text{ mA} + 7.13 \text{ mA}) = 89.8 \text{ mW}$$

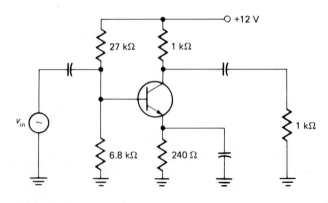

Fig. 9-19.
Calculating efficiency of
CE amplifier with
matched load.

The efficiency is

$$\eta = \frac{P_{L(max)}}{P_{CC}} \times 100 \text{ percent} = \frac{4.99 \text{ mW}}{89.8 \text{ mW}} = 5.56 \text{ percent}$$

As you see, the efficiency is less than 8.33 percent, the theoretical limit for an *RC*-coupled class A amplifier. No matter how you try to readjust the resistance values, you will always have an efficiency of less than 8.33 percent. In the next chapter, you will read about the class B push–pull amplifier, a special circuit often used at audio frequencies because its efficiency approaches 78.6 percent.

9-7
TRANSISTOR
POWER RATING

The temperature of the collector junction places a limit on the allowable power dissipation of a transistor. Depending on the transistor type, a junction temperature from 150 to 200°C will destroy the transistor. This is why a manufacturer specifies a maximum power rating for a transistor. If you keep the transistor's power dissipation less than the maximum power rating, the junction temperature will not reach destructive levels.

Derating Factor

Data sheets often specify the $P_{D(max)}$ of a transistor at an ambient temperature of 25°C. For instance, the 2N1936 has a $P_{D(max)}$ of 4 W for a T_A of 25°C. This means the 2N1936 can have a power dissipation as high as 4 W, provided the ambient temperature is 25°C or less.

What do you do if the ambient temperature is greater than 25°C? You have to *derate* (reduce) the power rating. Data sheets sometimes include a derating curve like Fig. 9-20. As you can see, the maximum power rating is 4 W from 0 to 25°C. Above 25°C the maximum power rating decreases linearly. If you want to use this transistor over an ambient temperature range of 0 to 100°C, the maximum allowable power dissipation would be 2 W, the worst case.

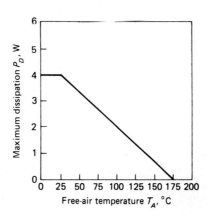

Fig. 9-20.
Derating curve for ambient temperature.

Some data sheets do not give a derating curve like Fig. 9-20. Instead, they list a derating factor D. The data sheet of a 2N1936 gives a derating factor of 26.7 mW/°C. This means that you have to subtract 26.7 mW for each degree the ambient temperature is above 25°C. In symbols,

$$\Delta P = D(T_A - 25°C) \tag{9-27}$$

where ΔP = decrease in power rating
  D = derating factor
  T_A = ambient temperature
For example, if the ambient temperature rises to 75°C, you have to reduce the power rating of a 2N1936 by

$$\Delta P = (26.7 \text{ mW})(75 - 25) = 1.34 \text{ mW}$$

Since the power rating is 4 W at 25°C, the new power rating is

$$P_{D(max)} = 4 \text{ W} - 1.34 \text{ W} = 2.66 \text{ W}$$

This agrees with the derating curve of Fig. 9-20.

Heat Sinks One way to increase the power rating of a transistor is to get rid of the heat faster. This is the purpose of a *heat sink* (a mass of metal). If we increase the surface area of the transistor case, we allow the heat to escape more easily into the surrounding air. For instance, Fig. 9-21*a* shows one type of heat sink. When this is pushed onto the transistor case, heat radiates more quickly because of the increased surface area of the fins.

Figure 9-21*b* is the outline of a power-tab transistor. The metal tab is fastened to the chassis, which acts like a massive heat sink. Because of this, heat escapes rapidly from the transistor, reducing the junction temperature. This means the transistor can dissipate more heat before reaching excessive junction temperatures.

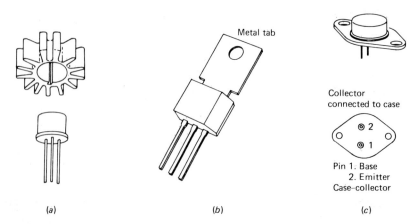

Metal tab

Collector
connected to case

⊚ 2

⊚ 1

Pin 1. Base
2. Emitter
Case–collector

(a)   (b)   (c)

Fig. 9-21.
(a) Attachable heat sink.
(b) Power-tab transistor.
(c) Power transistor.

Large power transistors like Fig. 9-21c have the collector connected directly to the case to let the heat escape as easily as possible. The case can then be placed in thermal contact with the chassis for heat-sinking. To prevent the collector from being shorted to ground, a thin mica washer and a heat-conductive grease is used between the case and the chassis.

Case Temperature When heat flows out of a transistor, it passes through the case of the transistor and into the heat sink, which then radiates the heat into the surrounding air. The temperature of the transistor case T_C will be slightly higher than the temperature of the heat sink T_S, which in turn is slightly higher than the ambient temperature T_A.

The data sheets of large power transistors give the derating curves for the case temperature rather than the ambient temperature. For instance, Fig. 9-22 shows the derating curve of a 2N5877. The power rating is 150 W for a case temperature of 25°C. Then it decreases linearly until it reaches zero at 200°C.

Sometimes you get a derating factor instead of a derating curve. In this case, you can use the following equation to calculate the reduction in power rating:

$$\Delta P = D(T_C - 25°C) \qquad (9\text{-}28)$$

Where ΔP = decrease in power rating
D = derating factor
T_C = case temperature

Thermal Analysis Thermal resistance is the resistance of a substance to the flow of heat. The quantity θ_{JA} is the symbol for the thermal resistance between the junction and the surrounding air. The larger the transistor case, the lower the thermal resistance, because heat can escape more easily to the surrounding air. For instance, the thermal resistance of a 2N3904 between the junction and the surrounding air is

$$\theta_{JA} = 357°C/W$$

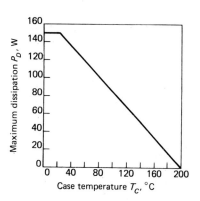

Fig. 9-22.
Derating curve for case temperature.

The 2N3719 has a larger case and radiates heat more easily. As you would expect, it has a smaller thermal resistance between the junction and the surrounding air:

$$\theta_{JA} = 175°C/W$$

Besides, θ_{JA}, there are two other thermal resistances worth knowing about: θ_{CS} is the resistance to heat flow between the transistor case and the heat sink, while θ_{SA} is the thermal resistance between the heat sink and the ambient air.

Courses in thermodynamics prove the following equation for the case temperature of a power transistor:

$$T_C = T_A + P_D(\theta_{CS} + \theta_{SA}) \qquad (9\text{-}29)$$

where T_C = case temperature
T_A = ambient temperature
P_D = power dissipation
θ_{CS} = thermal resistance between case and sink
θ_{SA} = thermal resistance between sink and ambient

With this formula, a designer can calculate the case temperature and the power rating using the derating factor or derating curve of the transistor. The following example shows how this is done.

EXAMPLE 9-12 A 2N5877 and a heat sink have the following thermal resistances: θ_{CS} = 0.5°C/W and θ_{SA} = 1.5°C/W. If the transistor has a power dissipation of 30 W, what is the case temperature when the ambient temperature is 70°C? What is the power rating at this temperature?

SOLUTION With Eq. (9-30), we can calculate the case temperature:

$$T_C = 70°C + (30 \text{ W})(0.5°C/W + 1.5°C/W) = 130°C$$

With the derating curve of Fig. 9-22, we can read a power rating of

$$P_D(\text{max}) = 60 \text{ W}$$

As you see, an ambient temperature of 70°C produces a case temperature of 130°C and a power rating of 60 W.

 **Summary** Every amplifier has two load lines: a dc load line and an ac load line. You use the dc equivalent circuit to work out the dc load line and the ac equivalent for the ac load line.

If you overdrive an amplifier, the output signal will be clipped on either or both peaks. Class A operation means no clipping on either peak. The ac compliance PP is the maximum unclipped peak-to-peak collector voltage. It equals $2V_{CEQ}$ or $2I_{CQ}r_L$, whichever is smaller.

When the Q point is centered on the dc load line, the ac compliance equals $2I_{CQ}r_L$. A convenient design for a voltage-divider biased amplifier is to use $V_E = 0.1V_{CC}$. This implies $R_C = 4R_E$.

To center the Q point on the ac load line, I_{CQ} must equal $V_{CC}/(R_C + R_E + .r_L)$. Then, the ac compliance equals either $2V_{CEQ}$ or $2I_{CQ}r_L$, because these expressions are equal. The maximum ac load power equals $PP^2/8R_L$.

The efficiency of a stage equals the maximum ac load power divided by the dc power from the supply. An *RC*-coupled class A amplifier has a maximum efficiency of 8.33 percent. You approach this efficiency when the load resistance is matched to the output impedance, and power losses in the biasing resistors are minimum. A transformer-coupled class A amplifier has a maximum efficiency of 50 percent.

The temperature of the collector junction places a limit on the allowable power dissipation of a transistor. Depending on the transistor type, a junction temperature from 150 to 200°C will destroy the transistor. One way to decrease the junction temperature is with heat sinks. They allow heat to escape more easily. The data sheets of smaller transistors provide either a derating curve or a derating factor with respect to ambient temperature. The data sheets of larger transistors give the derating curve or factor for case temperature.

Glossary

ac compliance The maximum peak-to-peak unclipped ac load voltage of a transistor amplifier.

ac load line A graph of I_C versus V_{CE} when a transistor is driven by an ac signal.

class A operation Operation of the transistor in the active region, equivalent to operating the transistor anywhere between ac saturation and ac cutoff.

compliance The voltage range of a current source.

cutoff clipping Clipping of the output signal when a large ac signal drives the transistor into cutoff.

dc compliance The dc collector voltage of a transistor amplifier.

derating factor The decrease in the power rating of a transistor for each degree rise in temperature.

efficiency The ratio of maximum ac load power to the dc power from the supply.

heat sink A mass of metal in thermal contact with the transistor case. The heat sink absorbs heat from the transistor more rapidly than the surrounding air.

saturation clipping Clipping of the output signal when a large ac input signal drives the transistor into saturation.

thermal resistance The resistance to the flow of heat between two points. With transistors, the most important thermal resistances are θ_{JA}, θ_{CS}, and θ_{SA}.

Review Questions

1. What is the difference between the ac and dc load line?
2. Define class A operation.
3. What is the ac compliance?
4. If $V_E = 0.1V_{CC}$ in a voltage-divider biased CE stage, how is R_C related to R_E?
5. What does I_{CQ} equal in an amplifier where the Q point is centered on the ac load line?
6. Define the efficiency of a stage.
7. What is the maximum efficiency of an RC-coupled class A amplifier?
8. What is a derating curve? A derating factor?
9. What are heat sinks and why are they used?

Problems

9-1. Draw the ac load line of Fig. 9-23 for an R_L of 2.4 kΩ.

9-2. Draw the ac load line of Fig. 9-24 for an R_L of 620 Ω.

9-3. Draw the ac load lines in Fig. 9-23 for these two values of R_L: 1 kΩ and 3.9 kΩ. What happens to the slope of the ac load line when R_L decreases?

9-4. What is the ac compliance of Fig. 9-23 for $R_L = 1.8$ kΩ?

9-5. What is the ac compliance of Fig. 9-24 for $R_L = 330$ Ω?

9-6. Calculate the ac compliance of Fig. 9-23 for each of the following values of R_L: 1 MΩ, 100 kΩ, 10 kΩ, 1 kΩ, and 100 Ω.

9-7. Calculate the ac compliance of Fig. 9-24 for each of the following values of R_L: 1 MΩ, 100 kΩ, 10 kΩ, 1 kΩ, and 100 Ω.

9-8. Redesign the circuit of Fig. 9-23 to meet these specifications: $V_E = 1.5$ V, $I_{CQ} = 1$ mA, $h_{FE} = 200$, and the Q point is centered on the dc load line. (Use a dc current through the voltage divider of 10 time the dc base current.)

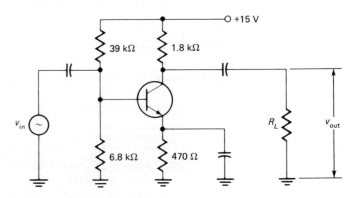

Fig. 9-23.
CE amplifier.

CHAPTER

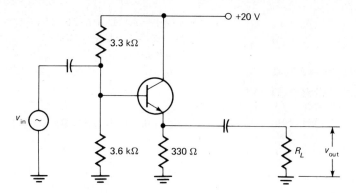

Fig. 9-24.
Emitter follower.

9-9. Redesign the emitter follower of Fig. 9-24 to meet these specifications: $V_E = 10$ V, $I_{CQ} = 2$ mA, $h_{FE} = 175$, and the Q point is centered on the dc load line. (Use a dc current through the voltage divider of 10 times the dc base current.)

9-10. We want to center the Q point of Fig. 9-23 on the ac load line for $R_L = 1.8$ kΩ. R_C and R_E are to remain the same. Select values of R_1 and R_2 using a dc current through the voltage divider of 10 times the dc base current. Assume $h_{FE} = 50$.

9-11. Calculate P_{DQ} and $P_L(\text{max})$ in Fig. 9-23 for $R_L = 3.3$ kΩ.

9-12. Calculate P_{DQ} and $P_L(\text{max})$ in Fig. 9-24 for $R_L = 680$ Ω.

9-13. What is the efficiency of Fig. 9-23 for an R_L of 1.8 kΩ? For an R_L of 1 kΩ? For an R_L of 3.3 kΩ?

9-14. What is the efficiency of Fig. 9-24 for an R_L of 330 Ω? For an R_L of 150 Ω? For an R_L of 680 Ω?

9-15. A 2N4401 has a power rating of 310 mW for an ambient temperature of 25°C. The derating factor is 2.81 mW/°C. What is the power rating for an ambient temperature of 100°C?

9-16. The 2N1936 has the derating curve of Fig. 9-20. What is the power rating for each of these ambient temperatures:
a. 50°C
b. 75°C
c. 125°C

9-17. A transistor has a power rating of 50 W for a case temperature of 25°C. If the derating factor is 0.285 W/°C, what is the power rating for a case temperature of 125°C?

9-18. A power transistor and its heat sink have the following thermal resistances: $\theta_{CS} = 0.45$°C/W and $\theta_{SA} = 1.65$°C/W. If the ambient temperature is 50°C, what is the case temperature when the transistor dissipates 25 W?

Other Power Amplifiers

Sylvester: This brilliant mathematician was highly inventive. He always did it his way because he found it intolerable and boring to master what others had done. He rediscovered some old theorems but found many new ones. His memory was so poor that he had difficulty remembering his own inventions and once even disputed that a certain theorem of his own could possibly be true.

The class B push-pull amplifier is a two-transistor circuit that is more efficient than a class A power amplifer. Another advantage of class B operation is less current drain from the power supply under no-signal conditions. This is important for battery-operated equipment because you don't want to waste battery power on a standby condition.

Class C operation is more efficient than class A or class B. But a class C amplifier has to be tuned to a resonant frequency. Because of this, class C amplifiers are narrow-band circuits used at radio frequencies (above 20 kHz).

Besides class A, B, and C operation, the transistor can be used as a switch. This means using only two points on the ac load line, typically cutoff and saturation. One important industrial application of switching operation is controlling large amounts of current through small load impedances.

10-1
CLASSES OF OPERATION

The transistor of a class A amplifier operates in the active region throughout the ac cycle. Therefore, collector current flows for 360°, as shown in Fig. 10-1a. At no point in the cycle is the transistor driven into saturation or cutoff.

In a class B amplifier, the transistor operates in the active region for only half the cycle. During the other half cycle, the transistor is cut off. As a result, collector current flows for 180°, as shown in Fig. 10-1b. Normally, a second transistor supplies the missing half cycle.

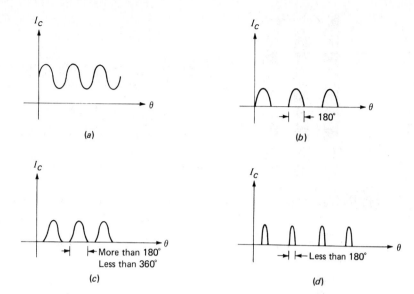

Fig. 10-1.
(a) Class A. (b) Class B.
(c) Class AB. (d) Class C.

Stated another way, two transistors operating class B form a *push-pull* amplifier where one transistor amplifies the positive half cycle and the other amplifies the negative half cycle.

Class AB operation is between class A and class B. The transistor of a class AB circuit operates in the active region for more than a half cycle but less than a whole cycle. For class AB operation, collector current flows for more than 180° but less than 360°, as shown in Fig. 10-1c.

Class C operation means collector current flows for less than 180°, as shown in Fig. 10-1d. In a practical class C circuit, the current exists for much less than 180° and looks like narrow pulses. When these narrow pulses drive a high-Q resonant tank circuit, the voltage across the tank is almost a perfect sine wave.

_____
10-2

**CLASS B PUSH-
PULL AMPLIFIER**

Class A is the common way to run a transistor in linear circuits because it leads to the simplest and most stable biasing circuits. But class A is not efficient because too much power is dissipated in the transistor rather than in the load resistance. In some applications like battery-powered systems, a class A power amplifier may cause too much current drain. For this reason, other classes of operation have evolved for power amplifiers.

**Definition of
Class B**

Class B operation of a transistor means that the collector current flows for only 180° of the ac cycle. With this type of operation, the Q point is located at cutoff on both the dc and ac load lines. The advantage of class B operation is less transistor dissipation, more load power, and greater stage efficiency.

Push-Pull Connection

When a transistor operates class B, it clips off half a cycle. To avoid distortion, you have to use two transistors in a *push-pull connection*. In this way, one transistor conducts during the positive half cycle and the other during the negative half cycle. With push-pull circuits, it is possible to build class B amplifiers that have high load power and low transistor dissipation.

Figure 10-2*a* shows one way to connect a class B push-pull emitter follower. On the positive half cycle of input voltage, the upper transistor conducts and the lower cuts off. The upper transistor acts like an emitter follower; so the output voltage across R_L approximately equals the positive half cycle of input voltage. On the negative half cycle of input voltage, the lower transistor acts like an emitter follower and produces the negative half cycle of output voltage. As you see, the upper transistor handles the positive half cycle and the lower transistor takes care of the negative half cycle.

It is possible to build class B push-pull CE amplifiers, but they have

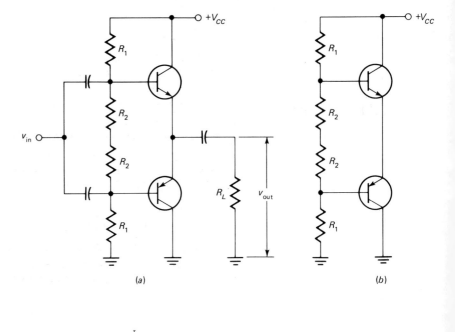

(a)   (b)

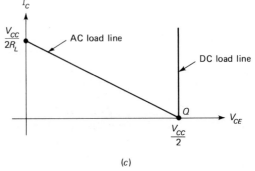

(c)

Fig. 10-2.
Class B push-pull emitter follower. (a) Circuit. (b) Dc equivalent circuit. (c) Dc and ac load lines.

too much distortion when the transistors are used. For this reason, you almost always see the class B push-pull amplifier as a combination of emitter followers like Fig. 10-2a because they have inherently low distortion. In this book, we will discuss only the class B push-pull emitter follower.

DC Equivalent Circuit Figure 10-2b shows the dc equivalent circuit. The designer selects biasing resistors to set the Q point at cutoff. This biases each emitter diode between 0.6 and 0.7 V. Ideally, the quiescent collector current is zero. In symbols,

$$I_{CQ} = 0$$

Notice the symmetry of the circuit. Because the transistors are in series, the voltage across each emitter diode is equal. As a result, half the supply voltage is dropped across each transistor, which means the quiescent collector-emitter voltage of each transistor is

$$V_{CEQ} = \frac{V_{CC}}{2}$$

DC Load Line When the transistors are open in Fig. 10-2b, half the supply voltage appears across each. This is why the lower end of the dc load line is located at $V_{CC}/2$ in Fig. 10-2c. On the other hand, when the transistors are shorted, the dc current tries to increase to infinity because there is no dc collector resistance. This is why the dc load line is vertical in Fig. 10-2c.

If a vertical dc load line seems dangerous, you are quite right. The most difficult thing about designing a class B push-pull amplifier is setting up a stable Q point at cutoff. Temperature variations and transistor tolerances can move the Q point up the dc load line to destructively high quiescent currents. Unless special precautions are taken, the transistors are easily destroyed in a class B push-pull amplifier. For now, we will assume the Q point is rock solid at cutoff, as shown in Fig. 10-2c.

AC Load Line The ac load line derived earlier still applies. The ac saturation current is

$$i_{c(\text{sat})} = I_{CQ} + \frac{V_{CEQ}}{r_L}$$

and the ac cutoff voltage is

$$v_{ce(\text{cutoff})} = V_{CEQ} + I_{CQ}r_L$$

In the class B push-pull emitter follower of Fig. 10-2a, $I_{CQ} = 0$ and $r_L = R_L$. Therefore, the ac saturation current and cutoff voltage simplify to

$$i_{c(\text{sat})} = \frac{V_{CEQ}}{R_L} \qquad \text{(10-1)}$$

and
$$v_{ce(\text{cutoff})} = V_{CEQ} \qquad \text{(10-2)}$$

Figure 10-2c shows the ac load line. When a transistor is conducting, its operating point swings upward along the ac load line; the operating point of the other transistor remains at cutoff. The voltage of the conducting transistor can swing all the way from cutoff to saturation. On the alternate half cycle, the other transistor does the same thing. Since the voltage swing in each direction equals $V_{cc}/2$, the ac compliance is

$$PP = 2V_{CEQ} = 2\frac{V_{CC}}{2}$$

or
$$PP = V_{CC} \qquad \text{(10-3)}$$

For instance, if $V_{CC} = 10$ V, then $PP = 10$ V. This means a class B push-pull emitter follower can produce an unclipped peak-to-peak load voltage of approximately 10 V.

Crossover Distortion

Ideally, the Q point of a class B push-pull amplifier is at cutoff. As a practical matter, the Q point should be slightly above cutoff to avoid distortion. Why? Suppose no bias at all is applied to the emitter diodes. On the positive half cycle the incoming ac voltage has to rise to about 0.7 V to overcome the barrier potential of the upper transistor (see Fig. 10-3a). Because of this, no collector current can flow when the ac input voltage is less than 0.7 V. The action on the negative half cycle is similar; the lower transistor does not turn on until the ac input voltage is more negative than -0.7 V.

For this reason, the output looks like Fig. 10-3b when no bias is applied to the emitter diodes. This voltage is distorted. It no longer is sinusoidal because of the clipping action between half cycles. Since

Fig. 10-3.
Slight forward bias
eliminates crossover
distortion.

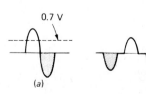

(a)   (b)

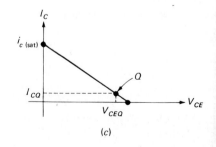

(c)

the clipping occurs between the time one transistor shuts off and the other one comes on, we call it *crossover distortion*.

To eliminate crossover distortion, we need to apply a slight forward bias to each emitter diode. This means locating the Q point slightly above cutoff, as shown in Fig. 10-3c. As a guide, an I_{CQ} from 1 to 5 percent of $i_c(\text{sat})$ is enough to eliminate crossover distortion in most class B push-pull amplifiers. For instance, if $i_c(\text{sat}) = 100$ mA, you have to set up an I_{CQ} between 1 and 5 mA to avoid cross distortion.

10-3

POWER FORMULAS

Figure 10-4a shows an *RC*-coupled class B push-pull amplifier. Q_1 conducts during the positive half cycle of input voltage, and Q_2 during the negative half cycle. Figure 10-4b shows the maximum unclipped current and voltage waveforms of the amplifier. The upper transistor

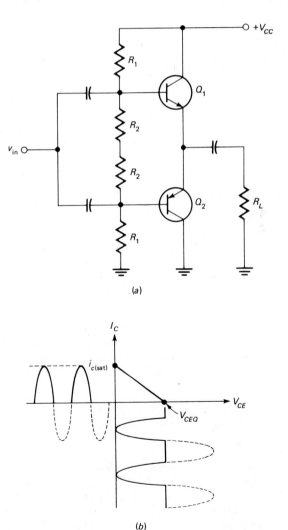

(a)

(b)

Fig. 10-4.
Class B push-pull emitter
follower and waveforms.

produces the solid half cycle, and the lower transistor produces the dashed half cycle.

Quiescent Transistor Dissipation Under no-signal or quiescent conditions, the transistors of a class B push-pull amplifier are idling because only a small trickle of current passes through them, whatever is needed to eliminate crossover distortion. You can calculate the quiescent power dissipation of each transistor with

$$P_{DQ} = V_{CEQ}I_{CQ} \qquad (10\text{-}4)$$

Since I_{CQ} is only 1 to 5 percent of $i_c(\text{sat})$, P_{DQ} is small.

Maximum Transistor Dissipation When there is an ac input signal, each transistor has large swings in current and voltage, which increases its power dissipation. In the worst case, the transistor power dissipation reaches a maximum value of

$$P_{D(\text{max})} = \frac{PP^2}{40R_L} \qquad (10\text{-}5)$$

(This can be derived with calculus.) Since $P_{D(\text{max})}$ is the worst-case power dissipation, each transistor in a class B amplifier must have a power rating greater than $PP^2/40R_L$.

Load Power As before, the ac load power is given by

$$P_L = \frac{V_{PP}{}^2}{8R_L}$$

Since PP represents the maximum unclipped load voltage, we can write

$$P_{L(\text{max})} = \frac{PP^2}{8R_L} \qquad (10\text{-}6)$$

In Fig. 10-4b, $PP = 2V_{CEQ}$. In Fig. 10-4a, $V_{CEQ} = V_{CC}/2$; therefore, $PP = V_{CC}$.

Efficiency In Fig. 10-4a, the maximum efficiency occurs when the current through the biasing resistors is small enough to ignore. In this case, the current drain is approximately equal to the current flowing through the upper transistor. For the maximum signal swing, the current in Q_1 is the half-wave rectified sine wave shown in Fig. 10-4b. Since it has a peak value of $i_{c(\text{sat})}$, it has a dc or average value of

$$I_{dc} = 0.318i_{c(\text{sat})}$$

Therefore, the dc current drain from the power supply is approximately

$$I_{CC} = 0.318i_{c(sat)} = 0.318\frac{V_{CEQ}}{R_L} = 0.318\frac{V_{CC}/2}{R_L} = 0.159\frac{V_{CC}}{R_L}$$

and the dc power from the supply is

$$P_{CC} = V_{CC}I_{CC} = 0.159\frac{V_{CC}^2}{R_L}$$

Now we are ready to calculate the maximum efficiency of a class B push-pull amplifier. Recall that the efficiency of an amplifier equals the maximum unclipped ac load power divided by the dc power from the supply. When the biasing current is negligible in Fig. 10-4a, the efficiency was a maximum value of

$$\eta = \frac{P_{L(max)}}{P_{CC}} = \frac{PP^2/8R_L}{0.159V_{CC}^2/R_L} = \frac{V_{CC}^2/8R_L}{0.159V_{CC}^2/R_L} \times 100\%$$

or
$$\eta = 78.6\% \tag{10-7}$$

Because the power losses in the biasing resistors are small, the efficiency of practical class B push-pull amplifiers approaches 78.6 percent, which is far more efficient than an RC-coupled class A amplifier whose maximum efficiency is only 8.33 percent. For this reason, many systems use RC-coupled class A amplifiers for the earlier stages and a class B push-pull amplifier for the last stage where the current drain is usually high.

EXAMPLE 10-1 A class B push-pull emitter follower like Fig. 10-4a has $V_{CC} = 15$ V and $R_L = 50 \, \Omega$. If I_{CQ} is 1 percent of $i_{c(sat)}$, what are the values of P_{DQ}, $P_{D(max)}$, and $P_{L(max)}$?

SOLUTION The ac saturation current is approximately

$$i_{c(sat)} = \frac{V_{CEQ}}{R_L} = \frac{7.5 \text{ V}}{50 \, \Omega} = 150 \text{ mA}$$

Therefore, the quiescent collector current is

$$I_{CQ} = 0.01i_{c(sat)} = 0.01(150 \text{ mA}) = 1.5 \text{ mA}$$

The standby power dissipation in each transistor is

$$P_{DQ} = V_{CEQ}I_{CQ} = (7.5 \text{ V})(1.5 \text{ mA}) = 11.3 \text{ mW}$$

The ac compliance is approximately

$$PP = 2V_{CEQ} = V_{CC} = 15 \text{ V}$$

When an input signal drives the amplifier, the worst-case transistor dissipation is

$$P_{D(max)} = \frac{PP^2}{40R_L} = \frac{(15 \text{ V})^2}{40(50 \text{ }\Omega)} = 113 \text{ mW}$$

The maximum ac load power is

$$P_{L(max)} = \frac{PP^2}{8R_L} = \frac{(15 \text{ V})^2}{8(50 \text{ }\Omega)} = 563 \text{ mW}$$

EXAMPLE 10-2 If the current through the biasing resistors of the preceding example is 2 percent of $i_{c(sat)}$, what is the efficiency of the amplifier with a maximum output signal?

SOLUTION The current through the biasing resistors is

$$I = 0.02i_{c(sat)} = 0.02(150 \text{ mA}) = 3 \text{ mA}$$

With a maximum output signal, the current through the upper transistor is a half-wave rectified sine wave with an average value of approximately

$$I_{dc} = 0.318i_{c(sat)} = 0.318(150 \text{ mA}) = 47.7 \text{ mA}$$

Therefore, the maximum current drain on the power supply is approximately

$$I_{CC} = I_1 + I_{dc} = 3 \text{ mA} + 47.7 \text{ mA} = 50.7 \text{ mA}$$

The dc power from the supply is

$$P_{CC} = V_{CC}I_{CC} = (15 \text{ V})(50.7 \text{ mA}) = 761 \text{ mW}$$

The efficiency is

$$\eta = \frac{P_{L(max)}}{P_{CC}} = \frac{563 \text{ mW}}{761 \text{ mW}} = 73.4 \text{ percent}$$

10-4

VOLTAGE-DIVIDER BIAS

Figure 10-5a shows voltage-divider bias for a class B push-pull amplifier. The two transistors have to be complementary, meaning that they have similar V_{BE} curves, maximum ratings, etc. For instance, a 2N3904 and 2N3906 are complementary, the first being an *npn* transistor and

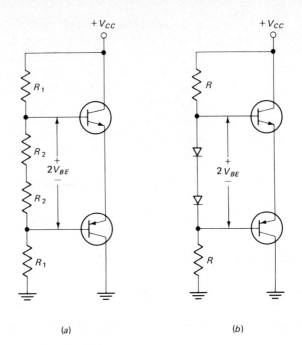

Fig. 10-5.
(a) Voltage-divider bias.
(b) Diode bias.

(a)

(b)

the second a *pnp*. These transistors have similar V_{BE} curves, maximum ratings, and so on. Complementary pairs like these are commercially available for almost any class B push-pull design.

Voltage-Divider Bias Not Stable

In a class A amplifier, voltage-divider bias produces a stable Q point, one that is fixed on the dc load line. But with a class B amplifier, voltage-divider bias does not have an emitter resistor in the dc equivalent circuit. For this reason, voltage-divider bias no longer produces a stable Q point.

The circuit of Fig. 10-5a has an unstable Q point for two reasons. First, the V_{BE} needed to set the Q point slightly above cutoff is typically between 0.6 and 0.7 V. The exact value varies from one transistor to the next. Therefore, the circuit needs an adjustable resistor to set the correct Q point. Second, when the ambient temperature increases, V_{BE} decreases approximately 2 mV per degree rise, which means that I_{CQ} increases. Therefore, even though you adjust the Q point at room temperature, I_{CQ} will increase when the temperature increases. This means the efficiency of the amplifier decreases at higher temperatures.

Diode Bias

Figure 10-5b shows *diode bias*. In this circuit, the diodes are called *compensating diodes* because they can compensate for temperature changes. The idea is to use the voltage across compensating diodes to produce the bias voltage for the emitter diodes. For this scheme to work, the curves of the compensating diodes must match the V_{BE} curves of the transistors. Also, the compensating diodes must have the same temperature coefficient as the emitter diodes. In other words, if the voltage across an emitter diode decreases 2 mV per degree rise, then the voltage across the compensating diode must also decrease 2 mV per degree rise. This way, when the temperature increases, the compensating diodes produce less voltage, as required by the emitter diodes.

For instance, assume that a bias voltage of 0.65 V sets up 2 mA of quiescent collector current. If the temperature rises 30°C, the voltage across such compensating diode decreases

$$(2 \text{ mV})(30) = 60 \text{ mV}$$

Since the required V_{BE} also decreases 60 mV, I_{CQ} remains at 2 mA.

In Fig. 10-5b, the base currents are much smaller than the diode currents. For this reason, the current through the biasing resistors is approximately

$$I = \frac{V_{CC} - 2V_{BE}}{2R}$$

When the diode curves match the V_{BE} curves of the transistors, the collector currents equal the diode currents. Therefore,

$$I_{CQ} = \frac{V_{CC} - 2V_{BE}}{2R} \qquad \textbf{(10-8)}$$

Figure 10-6 shows a class B push-pull amplifier with diode bias. The positive half cycle of input voltage is coupled through the capacitor and through the diodes to the bases. Because the diodes are forward-biased, their ac resistance is small, which allows most of the

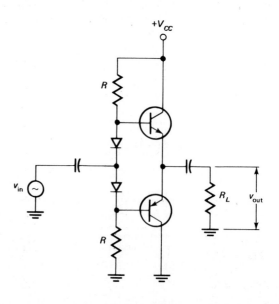

Fig. 10-6.
RC-coupled class B push-pull amplifier.

ac input voltage to reach the bases. The positive half cycle will turn on the upper transistor and shut off the lower one. The emitter voltage of the upper transistor now follows the base voltage, and the output capacitor couples the positive half cycle to the load resistance R_L. The action is complementary on the negative cycle. The final signal across the load is sinusoidal.

Darlington and Sziklai Pairs The Darlington pairs in Fig. 10-7a can increase the input impedance of a class B push-pull emitter follower. Since the Darlington beta is

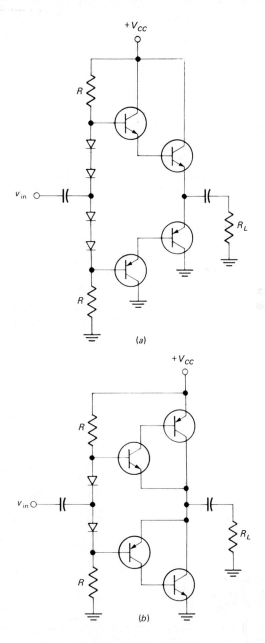

(a)

(b)

Fig. 10-7.
(a) Darlington pairs.
(b) Sziklai pairs.

much higher, the input impedance looking into each base is much higher. A circuit like this is useful with very small load resistors. Since each Darlington pair has two V_{BE} drops, we need to use four compensating diodes instead of two. This makes it more difficult to match the compensating diodes to the emitter diodes.

Figure 10-7b shows *Sziklai pairs*, sometimes called complementary Darlingtons. The upper Sziklai pair acts like a single *npn* transistor, while the lower Sziklai pair acts like a *pnp* transistor. Each Sziklai pair has an effective beta equal to the product of the individual betas, but produces only one V_{BE} drop. This is why we need only two compensating diodes instead of four.

Increases the input impedance of a class "B" push pull emitter follower.

EXAMPLE 10-3   Draw the dc and ac load lines for Fig. 10-8a.

SOLUTION   Since V_{CC} = 40 V, the ac saturation current is

$$i_{c(sat)} = \frac{V_{CEQ}}{R_L} = \frac{20 \text{ V}}{10 \text{ }\Omega} = 2 \text{ A}$$

The ac cutoff voltage is

$$V_{ce(cutoff)} = V_{CEQ} = 20 \text{ V}$$

Figure 10-8b shows the dc and ac load lines.

EXAMPLE 10-4   In Fig. 10-8a, calculate the value of P_{DQ}, $P_{D(max)}$, and $P_{L(max)}$.

SOLUTION   Assuming the diode curves match the V_{BE} curves, I_{CQ} equals the current through the compensating diodes:

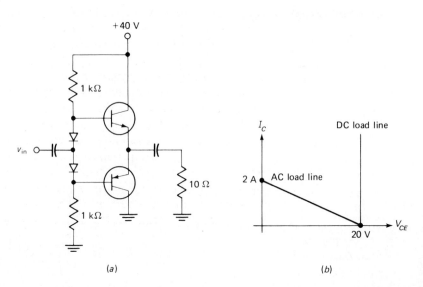

Fig. 10-8.
Drawing the dc and ac load lines of class B circuit.

(a)

(b)

260

$$I_{CQ} = \frac{V_{CC} - 2V_{BE}}{2R} = \frac{40 \text{ V} - 1.4 \text{ V}}{2 \text{ k}\Omega} = 19.3 \text{ mA}$$

Since V_{CEQ} is 20 V,

$$P_{DQ} = V_{CEQ}I_{CQ} = (20 \text{ V})(19.3 \text{ mA}) = 0.386 \text{ W}$$

The ac compliance is

$$PP = 2V_{CEQ} = 2(20 \text{ V}) = 40 \text{ V}$$

The maximum ac load power is

$$P_{L(max)} = \frac{PP^2}{8R_L} = \frac{(40 \text{ V})^2}{8(10 \text{ }\Omega)} = 20 \text{ W}$$

The worst-case power dissipation of each transistor is

$$P_{D(max)} = \frac{PP^2}{40R_L} = \frac{(40 \text{ V})^2}{40(10 \text{ }\Omega)} = 4 \text{ W}$$

10-5
CASCADED
STAGES

In the initial discussion of a class B push–pull emitter follower, capacitors were used to couple the ac signal into the amplifier. This is not the usual way to drive a class B emitter follower. As you may recall from Chap. 7, the usual way to drive an emitter follower is by direct coupling from the preceding stage.

Driver Stage

Figure 10-9 shows how to direct-couple into a class B push-pull emitter follower. Transistor Q_2 is a current source that sets up the dc current through the compensating diodes. By adjusting R_2, we can control the dc emitter current of Q_2. This means that Q_2 sources dc current through the compensating diodes. When the compensating diodes match the emitter diodes, the current through Q_3 and Q_4 approximately equals the current through R_C and R_E.

When an ac signal drives the input, Q_2 acts like a swamped amplifier. This stage is called a *driver stage*. The amplified and inverted ac signal at the Q_2 collector drives the class B push-pull emitter follower, as previously described. Because of the output coupling capacitor, the ac emitter voltage is coupled to the load resistance.

Since the driver stage is a swamped amplifier, you can use the following approximation for its voltage gain:

$$A = \frac{R_C \parallel z_{in}}{R_E + r'_e} \qquad \text{(10-9)}$$

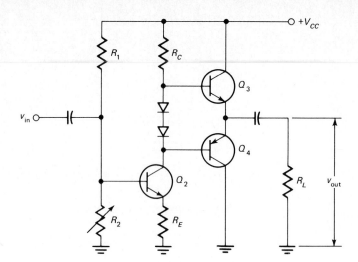

Fig. 10-9.
Direct-coupled driver stage
produces input voltage to
class B push-pull emitter
follower.

where z_{in} is the input impedance looking into the base of the conducting output transistor, and r'_e is the ac emitter resistance of the driver transistor. Often, z_{in} and r'_e are negligible and you can use

$$A = \frac{R_C}{R_E} \qquad\qquad \textbf{\textit{(10-10)}}$$

AC Compliance
Slightly Smaller

Ideally, a class B push-pull amplifier like Fig. 10-9 has an ac compliance of V_{CC}. But the driver stage reduces this ac compliance slightly. Here's why. On the positive half cycle, the base voltage of Q_3 can swing no higher than V_{CC}. Because of the V_{BE} drop in Q_3, the load voltage can swing no higher than $V_{CC} - V_{BE}$. This means we have lost approximately 0.7 V of ac compliance.

On the negative half cycle, the Q_2 collector voltage can swing no lower than the voltage across R_E. Usually, R_E is small, so that its voltage is close to zero. Nevertheless, the voltage across R_E prevents the negative voltage swing at the Q_2 collector from reaching zero. In addition to the voltage across R_E, there is a V_{BE} drop across Q_4, which further reduces the ac compliance.

Therefore, you can always expect the ac compliance of a class B push-pull emitter follower with a driver stage as shown in Fig. 10-9 to be a few volts less than V_{CC}.

Complete
Amplifier

Figure 10-10 is an example of a complete amplifier. It has three stages: a small-signal class A amplifier (Q_1), a driver stage (Q_2), and a class B push-pull emitter follower. The 1 kΩ in the driver stage is adjusted to produce 2.13 V at the base. Subtracting one V_{BE} drop gives 1.43 V at the emitter. The remaining dc voltages are listed for all nodes.

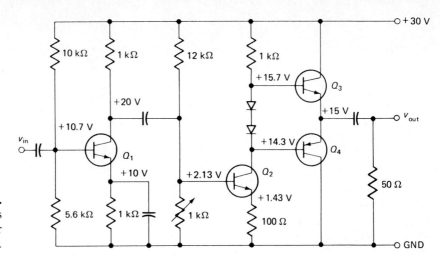

Fig. 10-10.
Cascaded stages drive class
B push-pull emitter
follower.

EXAMPLE 10-5   In Fig. 10-10, what does I_{CQ} equal in the output stage? If $\beta = 200$ for each output transistor, what is the voltage gain of the driver stage?

SOLUTION   The emitter resistor of the driver stage has 1.43 V across it. Therefore, the dc emitter current in Q_2 is,

$$I_E = \frac{V_E}{R_E} = \frac{1.43 \text{ V}}{100 \text{ } \Omega} = 14.3 \text{ mA}$$

Since this is the approximately dc current through the compensating diodes, the quiescent collector currents of Q_3 and Q_4 approximately equal 14.3 mA:

$$I_{CQ} = 14.3 \text{ mA}$$

The input impedance looking into the base of the conducting transistor is

$$z_{in} = \beta R_L = 200(100 \text{ } \Omega) = 20 \text{ k}\Omega$$

and the r'_e is

$$r'_e = \frac{25 \text{ mV}}{I_{CQ}} = \frac{25 \text{ mV}}{14.3 \text{ mA}} = 1.75 \text{ } \Omega$$

The voltage gain of the driver stage is

$$A = \frac{R_C \| z_{in}}{R_E + r'_e} = \frac{1 \text{ k}\Omega \| 20 \text{ k}\Omega}{100 \text{ } \Omega + 1.75 \text{ } \Omega} = 9.36$$

If you ignore Z_{in} and r'_e, you still get a fairly accurate voltage gain of

$$A = \frac{R_C}{R_E} = \frac{1\ k\Omega}{100\ \Omega} = 10$$

When troubleshooting, you can save time by ignoring z_{in} and r'_e.

EXAMPLE 10-6 Calculate the ideal values of PP and $P_{L(max)}$ for the output stage of Fig. 10-10.

SOLUTION Since $V_{CC} = 30$ V, the ac compliance is ideally

$$PP = 2V_{CEQ} = V_{CC} = 30\ V$$

The maximum unclipped ac load power is

$$P_{L(max)} = \frac{PP^2}{8R_L} = \frac{(30\ V)^2}{8(100\ \Omega)} = 1.13\ W$$

As previously discussed, the ac compliance will be a few volts less than V_{CC} because of the V_{BE} drops across the output transistors and the voltage across RE in the driver stage. For this reason, PP and $P_{L(max)}$ will be somewhat less than the ideal answers we have calculated.

EXAMPLE 10-7 You are troubleshooting the circuit of Fig. 10-10. Estimate the overall voltage gain.

SOLUTION Since 10 V is across 1 kΩ in the emitter of the first stage, you can mentally calculate an I_E of 10 mA and an r'_e of 2.5 Ω. Ignore the $z_{in(base)}$ of Q_2. Then, the z_{in} of the second stage is around 1 kΩ. This is in parallel with the 1 kΩ of collector resistance in the first stage. Therefore, an estimate for the voltage gain of the first stage is

$$A_1 = \frac{500\ \Omega}{2.5\ \Omega} = 200$$

The voltage gain of the second stage is approximately 10, found in Example 10-5. The voltage gain of the third stage is approximately 1 because of the emitter followers. The overall voltage gain is

$$A = A_1 A_2 A_3 = (200)(10)(1) = 2000$$

10-6

THE TUNED CLASS C AMPLIFIER

Figure 10-11a is an example of a class C amplifier. The input coupling capacitor, the base resistor, and the emitter diode form a negative clamper, so that only the peaks of the ac input voltage produce base current. As a result, the collector current is a train of narrow pulses, similar to Fig. 10-11b. This nonsinusoidal current contains a funda-

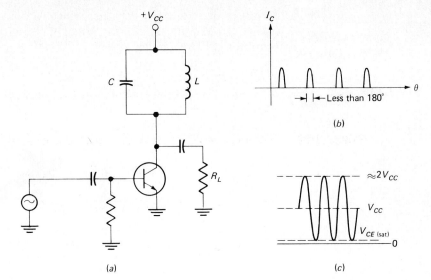

Fig. 10-11.
Class C tuned amplifier.
(a) Circuit. (b) Collector
current flows in pulses.
(c) Output voltage
is sinusoidal.

mental frequency plus harmonics. The fundamental frequency is given by

$$f = \frac{1}{T} \qquad \textbf{(10-11)}$$

where T is the period of the pulses in Fig. 10-11b. The harmonics are all multiples of the fundamental frequency, given by 2f, 3f, 4f, and so on. As an example, if f = 5 kHz, then the harmonics are 10 kHz, 15 kHz, 20 kHz, and so forth.

When the narrow current pulses drive the resonant collector circuit, all harmonics except the fundamental are filtered out, leaving only the fundamental frequency across the load resistor. As a result, the voltage at the collector is sinusoidal (see Fig. 10-11c). Since the quiescent collector voltage is V_{CC}, the maximum voltage swing is from approximately $V_{CE(\text{sat})}$ to 2V_{CC}. The tuned or resonant frequency is given by

$$f = \frac{1}{2\pi\sqrt{LC}} \qquad \textbf{(10-12)}$$

The class C amplifier is more efficient than class A or B because it can deliver more ac load power for the same transistor power dissipation. Typically, the efficiency is more than 90 percent. But a tuned class C amplifier has to use a resonant load with a high Q, which means that only a narrow band of frequencies is amplified. For this reason, class C amplifiers are normally used only in tuned RF applications such as radio and TV receivers.

A transistor is often used as a switch to control large load currents. Figure 10-12a illustrates the idea. When V_{in} is zero, the transistor operates at cutoff in Fig. 10-12b. In this case, no current flows through the load resistor R_L, which may be a lamp, relay, or other heavy load.

When V_{in} is large enough, the operating point switches from cutoff to saturation and the load current is approximately

$$I_{C(sat)} = \frac{V_{CC}}{R_L} \qquad \textbf{(10-13)}$$

The minimum input voltage that produces saturation is

$$V_{in} = I_{B(sat)}R_B + V_{BE} \qquad \textbf{(10-14)}$$

where $I_{B(sat)} = I_{C(sat)}/\beta_{dc}$. As long as the input voltage is greater than the V_{in} given by Eq. (10-15), the transistor acts like a closed switch.

Here are a few of the important quantities on the data sheet of a power-switching transistor:

I_{CEO} = cutoff current with base open (driven by current source)
I_{CES} = cutoff current with base shorted (driven by voltage source)
$V_{CE(sat)}$ = collector-emitter voltage at saturation
$V_{BE(sat)}$ = base-emitter voltage at saturation
t_{on} = time to switch from cutoff to saturation
t_{off} = time to switch from saturation to cutoff

As an example, a 2N5939 has the following values for $I_{C(sat)}$ = 5 A and $V_{CE(cutoff)}$ = 50 V, I_{CEO} = 2 mA, I_{CES} = 2 mA, V_{CE}(sat) = 0.6 V, $V_{BE(sat)}$ = 1.2 V, t_{on} = 135 ns, and t_{off} = 800 ns. Notice that it takes more time to turn a transistor off than it does to turn it on. This is because of the charge storage described in Chap. 3.

EXAMPLE 10-8

The 2N5939 has a $V_{CE(sat)}$ of 1 V maximum when $I_{C(sat)}$ is 10 A. What is the load current when the transistor is saturated in Fig. 10-13a? Also, what is the minimum base current that produces saturation if h_{FE} = 90?

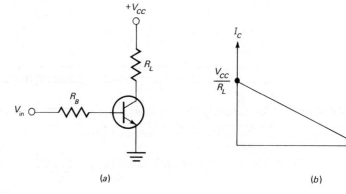

Fig. 10-12.
Transistor operating as
a switch. (a) Circuit.
(b) Load line.

(a)

(b)

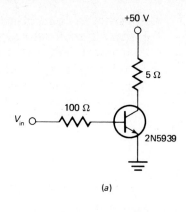

(a)

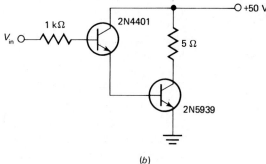

Fig. 10-13.
Switching circuits.

(b)

SOLUTION   The exact saturation current is

$$I_{C(sat)} = \frac{V_{CC} - V_{CE(sat)}}{R_L} = \frac{50 \text{ V} - 1 \text{ V}}{5 \text{ }\Omega} = 9.8 \text{ A}$$

The minimum base current that produces saturation is

$$I_{B(sat)} = \frac{I_{C(sat)}}{\beta_{dc}} = \frac{9.8 \text{ A}}{90} = 109 \text{ mA}$$

EXAMPLE 10-9   In the preceding example, it takes 109 mA of base current to control 9.8 A of collector current. By using an emitter follower as shown in Fig. 10-13b, we can reduce the required input current. The 2N4401 has an h_{FE} of 200. Calculate the minimum input voltage that produces saturation, using 0.7 and 1.6 V for the V_{BE} drops. (Note: power transistors have a larger V_{BE} drop because the base current is much larger and produces an additional drop across the base-spreading resistance.)

SOLUTION   As calculated in the preceding example, it takes 109 mA of base current to saturate the 2N5939. The $I_{B(sat)}$ of the 2N4401 is

$$I_{B(sat)} = \frac{I_{C(sat)}}{\beta_{dc}} = \frac{109 \text{ mA}}{200} = 0.545 \text{ mA}$$

Because of the current gain of the emitter follower, it now takes only 0.545 mA to control 9.8 A of load current. The required input voltage is

$$V_{in} = I_{B(sat)}R_B + V_{BE(1)} + V_{BE(2)}$$
$$= (0.545 \text{ mA})(1 \text{ k}\Omega) + 0.7 \text{ V} + 1.6 \text{ V} = 2.85 \text{ V}$$

Summary

Class B operation requires a push-pull connection of two transistors to avoid excessive distortion of the signal. It has the advantage of less quiescent power dissipation and better efficiency. To avoid crossover distortion, it is necessary to have an I_{CQ} from 1 to 5 percent of $i_{c(sat)}$. Ideally, a class B push-pull emitter follower has an ac compliance equal to the supply voltage.

Voltage-divider bias does not work well with class B operation because the exact V_{BE} needed to set the Q point at cutoff varies with transistors and with temperature. Diode bias is more common, the idea being to use compensating diodes to produce the required V_{BE} of the emitter diodes. Darlington and Sziklai pairs increase the input impedance of a class B push-pull amplifier. Usually, a direct-coupled driver stage produces the input voltage for a class B push-pull emitter follower.

Class C can deliver more ac output power than class A or class B, but class C has to be tuned to a resonant frequency. Because of this, class C is restricted to narrow-band RF applications.

A transistor can also be used as a switch to control the current through a load. When the input voltage is zero, the load current is zero. When the input voltage is large enough, the transistor saturates and the load current equals V_{CC}/R_L.

Glossary

compensating diode In a class B push-pull amplifier, this is a diode whose curve matches the transconductance curve of the emitter diode.

crossover distortion When unbiased, a class B push-pull amplifier has no output until the input voltage exceeds approximately 0.7 V. This results in clipping between positive and negative half cycles. To eliminate crossover distortion, a slight forward bias must be applied to the emitter diodes.

driver stage The stage supplying the signal drive for a class B push-pull emitter follower.

harmonics Any nonsinusoidal voltage is the superposition of sinusoidal voltages with frequencies of f, $2f$, $3f$, . . . , nf. Frequency f is called the fundamental or first harmonic, $2f$, is the second harmonic, $3f$ is the third harmonic, and so on. Frequency f equals $1/T$, where T is the period of the nonsinusoidal voltage.

switching operation Rather than swing the operating point sinusoidally over the ac load line, switching operation means forcing the operating point to jump suddenly from cutoff to saturation, or vice versa. In this type of operation, the transistor functions as a switch rather than a current source.

switching time The jump from cutoff to saturation, or vice versa, is fast but not instantaneous. It takes time to turn on a transistor (t_{on}) and more time to turn it off (t_{off}). Because of charge storage, t_{off} is always greater than t_{on}.

Review Questions

1. For what part of a cycle does a transistor operate with class B?
2. Where is the Q point ideally located for class B operation?
3. What is crossover distortion? How can you get rid of it?
4. Why is voltage-divider bias not suitable for class B operation?
5. Explain the idea behind diode bias.
6. How can you increase the input impedance of a class B push-pull emitter follower?
7. To avoid excessive distortion, what kind of load does a class C amplifier use?
8. When a transistor is used as a switch, how many points are used on the ac load line? What are these points called?

Problems

10-1. Draw the ac load line for Fig. 10-14.

10-2. Calculate $P_{D(max)}$ and $P_{L(max)}$ for Fig. 10-14.

10-3. Assuming the compensating diodes match the emitter diodes, what is the quiescent collector current in Fig. 10-14?

10-4. If the supply voltage changes from 20 to 30 V in Fig. 10-14, what does the quiescent collector current equal?

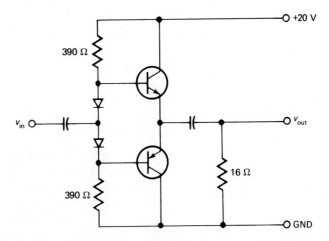

Fig. 10-14.

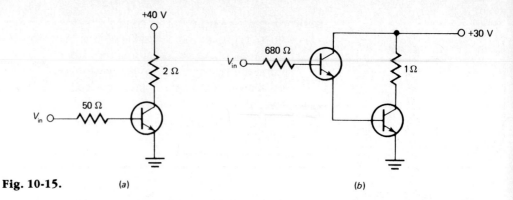

Fig. 10-15.   (a)   (b)

10-5. What is the efficiency of Fig. 10-14 for a maximum output signal?

10-6. If the supply voltage changes from 20 to 30 V in Fig. 10-14, what are the values of P_{DQ}, $P_{D(max)}$, and $P_{L(max)}$?

10-7. If V_{CC} changes from 20 to 30 V in Fig. 10-14, what are the values of P_{CC} and η?

10-8. A class C tuned amplifier has $L = 100\ \mu H$ and $C = 1000$ pF. What is the resonant frequency?

10-9. The transistor of Fig. 10-15a has a V_{BE} drop of 1.1 V at saturation. Ignoring $V_{CE(sat)}$, what is the maximum current through the load resistor? If $\beta_{dc} = 80$, what is the minimum input voltage that produces saturation?

10-10. If $V_{CE(sat)} = 1.2$ V in Fig. 10-15a, what is the value of collector current when the transistor is saturated?

10-11. The input transistor of Fig. 10-15b has an h_{FE} of 250, and the output transistor has an h_{FE} of 75. Ignoring $V_{CE(sat)}$, what is the maximum load current? What is the minimum input base current that produces saturation? If the input and output transistors have V_{BE} drops of 0.7 and 1.2 V, what is the minimum input voltage that produces maximum load current?

More Amplifier Theory

Fleming: In 1904, he saw what Edison missed—the commercial value of the Edison effect. He developed and patented a device that he called a "valve," because it allowed current to flow in one direction but not in the other. In the United States the Fleming valve became known as a diode, the first electronic device that was capable of converting alternating current to direct current.

There are three major topics in this chapter. First, we discuss various ways to cascade transistor stages, including inductive coupling, transformer coupling, and direct coupling. Second, we analyze the frequency response of an *RC*-coupled amplifier, covering topics like cutoff frequencies and bandwidth. Third, we examine *h* parameters, an exact method of analyzing transistor amplifiers.

11-1
RC COUPLING

Figure 11-1 illustrates *resistance-capacitance (RC)* coupling, the most widely used method for coupling a signal from one stage to the another. In ths approach, the signal developed across the collector resistor of each stage is coupled into the base of the next stage. In this way the *cascaded* (one after another) stages amplify the signal, and the overall gain equals the product of the individual gains.

The coupling capacitors transmit ac voltages but block dc voltages. Because of this, the stages are isolated as far as dc voltages are concerned. This is necessary to prevent dc interference between stages and shifting of *Q* points. The drawback to the approach is the decrease in voltage gain at lower frequencies because the coupling capacitors no longer act like ac shorts.

The bypass capacitors are needed because they short the emitters to ac ground. Without them, the voltage gain of each stage would be much less because of the swamping effect of R_E. These bypass capacitors limit the frequency response because their capacitive reactance increases at low frequencies.

If you are interested in amplifying ac signals with frequencies

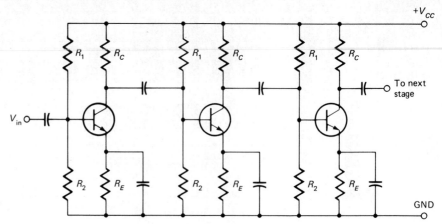

Fig. 11-1.
RC-coupled stages.

greater than 10 Hz, the *RC*-coupled amplifier is suitable. For discrete circuits it is the most convenient and least expensive way to build a multistage amplifier.

11-2
TWO-STAGE FEEDBACK

Earlier, we discussed single-stage feedback using a swamping resistor. It is also possible to use feedback around two stages, as shown in Fig. 11-2. The input signal v_{in} is amplified and inverted in the first stage. The output of this first stage is amplified and inverted again by the second stage. Part of this second-stage output is fed back to the first stage via the voltage divider formed by r_F and r_E. The feedback voltage v_F is applied to the emitter of the first stage. Because this voltage returns in phase with the input voltage, the base-emitter voltage decreases. This is called *negative feedback*.

Basic Idea

Negative feedback reduces the overall voltage gain of an amplifier. In exchange for this loss of gain, we get gain stability. Here is the idea. Suppose the r'_e of the second stage becomes smaller. The output voltage will increase, which means more voltage is fed back to the emitter of the first transistor. This reduces the base-emitter voltage and the output signal from the first stage. Since less voltage drives the second stage, the final output voltage is reduced. The overall effect is that the output voltage increases much less than it would have without the negative feedback.

Similarly, if the r'_e of the second stage increases, we get less output voltage. This means less feedback voltage to the first stage. The base-emitter voltage of the first stage therefore increases and produces a larger voltage to drive the second stage. With more input to the second stage, we get more output voltage, which almost completely offsets the original decrease in output voltage.

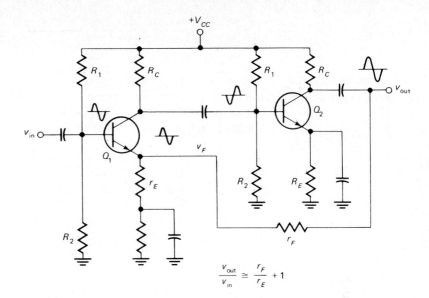

Fig. 11-2.
Two-stage feedback.

$$\frac{v_{out}}{v_{in}} \cong \frac{r_F}{r_E} + 1$$

Voltage Gain In Fig. 11-2, the ac voltage across the emitter diode is $v_{in} - v_F$. If the first stage has a gain of A_1 and the second stage a gain of A_2, the ac output voltage is

$$v_{out} = A_1 A_2 (v_{in} - v_F) \qquad\qquad \textbf{(11-1)}$$

The ac voltage being fed back to the first emitter is

$$v_F = \frac{r_E}{r_F + r_E} v_{out} \qquad\qquad \textbf{(11-2)}$$

This assumes that r_E is small compared with the output impedance of the Q_1 emitter. If Eq. (11-2) is substituted into Eq. (11.1), we can rearrange to get

$$\frac{v_{out}}{v_{in}} = \frac{A_1 A_2}{1 + A_1 A_2\, r_E/(r_F + r_E)}$$

In most circuits, the second term in the denominator is much greater than unity, and the equation simplifies to

$$\frac{v_{out}}{v_{in}} = \frac{r_F}{r_E} + 1 \qquad\qquad \textbf{(11-3)}$$

This tells us the overall voltage gain depends on the ratio of two discrete resistors. If precision resistors are used, we get a precise value of voltage gain. Furthermore, it means you can replace transistors without changing the voltage gain. This is ideal for mass production because we get a voltage gain that is almost constant despite transistor replacement or temperature change.

Open-Loop and Closed-Loop Gain

If we open the feedback resistor r_F of Fig. 11-2, the voltage gain will increase sharply because the negative feedback is lost. The gain with the feedback loop open is called the *open-loop voltage gain,* designated A_{OL}. In Fig. 11-2,

$$A_{OL} = A_1 A_2 \qquad \textbf{\textit{(11-4)}}$$

where $A_1 = r_{L1}/r'_{e1}$
$\quad\quad\ A_2 = r_{L2}/r'_{e2}$

For instance, if each stage has a voltage gain of 100, the open-loop gain is

$$A_{OL} = (100)(100) = 10,000$$

The voltage gain with the feedback loop closed is called the *closed-loop voltage gain,* symbolized A_{CL}. In other words, Eq. (11-3) may be written as

$$A_{CL} = \frac{r_F}{r_E} + 1 \qquad \textbf{\textit{(11-5)}}$$

EXAMPLE 11-1 If $r_F = 220\ \Omega$ and $r_E = 4.7\ \Omega$, what is the closed-loop voltage gain in Fig. 11-2?

SOLUTION The closed-loop voltage gain is

$$A_{CL} = \frac{r_F}{r_E} + 1 = \frac{220\ \Omega}{4.7\ \Omega} + 1 = 47.8$$

11-3

INDUCTIVE COUPLING

Occasionally, transistor stages will use *RF chokes* (inductors with a high reactance at radio frequencies) instead of collector resistors, as shown in Fig. 11-3. The idea is to prevent the loss of signal power that takes place in collector resistors. Such amplifiers are intended to amplify frequencies that are high enough for the RF chokes to appear as ac open circuits. In this case, the ac load resistance seen by the first stage is the input impedance of the second stage. The ac load resistance seen by the second stage is the final load resistance R_L.

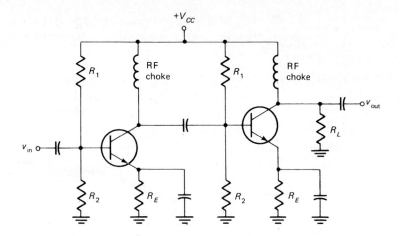

Fig. 11-3.
Inductive coupling.

11-4
TRANSFORMER COUPLING

Figure 11-4 shows transformer coupling to a load resistance such as a loudspeaker in a radio. With a transformer, the small load resistance (only 3.2 Ω for some loudspeakers) can be stepped up to a higher impedance level. This improves the voltage gain. Furthermore, since no signal power is wasted in a collector resistor, all the ac power is delivered to the final load resistor.

Transformer coupling was once popular at audio frequencies (20 Hz to 20 kHz). But the cost and bulkiness of audio transformers was a major disadvantage. When complementary transistors (matched *npn* and *pnp* transistors) became available, the class B push-pull emitter follower replaced the transformer-coupled output stage of most audio amplifiers.

The one area where transformer coupling has survived is *radio-frequency* (RF) amplifiers. Radio frequency means all frequencies greater than 20 kHz. In AM receivers, the RF signals have frequencies are from 535 to 1605 kHz. In TV receivers, the RF frequencies are from 54 to 216 MHz (channels 2 through 13). Transformer coupling is still used in RF amplifiers because RF transformers are much smaller and less expensive than audio transformers.

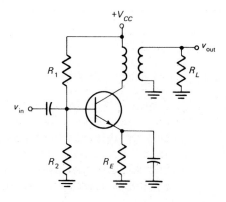

Fig. 11-4.
Transformer coupling.

Often, the stages of an RF amplifier are designed to amplify a narrow band of frequencies. This is how a radio or TV receiver separates one station from another. Figure 11-5a is an example of tuned transformer coupling. At resonance, the impedance of each *LC* tank is high. Above and below the resonant frequency the impedance decreases. Therefore, the voltage gain is maximum at resonance, as shown in Fig. 11-5b.

The formula for the resonant frequency is

$$f_r = \frac{1}{2\pi\sqrt{LC}} \tag{11-6}$$

The *cutoff frequencies* f_1 and f_2 of a tuned tank are the frequencies where the voltage gain decreases to 0.707 of the maximum voltage gain (see Fig. 11-5b). The *bandwidth* of a tuned tank is the difference between the cutoff frequencies:

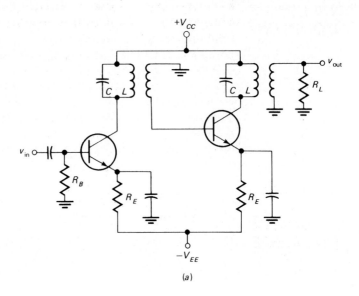

(a)

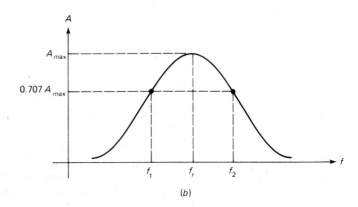

(b)

Fig. 11-5.
Tuned transformer
coupling. (a) Circuit.
(b) Frequency response.

$$B = f_2 - f_1 \qquad \textbf{(11-7)}$$

Basic courses in ac theory derive this alternative formula for bandwidth:

$$B = \frac{f_r}{Q} \qquad \textbf{(11-8)}$$

where Q is the circuit Q given by

$$Q = \frac{r_L}{X_L} \qquad \textbf{(11-9)}$$

Each collector current source drives a parallel LC tank. At resonance, the inductive reactance equals the capacitive reactance, leaving a purely resistive load on each collector. Above resonance, X_C becomes smaller than X_L, causing the voltage gain to decrease as shown in Fig. 11-5b. Below resonance, X_L is smaller than X_C, and the voltage gain again drops off.

Tuned transformer coupling like Fig. 11-5a produces a *bandpass amplifier,* one with maximum voltage gain at the resonant frequency. As we move away from the resonant frequency, the voltage gain decreases rapidly. Since the resonant frequency receives more gain than other frequencies, we are able to tune in a desired radio station or TV channel.

11-6
DIRECT
COUPLING

All amplifiers discussed so far have been capacitor-coupled or transformer-coupled between stages. This limits the low-frequency response. In other words, the voltage gain of the amplifiers decreases at lower frequencies. One way to avoid this is with *direct coupling,* which means providing a dc path between stages.

One-Supply Circuit

Figure 11-6 is a two-stage direct-coupled amplifier. No coupling or bypass capacitors are used. Because of this, dc voltage is amplified as well as ac voltage. With a quiescent input voltage of $+1.4$ V, about 0.7 V is dropped across the first emitter diode, leaving 0.7 V across the 680 Ω. (Note: the symbol $\approx$ stands for "approximately.") This sets up approximately 1 mA of collector current. This 1 mA produces a drop of 27 V across the collector resistor. Therefore, the first collector runs at about $+3$V with respect to ground.

Allowing 0.7 V for the second emitter diode, we get 2.3 V across the 2.4 kΩ. This results in approximately 1 mA of collector current, a drop of around 24 V across the collector resistor, and a final output voltage of approximately $+6$ V to ground. Therefore, a quiescent input voltage of $+1.4$ V produces a quiescent output voltage of $+6$ V.

Because of the high β_{dc}, we can ignore the loading effect of the second base upon the first collector. Ignoring r'_e, the first stage has a voltage gain of

$$A_1 = \frac{27,000 \ \Omega}{680 \ \Omega} = 40$$

The second stage has a voltage gain of

$$A_2 = \frac{24 \ k\Omega}{2.4 \ k\Omega} = 10$$

The overall voltage gain is

$$A = A_1 A_2 = (40)(10) = 400$$

The two-stage circuit will amplify any change in the input voltage by a factor of 400. For instance, if the input voltage changes by $+5$ mV, the final output voltage changes by

$$400(5 \ mV) = 2 \ V$$

Therefore, the total output voltage changes from $+6$ to $+8$ V.

Here is the main disadvantage of direct coupling. Transistor *parameters* (characteristics) like V_{BE} change with temperature. This causes quiescent collector currents and voltages to change. Because of the direct coupling, the voltage changes are coupled from one stage to the next, appearing at the final ouput as an amplified voltage change. This unwanted change is called *drift*. The trouble with drift is you cannot distinguish it from a genuine change produced by the input voltage.

Ground-Referenced Input

For the two-stage amplifier of Fig. 11-6 to work properly, we need a quiescent input voltage of $+1.4$ V. In typical applications, it is necessary to have a ground-referenced input, one where the quiescent input voltage is 0 V.

Figure 11-7 shows a ground-referenced input stage. This stage is a *pnp* Darlington with the input base returned to ground through the signal source. Because of this, the first emitter is approximately $+0.7$ V above ground, and the second emitter is about $+1.4$ V above ground. The $+1.4$ V biases the second stage, which operates as previously described.

The quiescent V_{CE} of the first transistor is only 0.7 V, and the quiescent V_{CE} of the second transistor is only 1.4 V. Nevertheless, both transistors are operating in the active region because the $V_{CE(sat)}$ of small-signal transistors is only about 0.1 V. Since the input signal is typically in millivolts, the input transistors continue to operate in the

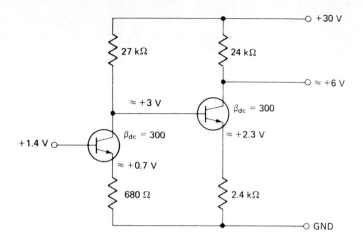

Fig. 11-6.
Direct-coupled stages.

active region. The *pnp* ground-referenced input is used a lot in audio integrated circuits.

Two-Supply Circuit

When positive and negative supply voltages are available, we can reference both the input and the output to ground. Figure 11-8 is an example. The first stage is emitter-biased with an I_E around 1 mA. This produces about +3 V at the first collector. Subtracting the V_{BE} drop of the second emitter diode leaves +2.3 V at the second emitter.

The emitter current in the second stage is around 1 mA. This flows through the collector resistor, producing about +6 V from the collector to ground. The final stage has +5.3 V across the emitter resistor, which gives about 1 mA of current. Therefore, the last collector has approximately +10 V to ground.

The output voltage divider references the output to ground. When the upper resistor is adjusted to 200 kΩ, the final output voltage is

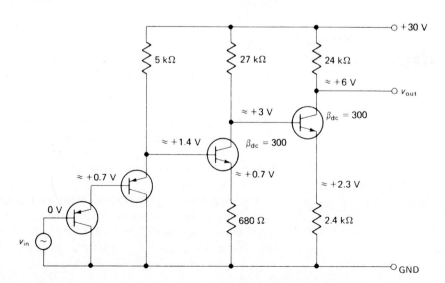

Fig. 11-7.
Direct coupling with ground-referenced input.

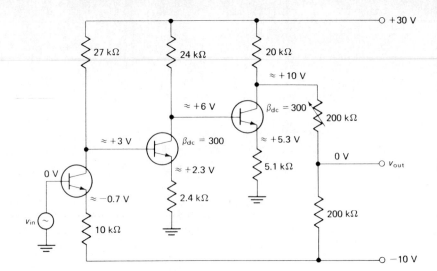

+30 V

27 kΩ   24 kΩ   20 kΩ

≈ +10 V

≈ +6 V $\beta_{dc} = 300$ 200 kΩ

≈ +3 V $\beta_{dc} = 300$ ≈ +5.3 V   0 V v_{out}

0 V   ≈ +2.3 V   5.1 kΩ

≈ −0.7 V   2.4 kΩ

Fig. 11-8.
Direct coupling with
ground-referenced input
and output.

v_{in} 10 kΩ   200 kΩ

−10 V

approximately 0 V. The adjustment allows us to eliminate errors caused by resistor tolerances, V_{BE} differences, etc.

What is the overall voltage gain? The first stage has a gain around 2.7, the second stage about 10, the third stage approximately 4, and the voltage divider around 0.5. Therefore,

$$A = (2.7)(10)(4)(0.5) = 54$$

Review of Direct Coupling

This gives you the idea behind direct coupling. Capacitors are not used, allowing dc voltage as well as ac voltage to reach the next stage. The main advantage of direct coupling is that the amplifier has no lower frequency limit because it amplifies all frequencies down to zero or dc voltage. But the disadvantage of direct coupling is that all dc voltage changes are amplified, including those caused by changes in supply voltage, transistor variations, etc.

There is a way to reduce drift with a direct-coupled circuit called a *differential amplifier.* This special kind of amplifier is the key circuit used in *operational amplifiers.* Chapter 15 discusses differential amplifiers and operational amplifiers in detail.

11-7

RESPONSE OF AN RC-COUPLED AMPLIFIER

The *RC*-coupled amplifier is the most common type of discrete amplifier. Figure 11-9 shows the *frequency response,* a graph of ac output voltage versus frequency for a fixed ac input voltage. For very low or very high frequencies, the output voltage decreases. In the middle of the frequency range, however, the output voltage is constant. It is in this middle range of frequencies that an *RC*-coupled amplifier normally operates.

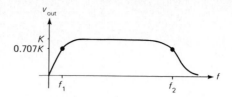

Fig. 11-9.
Frequency response of
RC-coupled amplifier.

Cutoff Frequencies In Fig. 11-9, the output voltage equals K in the mid-frequency range. If we increase or decrease the frequency, we reach a point where the output voltage equals $0.707K$. The frequencies for these $0.707K$ points are called the *cutoff frequencies* (also known as the half-power frequencies, break frequencies, and corner frequencies).

Figure 11-9 shows two cutoff frequencies f_1 and f_2. Coupling and bypass capacitors are responsible for the lower cutoff frequency f_1. Stray-wiring and internal transistor capacitances are the cause of the upper cutoff frequency f_2. All frequencies between f_1 and f_2 are the *passband* of the amplifier. Outside the passband, the voltage gain drops off rapidly.

Bandwidth The bandwidth B is important when discussing amplifiers. It is defined as the width of the passband:

$$B = f_2 - f_1 \qquad \textbf{(11-10)}$$

For instance, if an audio amplifier has $f_1 = 20$ Hz and $f_2 = 15$ kHz, then it has a bandwith of

$$B = 15,000 \text{ Hz} - 20 \text{ Hz} = 14,980 \text{ Hz} \cong 15 \text{ kHz}$$

In a direct-coupled amplifier, there is no lower cutoff frequency, so that $f_1 = 0$. In this case, the bandwidth equals the upper cutoff frequency:

$$B = f_2 \qquad \textbf{(11-11)}$$

11-8

LOWER CE CUTOFF FREQUENCIES In a CE amplifier like Fig. 11-10, the coupling and bypass capacitors produce lower cutoff frequencies. What happens is this. As the frequency decreases, the capacitive reactance increases. Eventually, the coupling capacitors drop a significant amount of the ac voltage, and the bypass capacitor no longer ac grounds the emitter.

Input Coupling Capacitor In Fig. 11-10, the input impedance of the stage is

$$z_{in} = R_1 \parallel R_2 \parallel z_{in(base)}$$

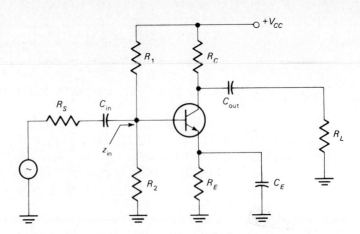

Fig. 11-10.
CE amplifier.

The input coupling capacitor C_{in} is in series with R_S and z_{in}. As discussed in basic circuit theory, the ac current is down to the 0.707 point when the capacitive reactance equals the total series resistance:

$$X_C = R$$

or

$$\frac{1}{2\pi f C_{in}} = R_S + z_{in}$$

$R_S + Z_{in} = R$

Solving for frequency, we get

$$f = \frac{1}{2\pi(R_S + z_{in})C_{in}}$$

To distinguish this cutoff frequency from others, we will add the subscript "in" as follows:

$$f_{in} = \frac{1}{2\pi(R_S + z_{in})C_{in}} \qquad (11\text{-}12)$$

Use this formula when you want to calculate the cutoff frequency produced by the input coupling capacitor.

Output Coupling Capacitor On the output side, coupling capacitor C_{out} is in series with the output impedance R_C and the load resistance R_L. The ac output current is down to the 0.707 point when

$$X_C = R$$

or

$$\frac{1}{2\pi f C_{out}} = R_C + R_L$$

Solving for frequency gives

$$f = \frac{1}{2\pi(R_C + R_L)C_{\text{out}}}$$

To distinguish this from other cutoff frequencies, we add the subscript "out" to get

$$f_{\text{out}} = \frac{1}{2\pi(R_C + R_L)C_{\text{out}}} \qquad \textbf{(11-13)}$$

This is the cutoff frequency produced by the output coupling capacitor.

Emitter Bypass Capacitor

In Fig. 11-10, the bypass capacitor sees a resistance of R_E in parallel with the output impedance of the emitter. This output impedance equals the ac resistance of the emitter diode plus the base resistance divided by the current gain. As a formula,

$$z_{\text{out}} = r'_e + \frac{R_S \parallel R_1 \parallel R_2}{\beta} \qquad \textbf{(11-14)}$$

This is the impedance looking back into the emitter from the bypass capacitor. The equivalent resistance in parallel with the bypass capacitor is $R_E \parallel z_{\text{out}}$.

When the capacitive reactance equals the total parallel resistance, the voltage gain drops to the 0.707 point. In other words, the cutoff frequency occurs when

$$X_C = R$$

or

$$\frac{1}{2\pi fC_E} = R_E \parallel z_{\text{out}}$$

or

$$\frac{1}{2\pi fC_E} = R_E \parallel z_{\text{out}}$$

Solving for frequency gives

$$f = \frac{1}{2\pi(R_E \parallel z_{\text{out}})C_E}$$

To distinguish this from other cutoff frequencies, we add the subscript "E" to get

$$f_E = \frac{1}{2\pi(R_E \parallel z_{\text{out}})C_E} \qquad \textbf{(11-15)}$$

This it the cutoff frequency produced by the emitter bypass capacitor.

Which One to Use The three capacitors produce three different cutoff frequencies. The highest one is the most critical because it is frequency where the voltage gain first decreases from the passband value. For instance, if the three cutoff frequencies are 10, 50, and 200 Hz, then the 200 Hz is the most important. Usually, the three cutoff frequencies are different; only by a coincidence would they all be the same. Therefore, when analyzing an amplifier, use the highest of the lower cutoff frequencies because this is the one that places a limit on amplifier performance.

11-9
TRANSISTOR CUTOFF FREQUENCIES

As the frequency increases, certain things happen inside a transistor that reduce the voltage gain. Bear in mind that a transistor has internal capacitances, charge storage, and other effects that can alter its high-frequency response.

Alpha Cutoff Frequency

The *ac alpha* of a transistor is the ratio of the ac collector current to the ac emitter current:

$$\alpha = \frac{i_c}{i_e}$$

At low frequencies, α approaches unity. But as the frequency increases, we eventually reach a point where charge storage decreases the value of α. The higher the frequency, the lower the value of α.

The *alpha cutoff frequency* f_α is the frequency where α has dropped to 0.707 of its low-frequency value. For instance, suppose a transistor has an α of 0.98 at low frequencies and an f_α of 300 MHz. If you try to operate this transistor at 300 MHz, the value of α will be

$$\alpha = (0.707)(0.98) = 0.693$$

This means that at 300 MHz the ac collector current is only 0.693 times the ac emitter current.

The f_α is one of the limitations on the upper-frequency response of a CB amplifier. When possible, we select a transistor whose f_α is much higher than the highest operating frequency of the CB amplifier.

Beta Cutoff Frequency

The *beta cuttoff frequency* f_β is another important transistor parameter or characteristic. It is the frequency where the β of the transistor has decreased to 0.707 of its low-frequency value. For instance, if a transistor has a low-frequency β of 250 and an f_β of 2 MHz, then at 2 MHz the β of the transistor is

$$\beta = (0.707)(250) = 177$$

Current-Gain Bandwidth Product

The *current-gain bandwidth product* f_T is the frequency where β equals unity. When the frequency increases beyond f_β, β keeps decreasing until eventually it equals unity. The frequency at this point is f_T, which is called the current-gain bandwidth product.

Relationships

The f_T of a transistor is much higher than the f_β. The relation between these two frequencies is

$$f_\beta = \frac{f_T}{\beta} \qquad (11\text{-}16)$$

where β is the low-frequency value of β. If a data sheet lists an f_T of 100 MHz and a low-frequency β of 50, then,

$$f_\beta = \frac{100\ \text{MHz}}{50} = 2\ \text{MHz}$$

The f_α and f_T are also related. As a rough approximation, $f_\alpha = f_T$. In actuality, f_T is less than f_α. For junction transistors, a better approximation is

$$f_T \cong \frac{f_\alpha}{1.2} \qquad (11\text{-}17)$$

The various transistor cutoff frequencies are important in high-frequency analysis. The f_α is one of the limitations of a CB amplifier. The f_β and f_T are limitations of a CE amplifier. The analysis of transistor amplifiers at high frequencies is very complicated and is discussed elsewhere.* As a rough guideline, the f_β of a transistor in a CE amplifier should be higher than the highest frequency you are trying to amplify. Likewise, the f_α of a CB amplifier should be higher than the highest frequency being amplified.

11-10 HYBRID PARAMETERS

Hybrid (*h*) parameters are easy to measure. This is the reason some transistor data sheets specify low-frequency characteristics in terms of four *h* parameters. This section tells you what *h* parameters are, and how they are related to the *r* parameters we have been using.

What They Are

The four *h* parameters of the CE connection are

h_{ie} = imput impedance with an r_L of zero
h_{fe} = ac current gain with an r_L of zero

---

*See Malvino, Albert P.: *Electronic Principles,* Third Edition, McGraw-Hill Book Company, New York, 1984, pages 398–410.

h_{re} = reverse voltage gain with an R_s of infinity
h_{oe} = output admittance with an R_s of infinity

The first two parameters are specified for an ac load resistance of zero, equivalent to a shorted output. The next two parameters are specified for a source resistance of infinity, equivalent to an open circuit.

Figure 11-11a shows how to measure h_{ie} and h_{fe}. To begin with, we use an ac short across the output. This prevents loading effects and ensures unambiguous values for h_{ie} and h_{fe}. In other words, the input impedance and current gain will vary if the load resistance is too large. To avoid this, we short the output terminals. In Fig. 11-11a, the ratio of input voltage to input current is the input impedance:

$$h_{ie} = \frac{v_b}{i_b} \quad \text{(for } v_{ce} = 0\text{)}$$

The ratio of output current to input current is the current gain:

$$h_{fe} = \frac{i_c}{i_b} \quad \text{(for } v_{ce} = 0\text{)}$$

To get the other two parameters, we use the ac equivalent circuit of Fig. 11-11b. Here we drive the collector with a voltage source v_c. This forces a collector current i_c to flow. The base is left open. The ratio of ac base voltage to ac collector voltage is the reverse voltage gain:

$$h_{re} = \frac{v_b}{v_c} \quad \text{(for } i_b = 0\text{)}$$

The fourth parameter is the ratio of ac collector current to ac collector voltage:

$$h_{oe} = \frac{i_c}{v_c} \quad \text{(for } i_b = 0\text{)}$$

Because this is a ratio of current to voltage, it is known as the output admittance.

Fig. 11-11.
Ac equivalent circuits for measuring h parameters. (a) h_{ie} and h_{fe}. (b) h_{re} and h_{oe}.

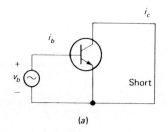

(a)

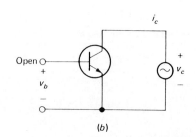

(b)

Relation to r Parameters   The r parameters (r'_e, r'_b, r'_c, α, and β) are the easiest to work with, but the h parameters are the easiest to measure. Therefore, we need to know how to convert from h parameters to r parameters. This way, when a data sheet specifies the h parameters of a transistor, we can convert to the r parameters and analyze transistor amplifiers with the simple methods of earlier chapters.

Table 11-1 shows the approximate relations between r and h parameters. As indicated, $\beta = h_{fe}$, $\alpha = h_{fe}/(h_{fe} + 1)$, and so on. Most data sheets list the CE h parameters; so with Table 11-1 you can convert to r parameters.

The only unusual entry in Table 11-1 is

$$r'_b = \frac{h_{rb}}{h_{ob}}$$

This is the ratio of the reverse voltage gain and the output admittance of a CB circuit. The easiest and most reliable way to measure r'_b is with a CB circuit. Reverse voltage gain h_{rb} divided by output admittance h_{ob} gives the value of base spreading resistance r'_b.

TABLE 11-1. Approximate Relations

r parameter	h parameter
β	h_{fe}
α	$h_{fe}/(h_{fe} + 1)$
r'_e	h_{ie}/h_{fe}
r'_c	h_{fe}/h_{oe}
r'_b	h_{rb}/h_{ob}

EXAMPLE 11-2   The data sheet of a 2N3904 shows the following typical values at $I_C = 1$ mA:

$$h_{ie} = 3.5 \text{ k}\Omega$$
$$h_{fe} = 120$$
$$h_{re} = 1.3(10^{-4})$$
$$h_{oe} = 8.5 \text{ }\mu\text{S}$$

What are the values of α, β, r'_e, and r'_c?

SOLUTION

$$\beta = h_{fe} = 120$$

$$\alpha = \frac{h_{fe}}{h_{fe} + 1} = \frac{120}{120 + 1} = 0.992$$

$$r'_e = \frac{h_{ie}}{h_{fe}} = \frac{3500 \ \Omega}{120} = 29 \ \Omega$$

$$r'_c = \frac{h_{fe}}{h_{oe}} = \frac{120}{8.5(10^{-6})} = 14.1 \ \text{M}\Omega$$

EXAMPLE 11-3   The 2N1975 data sheet specifies $h_{rb} = 1.75(10^{-4})$ and $h_{ob} = 1 \ \mu\text{S}$. What is the value of r'_b?

SOLUTION

$$r'_b = \frac{h_{rb}}{h_{ob}} = \frac{1.75(10^{-4})}{10^{-6} \ \text{S}} = 175 \ \Omega$$

11-11
EXACT HYBRID FORMULAS

The ideal transistor approximation is adequate for most troubleshooting and design. When more accurate answers are needed, you can use the h-parameter formulas of this section. These formulas are complicated because they take everything into account. Normally, you would not use these formulas unless a computer was available.

In deriving exact formulas based on h parameters, it is helpful to use the numerical parameters h_{11}, h_{12}, h_{21}, and h_{22} shown in Fig. 11-12. These parameters have the following meaning for any transistor connection:

h_{11} = input impedance with a shorted output
h_{12} = reverse voltage gain with an open input
h_{21} = current gain with a shorted output
h_{22} = output admittance with an open input

Table 11-2 shows how these parameters are related to the h parameters of each transistor connection. For instance, for a CE amplifier the relations are

TABLE 11-2. Hybrid Relations

Numerical	CE	CC	CB
h_{11}	h_{ie}	h_{ic}	h_{ib}
h_{12}	h_{re}	h_{rc}	h_{rb}
h_{21}	h_{fe}	h_{fc}	h_{fb}
h_{22}	h_{oe}	h_{oc}	h_{ob}

$$h_{11} = h_{ie}$$
$$h_{12} = h_{re}$$
$$h_{21} = h_{fe}$$
$$h_{22} = h_{oe}$$

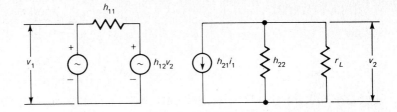

Fig. 11-12.
Ac equivalent circuit of amplifier with h parameter model.

Table 11-3 lists the formulas for an amplifier. All you have to do is substitute the parameters for the particular connection. For example, suppose you want the voltage gain of a CE amplifier. With Tables 11-2 and 11-3, you can write

$$A = \frac{h_{fe} r_L}{h_{ie}(1 + h_{oe} r_L) - h_{re} h_{fe} r_L}$$

The exact formulas of Table 11-3 give exact answers, provided you have the exact h parameters of the transistor being used. This is where a practical problem arises. The tolerance in h parameters is huge. As an example, the data sheet of a 2N3904 gives these h parameters:

$$h_{ie} = 1 \text{ to } 10 \text{ k}\Omega$$
$$h_{re} = 0.5(10^{-4}) \text{ to } 8(10^{-4})$$
$$h_{fe} = 100 \text{ to } 400$$
$$h_{oe} = 1 \text{ to } 40 \text{ } \mu\text{S}$$

With variations like these, exact formulas lose much of their appeal.

TABLE 11-3 Hybrid Formulas

	Exact	Approximate
A_v	$\dfrac{h_{21} r_L}{h_{11}(1 + h_{22} r_L) - h_{12} h_{21} r_L}$	$\dfrac{h_{21} r_L}{h_{11}}$
A_i	$\dfrac{h_{21}}{1 + h_{22} r_L}$	h_{21}
Z_{in}	$h_{11} - \dfrac{h_{12} h_{21}}{h_{22} + 1/r_L}$	h_{11}
Z_{out}	$\dfrac{R_S + h_{11}}{(R_S + h_{11})h_{22} - h_{12} h_{21}}$	$\dfrac{1}{h_{22}}$

EXAMPLE 11-4   A 2N3904 is used in a CE amplifier with $r_L = 2$ kΩ. With the h parameters of Example 11-2, calculate the exact voltage gain.

SOLUTION

$$A = \frac{h_{fe}r_L}{h_{ie}(1 + h_{oe}r_L) - h_{re}h_{fe}r_L}$$

$$= \frac{120(2000)}{3500[1 + 8.5(10^{-6})2000] - 1.3(10^{-4})120(2000)}$$
$$= 68$$

Summary

RC coupling between stages is the most convenient and least expensive way to build a multistage amplifier. Negative feedback around two stages stabilizes the overall voltage gain against transistor and temperature changes.

Other types of coupling include inductive coupling, transformer coupling, and direct coupling. With inductive coupling, only high frequencies are coupled. With transformer coupling, only a band of frequencies is coupled. With direct coupling, both dc and ac voltage are coupled. Direct coupling has a disadvantage called drift, a change in the output voltage produced by changes in supply voltages, transistors, etc.

The frequency response of an *RC*-coupled amplifier is a graph of ac output voltage versus frequency for a fixed ac input voltage. The cutoff frequencies are the frequencies where the voltage gain decreases to 0.707 of its maximum value. The bandwidth of an amplifier is the difference between the upper and lower cutoff frequencies. Coupling and bypass capacitors produce the lower cutoff frequency, while stray-wiring capacitance and internal transistor capacitances produce the upper cutoff frequency.

The alpha cutoff frequency f_α is the frequency where α has decreased to 0.707 of its low-frequency value. The beta cutoff frequency f_β is the frequency where β has decreased to 0.707 of its low-frequency value. The current-gain bandwidth product f_T is the frequency where β equals unity. These cutoff frequencies limit the upper frequency where β equals unity. These cutoff frequencies limit the upper frequency response of an amplifier.

The *h* parameters are easily and accurately measured. For this reason, they are used on data sheets to specify transistor characteristics. The *r* parameters are easier to understand and work with. One approach is to convert *h* parameters to *r* parameters to get answers that are adequate for most troubleshooting and design. Another approach is use the more complicated *h*-parameter formulas to get accurate answers. The main drawback with *h* parameters is their large tolerances.

Glossary

audio frequency Any frequency between 20 Hz and 20 kHz.

bandwidth The difference between the upper and lower cutoff frequencies.

cutoff frequency The frequency where the voltage gain is down to 0.707 of its maximum value.

direct coupling A circuit where the dc outpt voltage of one stage is coupled to the next stage.

drift This is a change in the dc output voltage caused by temperature, power-supply variations, transistor changes, etc. Drift is indistinguishable from a change produced by the input voltage.

negative feedback Returning part of the amplified output voltage to the input with a phase that opposes the input voltage. Although this decreases the voltage gain, negative feedback is widely used because it stabilizes the voltage gain against transistor and temperature changes.

RF chokes Inductors that appear open to RF frequencies.

Review Questions

1. What is the most widely used method of coupling between the stages of a discrete amplifier?
2. What is negative feedback?
3. What is the difference between the open-loop voltage gain and the closed-loop voltage gain?
4. What kind of coupling uses RF chokes instead of collector resistors?
5. What is the disadvantage of transformer coupling at audio frequencies?
6. Define the cutoff frequencies of an amplifier. How is the bandwidth related to these cutoff frequencies?
7. What is the advantage of direct coupling? The disadvantage?
8. In an *RC*-coupled amplifier, which capacitors produce the lower cutoff frequencies?
9. Which are the three transistor cutoff frequencies that affect the upper cutoff frequencies of an amplifier?
10. Name the four *h* parameters for the CE connection.
11. How are the *h* parameters measured?
12. What is the main drawback of *h* parameters?

Problems

11-1. A two-stage negative-feedback amplifier has $r_F = 1$ kΩ and $R_E = 18$ Ω. What is the closed-loop voltage gain?

11-2. Suppose the RF chokes of Fig. 11-3 are 100 mH each. What is the inductive reactance of each choke at 1 MHz?

11-3. The transformer of Fig. 11-4 has a 10:1 turns ratio. If $R_L = 100$ Ω, what is the load resistance reflected into the primary winding? If $r'_e = 40$ Ω, what is the voltage gain from the base to the collector? The voltage gain from the base to the secondary winding?

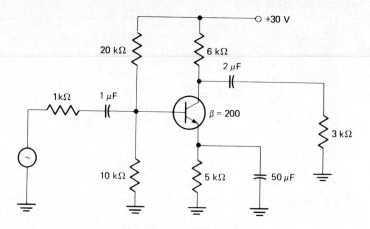

Fig. 11-13.

11-4. In Fig. 11-5, $L = 200\ \mu H$ and $C = 500$ pF. What is the resonant frequency of each tuned circuit? If $r_L = 12$ kΩ, what are the Q and bandwidth of each tuned tank?

11-5. An amplifier has a maximum voltage gain of 2000. If the cutoff frequencies are 30 Hz and 12 kHz, what is the bandwidth? The voltage gain at each cutoff frequency?

11-6. In Fig. 11-13, calculate the cutoff frequencies produced by the input coupling capacitor, output coupling capacitor, and emitter bypass capacitor.

11-7. A data sheet lists an f_T of 250 MHz. What does f_α equal? If h_{fe} is 175, what does f_β equal?

11-8. A 2N4401 has the following minimum and maximum values for its h parameters at $I_C = 1$ mA:

$$\text{Minimum:}\quad h_{ie} = 1\text{ k}\Omega$$
$$h_{fe} = 40$$
$$h_{re} = 1(10^{-4})$$
$$h_{oe} = 1\ \mu S$$

$$\text{Maximum:}\quad h_{ie} = 15\text{ k}\Omega$$
$$h_{fe} = 500$$
$$h_{re} = 8(10^{-4})$$
$$h_{oe} = 30\ \mu S$$

Calculate the values of α, β, r'_e, and r'_c for the minimum and maximum values of h parameters.

11-9. A CE amplifier has an R_s of 600 Ω and an r_L of 1.2 kΩ. What are the exact values of A_v, A_i, z_{in}, and z_{out} using the maximum h parameters of Prob. 11-8?

JFETS

Leibnitz: Acclaimed the universal genius of his age, he knew more law, history, politics, religion, literature, metaphysics, and philosophy than most experts. He also displayed mathematical ability when he invented calculus, independently and shortly after Newton. To Leibnitz, a method of analysis was perfect if "we can foresee from the start that following the method we shall attain our goal."

The *bipolar* transistor is the backbone of linear electronics. Its operation relies on two types of charge, holes and electrons. This is why it is called bipolar. For most linear applications, the bipolar transistor is the best choice. But there are some applications in which the *unipolar* transistor is better suited. The operation of a unipolar transistor depends on only one type of charge, either holes or electrons.

The junction field-effect transistor, abbreviated JFET, is an example of a unipolar transistor. This chapter discusses JFET fundamentals, biasing, amplifiers, and analog switches. JFETs are voltage-controlled devices, in contrast to bipolar transistors, which are current-controlled.

12-1
BASIC IDEAS

Figure 12-1*a* shows part of a JFET. The lower end is called the *source* and the upper end is the *drain*. The piece of semiconductor between the source and the drain is known as the *channel*. Since *n* material is used for the JFET in Fig. 12-1*a*, the majority carriers are free electrons.

By doping two *p* regions in the sides of the channel, we get the *n*-channel JFET of Fig. 12-1*b*. Each of these *p* regions is called a *gate*. When the manufacturer connects a separate external lead to each gate, the device is called a *dual-gate* JFET. The main use of a dual-gate JFET is with a frequency mixer, a circuit used in communication electronics.

This chapter concentrates on the *single-gate* JFET, a device whose gates are internally connected by the manufacturer. A single-gate JFET

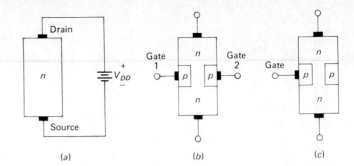

Fig. 12-1.
(a) Channel. (b) Dual-gate
JFET. (c) Single-gate JFET.

(a)   (b)   (c)

has only one external gate lead, as shown in Fig. 12-1c. When you see this symbol, remember the two p regions have the same potential because they are internally connected.

Biasing the JFET

Figure 12-2a shows the normal polarities for biasing an n-channel JFET. The idea is to apply a negative voltage between the gate and the source. Since the gate is reverse-biased, only a very small reverse current flows in the gate lead. To a first approximation, the gate current is zero.

The name *field effect* is related to the depletion layers around each *pn* junction. Figure 12-2b shows these depletion layers. Free electrons moving between the source and the drain must flow through the narrow channel between depletion layers. The size of these depletion layers determines the width of the conducting channel. The more negative the gate voltage is, the narrower the conducting channel becomes, because the depletion layers get closer to each other. Therefore, the gate voltage controls the current that flows between the source and the drain. The more negative the gate voltage is, the smaller the current.

The key difference between a JFET and a bipolar transistor is this. the gate is reverse-biased, whereas the base is forward-biased. This crucial difference means the JFET is a voltage-controlled device because the input voltage alone controls the output current. In a bipolar transistor, the input current controls the output current.

The input resistance of a JFET approaches infinity because it is typically greater than 10 MΩ. Therefore, in applications where a high

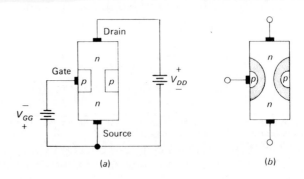

Fig. 12-2.
(a) Normal bias voltages.
(b) Depletion layers.

(a)   (b)

294

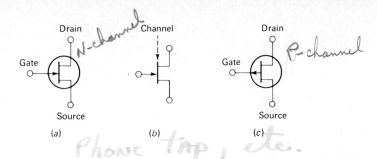

Drain — N-channel — Channel   Drain — P-channel

Gate   Gate

Source   Source

(a)   (b)   (c)

Phone "tap", etc.

Fig. 12-3.
Schematic symbols.

input resistance is needed, the JFET is preferred to the bipolar transistor. For instance, the input stage of a measuring instrument like an oscilloscope or electronic voltmeter often uses a JFET rather than a bipolar transistor.

But the JFET is less sensitive to changes in input voltage than a bipolar transistor. In almost any JFET, an input voltage change of 0.1 V produces an output current change of less than 10 mA. In a bipolar transistor, a change of 0.1 V in V_{BE} can easily produce a change of more than 10 mA in collector current. This implies that a JFET produces less voltage gain than a bipolar transistor.

Schematic Symbol

Figure 12-3a shows the schematic symbol of a JFET. As a memory aid, visualize the thin vertical line in Fig. 12-3b as the channel. The source and drain connect to this line. Also, the gate arrow points to the n material, similar to an ordinary diode. This should remind you that you are looking at an n-channel JFET.

The schematic symbol of a p-channel JFET is identical, except the arrow points in the opposite direction, as shown in Fig. 12-3c. Since a p-channel JFET is the complement of an n-channel JFET, holes are the majority carriers instead of free electrons.

Shorted-Gate Condition

Figure 12-4a shows a JFET with normal biasing voltages. Notice that the gate supply voltage is negative. This reverse-biases the gate and sets up the depletion layers as previously described. If the gate voltage is reduced to zero, the gate is effectively shorted to the source. This is called the *shorted-gate* condition.

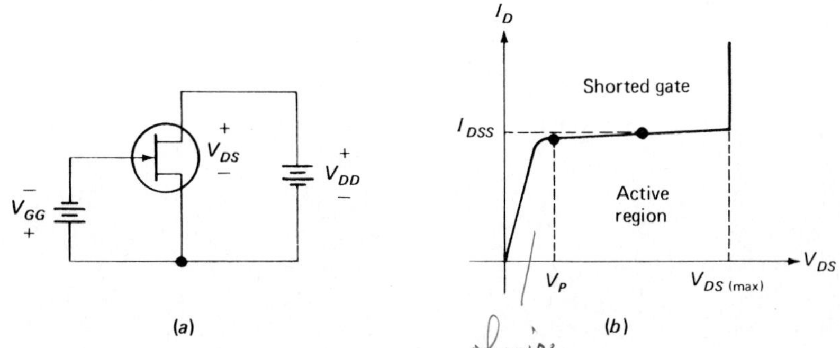

Fig. 12-4.
(a) Biased JFET.
(b) Shorted-gate drain curve.

(a)

(b)

ohmic Region

Figure 12-4*b* is a graph of drain current versus drain voltage for the shorted-gate condition. Notice the similarity to a collector curve. The drain current rises rapidly in the saturation region but then levels off in the active region. Between voltages V_P and $V_{DS(max)}$, the drain current is almost constant. When the drain voltage is too large, the JFET breaks down as shown. As with a bipolar transistor, the active region is along the almost horizontal part of the curve. In this region the JFET acts like a current source.

Pinch-off Voltage

The *pinch-off voltage* V_P is the drain voltage above which drain current becomes almost constant for the shorted-gate condition. When the drain voltage equals V_P, the conducting channel becomes extremely narrow and the depletion layers almost touch. Because of the small passage between the depletion layers, further increases in drain voltage produce only the slightest increase in drain current. In Fig. 12-5, $V_P = 4$ V.

Shorted-Gate Drain Current

In Fig. 12-4*b*, the subscripts of I_{DSS} stand for *Drain* to *Source* with *Shorted* gate. Data sheets specify I_{DSS} for a drain voltage in the active region, typically between 10 and 20 V. The important idea to remember is this: because the curve is almost flat in the active region, I_{DSS} is a close approximation for the drain current anywhere in the active region for the shorted-gate condition. Furthermore, because it applies to the shorted-gate condition, I_{DSS} is the maximum drain current you can get with normal operation of a JFET. All other gate voltages are negative and result in less drain current.

Gate-Source Cutoff Voltage

Drain curves resemble collector curves. For instance, Fig. 12-5 shows the drain curves of a typical JFET. The highest curve is for $V_{GS} = 0$, the shorted-gate condition. The pinch-off voltage is approximately 4 V, and the breakdown voltage is 30 V. As you see, I_{DSS} is 10 mA.

When $V_{GS} = V_{GS(off)}$, the depletion layers touch, cutting off the drain current. Since V_P is the drain voltage that pinches off current for the shorted-gate condition,

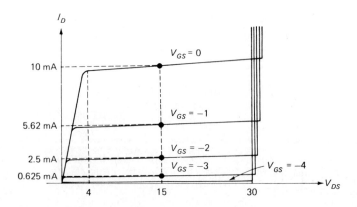

Fig. 12-5.
Drain curves.

$$V_P = -V_{GS(\text{off})} \qquad \textbf{(12-1)}$$

Some data sheets do not list V_P, but they almost always list $V_{GS(\text{off})}$, which is equivalent. For instance, if you see $V_{GS(\text{off})} = -4$ V on a data sheet, you will immediately know $V_P = 4$ V.

Transconductance Curve

The JFET transconductance curve is a graph of I_D versus V_{GS}. By reading the value of I_D and V_{GS} in Fig. 12-5, we can plot the transconductance curve of Fig. 12-6a. The transconductance curve of Fig. 12-6a is part of a parabola. By a calculus derivation, it is possible to prove that it has an equation of

$$I_D = I_{DSS} \left[1 - \frac{V_{GS}}{V_{GS(\text{off})}} \right]^2 \qquad \textbf{(12-2)}$$

This is an ideal formula that can be used as an approximation for any JFET. *Square law* is another word for parabolic. This is why JFETs are sometimes called square-law devices.

Normalized Transconductance Curve

We can rearrange Eq. (12-2) to get

$$\frac{I_D}{I_{DSS}} = \left[1 - \frac{V_{GS}}{V_{GS(\text{off})}} \right]^2 \qquad \textbf{(12-3)}$$

By substituting 0, 1/4, 1/2, 3/4, and 1 for $V_{GS}/V_{GS(\text{off})}$, we can calculate corresponding values of 1, 9/16, 1/4, 1/16, and 0 for I_D/I_{DSS}. Figure 12-6b summarizes these results in a normalized transconductance curve. This curve applies to all JFETs.

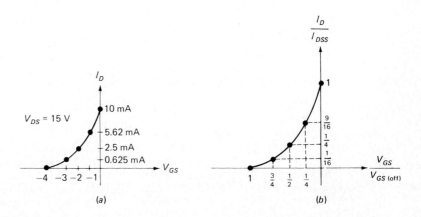

Fig. 12-6.
Transconductance curves.

(a)

(b)

12-2
GATE BIAS

Figure 12-7a is an example of *gate bias* (similar to base bias of a bipolar transistor). A fixed voltage V_{GG} is applied to the gate through a biasing resistor R_G. This is the worst way to set up the Q point of a linear JFET amplifier because of the variation between the minimum and maximum values of JFET parameters.

For example, here are the parameters of a 2N5459:

Parameter	Minimum	Maximum
I_{DSS}	4 mA	16 mA
$V_{GS(off)}$	−2 V	−8 V

This implies that the minimum and maximum transconductance curves are displaced as shown in Fig. 12-7b. If $V_{GS} = -1$ V, the Q point can be as high as Q_1 or as low as Q_2. In mass production, it means the Q point is unstable because it may lie anywhere between Q_1 and Q_2.

12-3
SELF-BIAS

Figure 12-8a shows self-bias, another way to bias a JFET. Only a drain supply is used; there is no gate supply. The idea is to use the voltage across the source resistor R_S to produce the gate-source reverse voltage. This form of negative feedback stabilizes the drain current against changes in temperature and JFET replacement. If the drain current increases, the voltage drop across R_S increases because the $I_D R_S$ voltage increases. This increases the gate-source reverse voltage, which makes the channel narrower and reduces the drain current. The overall effect is to partially offset the original increase in drain current.

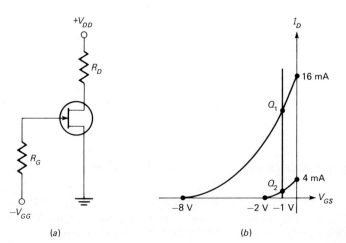

Fig. 12-7.
(a) Gate bias. (b) Q point
is unstable.

(a)

(b)

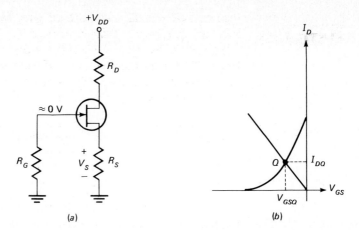

Fig. 12-8.
Self bias.

(a)

(b)

Similarly, if the drain current decreases, the gate-source reverse voltage decreases and the channel gets wider. This allows more free electrons through, and the drain current increases. This partially offsets the original decrease in drain current.

Gate-Source Voltage   Since the gate is reverse-biased in Fig. 12-8a, negligible gate current flows through R_G. Therefore, the gate voltage with respect to ground is zero:

$$V_G = 0$$

The source voltage to ground equals the product of drain current and source resistance:

$$V_S = I_D R_S$$

The gate-source voltage is the difference between the gate voltage and the source voltage:

$$V_{GS} = V_G - V_S = 0 - I_D R_S$$

or $$V_{GS} = -I_D R_S \qquad \textbf{(12-4)}$$

The greater the drain current, the more negative the gate-source voltage becomes.

Self-Bias Line   Equation (12-4) is a linear equation that passes through the origin when plotted as shown in Fig. 12-8b. The intersection of this line and the transconductance curve gives us the Q point, whose coordinates are V_{GSQ} and I_{DQ}. By rearranging Eq. (12-4), we get the following useful formula for the source resistor:

$$R_S = \frac{-V_{GS}}{I_D}$$

or

$$R_S = \frac{-V_{GSQ}}{I_{DQ}} \qquad \qquad (12\text{-}5)$$

For instance, if the desired Q point has $I_{DQ} = 5$ mA and $V_{GSQ} = -2.35$ V, then the required source resistance for self-bias is

$$R_S = \frac{-(-2.35 \text{ V})}{5 \text{ mA}} = 470 \ \Omega$$

Optimum Q Point Figure 12-9a shows the effect of different source resistors. When R_S is too large, the Q point is too far down on the transconductance curve and the drain current is too small. On the other hand, if R_S is very small, the Q point is too near I_{DSS}. Finally, there is a correct value of R_S that sets up a Q point near the middle of the current range. Ideally we would like a quiescent drain current of $I_{DSS}/2$ to allow maximum swing in both directions. This is not a hard-and-fast rule, only a guideline.

One way to get the correct size of R_S is using Eq. (12-5) and the transconductance curve given on the data sheet of a JFET. Suppose a data sheet gives a transconductance curve with an I_{DSS} of 16 mA as shown in Fig. 12-9b. Then draw a line from the origin through 8 mA as shown. Next, read the corresponding value of V_{GS}, which is -2 V. Then, use Eq. (12-5) to get

$$R_S = \frac{-(-2 \text{ V})}{8 \text{ mA}} = 250 \ \Omega$$

If the data sheet does not give the transconductance curve, you can use this formula to calculate the source resistance:

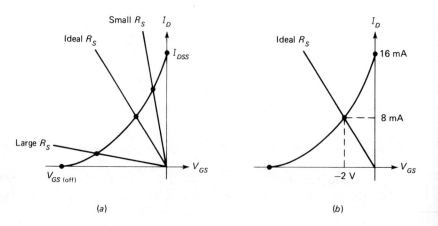

Fig. 12-9.
Effect of different source resistances.

(a)

(b)

$$R_S = \frac{-V_{GS(off)}}{I_{DSS}} \qquad \textbf{(12-6)}$$

The derivation for this is similar to that for Eq. (12-5). The value of R_S given by Eq. (12-6) automatically sets the Q point near the middle of the transconductance curve. As an example, if a data sheet lists $V_{GS(off)} = -6$ V and $I_{DSS} = 10$ mA, then Eq. (12-6) gives

$$R_S = \frac{-(-6\text{ V})}{10\text{ mA}} = 600\ \Omega$$

12-4
VOLTAGE-DIVIDER AND SOURCE BIAS

Self-bias is one way to stabilize the Q point. In this section, we discuss two more biasing methods, both of which are similar to methods used with bipolar transistors.

Voltage-Divider Bias

Figure 12-10*a* shows one of the better ways to bias a JFET. The idea is similar to voltage-divider bias used with a bipolar transistor. The voltage applied to the gate is

$$V_G = \frac{R_2}{R_1 + R_2} V_{DD} \qquad \textbf{(12-7)}$$

This is the dc voltage from the gate to ground. Because of V_{GS}, the voltage from the source to ground is

$$V_S = V_G - V_{GS} \qquad \textbf{(12-8)}$$

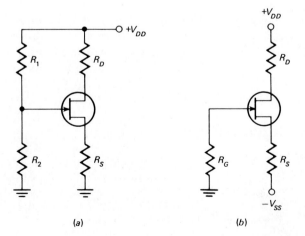

Fig. 12-10.
(a) Voltage-divider bias.
(b) Source bias.

(a)

(b)

Therefore, the drain current equals

$$I_D = \frac{V_G - V_{GS}}{R_S} \qquad \textbf{(12-9)}$$

and the dc voltage from the drain to ground is

$$V_D = V_{DD} - I_D R_D \qquad \textbf{(12-10)}$$

If V_G is much larger than V_{GS} in Eq. (12-9), the drain current is approximately constant for any JFET.

But there is a problem. In a bipolar transistor, V_{BE} is approximately 0.7 V, with only minor variations from one transistor to the next. In a JFET, however, V_{GS} can vary several volts from one JFET to the next. With typical supply voltages, it is difficult to make V_G large enough to swamp out V_{GS}. For this reason, voltage-divider bias is less effective with JFETs than with bipolars.

Source Bias

Figure 12-10*b* shows source bias, similar to emitter bias. The idea is to swamp out the variations in V_{GS} by making V_{SS} much larger than V_{GS}. Since most of V_{SS} appears across R_S, the drain current is roughly equal to V_{SS}/R_S. The exact value is given by

$$I_D = \frac{V_{SS} - V_{GS}}{R_S} \qquad \textbf{(12-11)}$$

For source bias to work well, V_{SS} must be much greater than V_{GS}. However, a typical range for V_{GS} is from -1 to -5 V, so that effective swamping is not possible with typical supply voltages, which may be only -10 to -15 V.

12-5

CURRENT-SOURCE BIAS

There is a way to get a solid Q point with JFETs. We need to produce a drain current that is independent of V_{GS}. Voltage-divider bias and source bias attempt to do this by swamping out the variations in V_{GS}. But it is difficult to swamp out V_{GS} with typical supply voltages because V_{GS} may be several volts. This section discusses two circuits that truly swamp out V_{GS}.

Two Supplies

When positive and negative supplies are available, you can use *current-source bias,* shown in Fig. 12-11*a*. Since the bipolar transistor is emitter-biased, its collector current is given by

$$I_C \cong \frac{V_{EE} - V_{BE}}{R_E} \qquad \textbf{(12-12)}$$

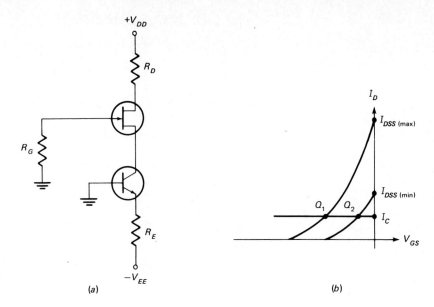

Fig. 12-11.
Current-source bias.

(a) (b)

Because the bipolar transistor acts like a dc current source, it forces the JFET drain current to equal the bipolar collector current:

$$I_D = I_C$$

Figure 12-11*b* illustrates how effective current-source bias is. Since I_C is constant, both Q points have the same value of drain current. The current source effectively wipes out the influence of V_{GS}. Although V_{GS} is different for each Q point, it no longer influences the value of drain current.

One Supply When only a positive supply is available, you can use a circuit like Fig. 12-12 to set up a constant drain current. In this case, the bipolar transistor is voltage-divider-biased. As a result, the emitter and collector currents are constant for all bipolar transistors. This forces the JFET drain current to equal the bipolar collector current.

EXAMPLE 12-1 If $V_{GS} = -2$ V in Fig. 12-12*b*, what is the drain voltage to ground? The source voltage to ground?

SOLUTION The voltage divider produces a base voltage:

$$V_B = \frac{R_2}{R_1 + R_2} V_{DD} = \frac{10 \text{ k}\Omega}{30 \text{ k}\Omega} 30 \text{ V} = 10 \text{ V}$$

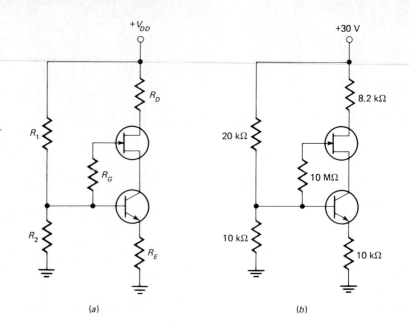

Fig. 12-12.
Single-supply current-
source bias.

(a)   (b)

Therefore, the emitter current is

$$I_E = \frac{V_B - V_{BE}}{R_E} = \frac{10\ V - 0.7\ V}{10\ k\Omega} = 0.93\ mA$$

So, the drain current is approximately 0.93 mA. The dc voltage from the drain to ground is

$$V_D = V_{DD} - I_D R_D = 30\ V - (0.93\ mA)(8.2\ k\Omega) = 22.4\ V$$

Because $V_{GS} = -2\ V$, the source voltage to ground is

$$V_S = V_G - V_{GS} = 10\ V - (-2\ V) = 12\ V$$

12-6
AC MODEL OF
JFET

The ac analysis of a JFET amplifier is similar to that given earlier for a bipolar amplifier. There are three basic connections: common-source (CS), common-drain (CD), and common-gate (CG). Before discussing these JFET amplifiers, we need some background information.

Transconductance

Every JFET has an ac characteristic called *transconductance*, designated g_m. Mathematically, it is defined as

$$g_m = \frac{\Delta I_D}{\Delta V_{GS}}$$

(12-13)

where ΔI_D = change in drain current
ΔV_{GS} = change in gate-source voltage

This says that transconductance equals the change in drain current divided by the corresponding change in gate-source voltage for a constant drain-source voltage. As an example, if a change in gate voltage of 0.1 V produces a change in drain current of 0.2 mA, then

$$g_m = \frac{0.2\ \text{mA}}{0.1\ \text{V}} = 2(10^{-3})\ \text{S} = 2000\ \mu\text{S}$$

(The symbol S stands for the unit *siemens,* also referred to as the mho. This unit represents conductance, the ratio of current to voltage.)

Most data sheets continue to use mho instead of siemens. They may also use the symbol g_{fs} for g_m. As an example, the data sheet of a 2N5451 lists a typical g_{fs} of 2000 μmhos for a drain current of 1 mA. This is identical to saying that it has a typical g_m of 2000 μS at 1 mA.

Graphical Meaning of Transconductance

Figure 12-13 brings out the meaning of g_m in terms of the transconductance curve. As you see, g_m is equivalent to the slope of the transconductance curve; the steeper the graph, the higher the g_m. To calculate g_m at any quiescent point, select two nearby points that straddle the Q point, such as A and B. The change in I_D divided by the change in V_{GS} equals the value of g_m. If the Q point is further up the transconductance curve between C and D, we get more change in I_D for the same change in V_{GS}. Therefore, g_m is larger. In a nutshell, g_m tells us how much control gate voltage has over drain current. The higher g_m is, the more effective gate voltage is in controlling drain current.

Data sheets for JFETs usually include a graph that shows how g_m varies with drain current. Therefore, once we figure out how much

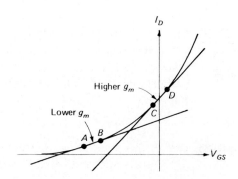

Fig. 12-13.
Transconductance increases
at higher drain currents.

dc drain current a JFET amplifier has, we can look up the value of g_m for this value of quiescent drain current.

If the data sheet does not show a graph of g_m versus I_D, then you can use the following approximation:

$$g_m = g_{m0}\left[1 - \frac{V_{GS}}{V_{GS(\text{off})}}\right]$$ (12-14)

or its equivalent from

$$g_m = \frac{2I_{DSS}}{-V_{GS(\text{off})}}\left[1 - \frac{V_{GS}}{V_{GS(\text{off})}}\right]$$

where g_m = transconductance at any Q point
g_{m0} = transconductance for zero gate voltage

To derive the foregoing equation, you need to apply calculus to Eq. (12-2). Another way to find the value of g_m at different Q points is with Fig. 12-14, whose derivation is also based on a calculus.

Simple JFET Model Figure 12-15 shows a simple ac equivalent circuit for a JFET. A very high resistance R_{GS} is between the gate and the source. This is well into the tens or hundreds of megohms. The drain of a JFET acts like a current source with a value of $g_m v_{gs}$, where v_{gs} is the ac voltage between the gate and the source. This model is an ideal approximation because it does not include the resistance of the current source, internal capacitances, etc. At low frequencies, we can use this simple ac model for troubleshooting and preliminary design.

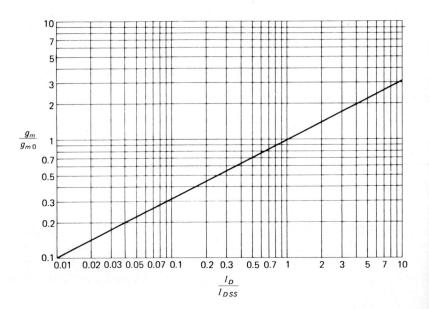

Fig. 12-14.
Variation of
transconductance with
operating point.

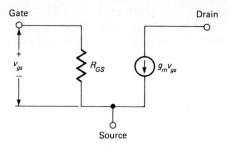

Gate   Drain

$+$
V_{gs}
$-$

R_{GS} $g_m v_{gs}$

Source

Fig. 12-15.
Ac model of JFET.

EXAMPLE 12-2   A data sheet lists the following quantities: $g_{mo} = 3000 \ \mu$mhos, $V_{GS(off)} = -4$ V, and $I_{DSS} = 10$ mA. Calculate the transconductance for each of these:
a. $V_{GS} = -1$ V
b. $I_D = 2.5$ mA

SOLUTION   a. Since V_{GS} is specified, we can use Eq. (12-14):

$$g_m = (3000 \ \mu S) \left(1 - \frac{-1 \ V}{-4 \ V} \right) = 2250 \ \mu S$$

b. Since I_D is given, proceed as follows. Calculate the current ratio:

$$\frac{I_D}{I_{DSS}} = \frac{2.5 \ mA}{10 \ mA} = 0.25$$

Next use Fig. 12-14 to read

$$\frac{g_m}{g_{mo}} = 0.5$$

or $g_m = 0.5 g_{mo} = 0.5(3000 \ \mu S) = 1500 \ \mu S$

12-7
JFET AMPLIFIERS

The main advantage of JFET amplifiers is high input resistance, while the main disadvantage is low voltage gain. What follows is an analysis of the three types of connections.

Common-Source Amplifier

Figure 12-16a shows a CS amplifier with voltage-divider bias. When a small ac signal is coupled into the gate, it produces variations in gate-source voltage. This produces a sinusoidal drain current. Since an alternating current flows through the drain resistor, we get an amplified ac voltage at the output.

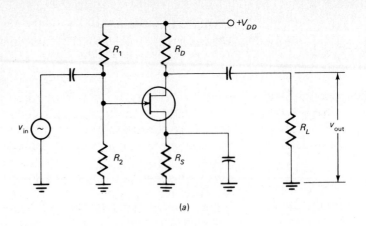

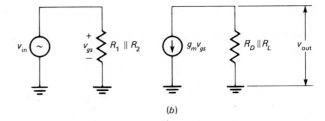

Fig. 12-16.
(a) Common-source
amplifier. (b) Ac equivalent
circuit.

An increase in gate-source voltage produces more drain current, which means drain voltage is decreasing. Since the positive half cycle of input voltage produces the negative half cycle of output voltage, we get phase inversion in a CS amplifier, similar to a CE amplifier.

Figure 12-16b shows the ac equivalent circuit of a CS amplifier. On the input side, R_1 is in parallel with R_2. Because the internal resistance R_{GS} is large enough to ignore, the input impedance at low frequencies is approximately

$$z_{in} = R_1 \parallel R_2 \qquad\qquad (12\text{-}15)$$

On the output side, the ac resistance loading the drain is

$$r_D = R_D \parallel R_L \qquad\qquad (12\text{-}16)$$

Since the current through r_D is $g_m v_{gs}$, the magnitude of ac output voltage is

$$v_{out} = g_m v_{gs} r_D = g_m v_{in} r_D$$

or $$A = \frac{v_{out}}{v_{in}} = g_m r_D \qquad\qquad (12\text{-}17)$$

The analysis is similar for the other types of bias. Because of this, the voltage gain is the same. The only difference is the input impedance. For instance, a self-biased circuit still has a voltage gain of $g_m r_D$, but its input impedance becomes R_G, the resistance of the gate resistor.

Common-Drain Amplifier

Figure 12-17a shows a CD amplifier. It is similar to an emitter follower. An ac signal drives the gate, producing an ac drain current. This flows through the unbypassed source resistor and produces an ac output voltage that is approximately equal to and in phase with the ac input voltage. For this reason, the circuit is called a *source follower*. Because of its high input impedance at low frequencies, a source follower is often used at the front end of measuring instruments like electronic voltmeters and oscilloscopes.

Figure 12-17b is the ac equivalent circuit. The equivalent ac resistance loading the source is

$$r_S = R_S \| R_L$$

To derive a formula for the voltage gain, start by summing voltages

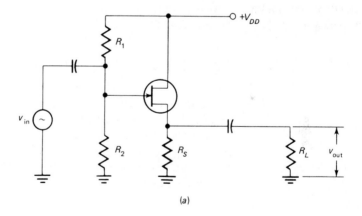

(a)

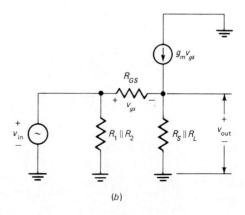

(b)

Fig. 12-17.
(a) Source follower. (b) Ac equivalent circuit.

around the input loop:

$$v_{gs} + g_m v_{gs} r_s - v_{in} = 0$$

or
$$v_{in} = (1 + g_m r_s)v_{gs}$$

Since R_{GS} approaches infinity, all of the current flows through the ac load resistance and the output voltage is

$$v_{out} = g_m v_{gs} r_s$$

Taking the ratio of output to input voltage gives

$$\frac{v_{out}}{v_{in}} = \frac{g_m r_s}{1 + g_m r_s}$$

or
$$A = \frac{g_m r_s}{1 + g_m r_s} \qquad (12\text{-}18)$$

When $g_m r_s$ is much greater than 1, A approaches unity.

Common-Gate Amplifier Figure 12-18a is a CG amplifier, and Fig. 12-18b is the ac equivalent circuit. The ac output voltage is

$$v_{out} = g_m v_{gs} r_D$$

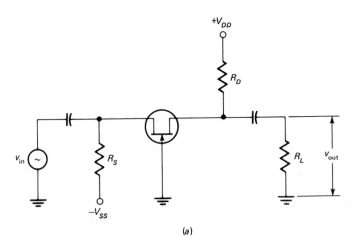

(a)

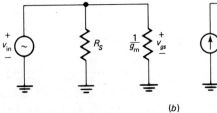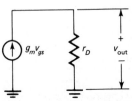

(b)

Fig. 12-18.
(a) Common-gate amplifier.
(b) Ac equivalent circuit.

and the ac input voltage is

$$v_{in} = v_{gs}$$

The ratio of output to input gives

$$A = \frac{v_{out}}{v_{in}} = g_m r_D \qquad\qquad (12\text{-}19)$$

The ac input current to the JFET is

$$i_{in} = i_d = g_m v_{gs}$$

which can be rearranged as

$$\frac{v_{gs}}{i_{in}} = \frac{1}{g_m}$$

Since $v_{gs} = v_{in}$,

$$z_{in} = \frac{1}{g_m} \qquad\qquad (12\text{-}20)$$

For instance, if $g_m = 2000 \ \mu S$,

$$z_{in} = \frac{1}{2000 \ \mu S} = 500 \ \Omega$$

Therefore, the input impedance of a CG amplifier is low. This differs from the CS and CD amplifiers, where the input impedance approaches infinity at low frequencies. Because of its small input impedance, the CG amplifier has only a few applications (such as a cascode amplifier or a differential amplifier).

EXAMPLE 12-3   A CS amplifier has $g_m = 2000 \ \mu S$, $R_D = 4.7 \ K\Omega$, and $R_L = 10 \ k\Omega$. If $v_{in} = 2 \ mV$, what does v_{out} equal?

SOLUTION   The voltage gain is

$$A = g_m r_D = (2000 \ \mu S)(4.7 \ k\Omega \| 10 \ k\Omega) = 7.83$$

The output voltage is

$$v_{out} = A v_{in} = 7.83(2 \ mV) = 15.7 \ mV$$

EXAMPLE 12-4   A source follower has the following values: $g_m = 2500 \ \mu S$, $R_S = 7.5 \ k\Omega$, and $R_L = 3 \ k\Omega$. What does the voltage gain equal?

SOLUTION   The ac load resistance is

$$r_s = R_D \parallel R_L = 7.5 \text{ k}\Omega \parallel 3 \text{ k}\Omega = 2.14 \text{ k}\Omega$$

The voltage gain is

$$A = \frac{g_m r_s}{1 + g_m r_s} = \frac{(2500 \text{ }\mu\text{S})(2.14 \text{ k}\Omega)}{1 + (2500 \text{ }\mu\text{S})(2.14 \text{ k}\Omega)} = 0.843$$

12-8

**THE JFET
ANALOG SWITCH**

The *analog switch* is one of the main applications of a JFET. The word "analog" means continuously variable, or able to take on any value in a given interval. The sine wave is an example of an analog voltage because its instantaneous value is continuously varying between the negative and positive peaks. An analog switch either transmits or blocks an analog input voltage. This analog voltage may be as simple as a sine wave or as complex as voice and music. The idea behind a JFET analog switch is to use only two points on the load line: cutoff and saturation. When the JFET is cut off, it's like an open switch. When it's saturated, it's like a closed switch.

Load Line

Figure 12-19*a* shows a JFET with a grounded source. Since the dc load resistance equals the ac load resistance, the dc and ac load lines are identical. When V_{GS} is zero, the JFET is saturated and operates at the upper end of the load line in Fig. 12-19*b*. When V_{GS} is equal to or more negative than $V_{GS(\text{off})}$, the JFET is cut off and operates at the lower end of the load line. Ideally, the JFET acts like a closed switch when saturated and like an open switch when cut off.

The region between saturation and cutoff is not used in switching operation. In other words, only two points on the load line are used: saturation and cutoff. We get this two-state operation by applying either a zero gate voltage or a large negative gate voltage.

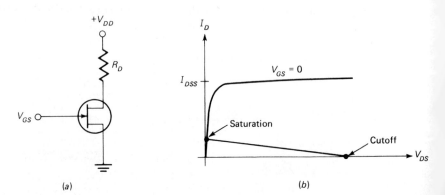

Fig. 12-19.
JFET switching action.
(a) Circuit. (b) Load line.

(a)

(b)

Dc On-State Resistance

The *dc on-state resistance* is defined as the ratio of total drain voltage to total drain current when the JFET is operating in the saturation region:

$$r_{DS(on)} = \frac{V_{DS}}{I_D} \qquad (12\text{-}21)$$

For instance, if the saturation point in Fig. 12-19b has $V_{DS} = 0.1$ V and $I_D = 0.8$ mA, then

$$r_{DS(on)} = \frac{0.1 \text{ V}}{0.8 \text{ mA}} = 125 \ \Omega$$

This means the JFET has a dc resistance of 125 Ω between its drain and its source.

Dc resistance $r_{DS(on)}$ is useful when you are testing a JFET with an ohmmeter. Remember to connect a jumper wire between the gate and the source to ensure that V_{GS} is zero. As an example, the data sheet of a 2N5951 indicates a minimum $r_{DS(on)}$ of 100 Ω and a maximum $R_{DS(on)}$ of 500 Ω for $V_{GS} = 0$. An ohmmeter connected between the drain and source of a 2N5951 should read between 100 and 500 Ω when the gate is shorted to the source.

AC Input Voltage Replaces Supply Voltage

When a JFET is used as an analog switch, the ac input voltage v_{in} replaces the drain supply voltage V_{DD} as shown in Fig. 12-20a. This input voltage is sinusoidal, swinging both positive and negative. Furthermore, v_{in} is deliberately kept small, typically 100 mV peak or less.

Figure 12-20b shows a magnified view of the $V_{GS} = 0$ drain curve of a 2N5951. Because of the small drain voltage, the drain curve is linear near the origin; any increase in drain voltage produces a proportional increase in drain current. This implies that a JFET acts like a linear resistance for small ac signals.

Unlike a bipolar transistor, the drain curve of a JFET extends into the third quadrant. In other words, when V_{DS} becomes negative, the drain current reverses direction. This bilateral property allows us to use a small sinusoidal signal to drive the drain of an analog switch.

Since v_{in} has replaced V_{DD} in Fig. 12-20a, the upper end of the load line has a saturation current of

$$i_{d(sat)} = \frac{v_{in}}{R_D}$$

and the lower end has a cutoff voltage of

$$v_{ds(cutoff)} = v_{in}$$

Figure 12-20c shows this ac load line.

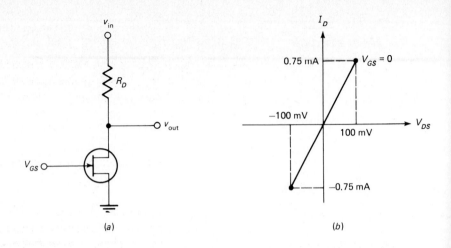

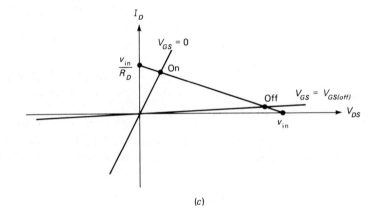

Fig. 12-20.
(a) Input voltage supplies
drain. (b) Ohmic region.
(c) Switching between on
and off points.

When $V_{GS} = 0$ in Fig. 12-20a, the JFET operates at the "on" point in Fig. 12-20c. When $V_{GS} = V_{GS(off)}$, the JFET operates at the "off" point. Incidentally, the drain curve for $V_{GS} = V_{GS(off)}$ is extremely close to the horizontal axis, so that the "off" point almost superimposes the cutoff point on the ac load line.

AC On-State Resistance   When v_{in} is a sinusoidal voltage, the cutoff point on the ac load line of Fig. 12-20c moves sinusoidally along the V_{DS} axis, while the slope of the ac load line remains the same. When $V_{GS} = 0$, the instantaneous operating point moves up and down along the drain curve for $V_{GS} = 0$. Since the instantaneous operating point of Fig. 12-20c moves along a linear drain curve, the JFET acts like an ac resistance given by

$$r_{ds(on)} = \frac{\Delta V_{DS}}{\Delta I_D} \qquad (12\text{-}22)$$

where ΔV_{DS} = change in drain voltage
ΔI_D = change in drain current

This resistance, known as the *small-signal on-state resistance,* equals the reciprocal slope of the $V_{GS} = 0$ drain curve. The steeper the saturation region, the smaller the ac resistance.

For small-signal operation of a JFET near the origin, the drain curve is linear and $r_{ds(on)} = r_{DS(on)}$. This means you can use total voltages and currents to calculate $r_{ds(on)}$. As an example, the 2N5951 whose typical drain curve is given in Fig. 12-20b has a small-signal on-state resistance of

$$r_{ds(on)} = \frac{100 \text{ mV}}{0.75 \text{ mA}} = 133 \ \Omega$$

Shunt Switch   Figure 12-21a is a *shunt switch.* The idea is to either transmit or block the ac input signal. Voltage v_{in} is small, typically less than 100 mV. When the control voltage V_{con} is zero, the JFET is saturated. Since the JFET is equivalent to a closed switch, v_{out} is ideally zero. When V_{con} is equal to or more negative than $V_{GS(off)}$, the JFET is like an open switch and v_{out} equals v_{in}.

The switching is not perfect. Because of the $r_{ds(on)}$ of a saturated JFET, the ac equivalent circuit appears as shown in Fig. 12-21b. When the switch is open, all the input voltage reaches the output. But when the switch is closed, a small amount of the input voltage still reaches the output because $r_{ds(on)}$ is not zero. For the switching to be effective, $r_{ds(on)}$ must be much smaller than R_D. For instance, a 2N3970 has an $r_{ds(on)}$ of 30 Ω. If R_D is 4.7 kΩ, then

$$v_{out} = \frac{30 \ \Omega}{4730 \ \Omega} v_{in} = 0.00634 v_{in}$$

This means that less than 1 percent of the input signal reaches the output when the JFET is saturated.

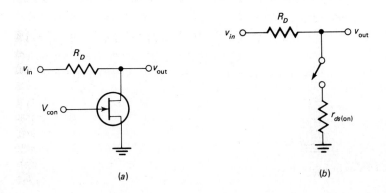

Fig. 12-21.
(a) Shunt switch. (b) Ac equivalent circuit.

(a)

(b)

Series Switch   Figure 12-22a is a *series switch*. When v_{con} is zero, the JFET is equivalent to a closed switch. In this case, the output approximately equals the input. When V_{con} is equal to or more negative than $V_{GS(off)}$, the JFET is like an open switch and v_{out} is approximately zero.

Figure 12-22b is the ac equivalent circuit. When the switch is open, no signal reaches the output. When the switch is closed, most of the signal reaches the output, provided $r_{ds(on)}$ is much smaller than R_D. You will see both kinds of JFET switches used in industry. The series switch is used more often than the shunt switch because it has a better on-off ratio. The series switch is another example of an analog switch, a device that either transmits or blocks an ac signal.

12-9

THE VOLTAGE-VARIABLE RESISTANCE

The saturation region of a JFET is also called the *ohmic* region because a saturated JFET acts like a resistance instead of a current source. In the ohmic region, $r_{ds(on)}$ can be controlled by V_{GS}. The more negative V_{GS} is, the larger $r_{ds(on)}$ becomes.

Figure 12-23 shows the typical drain curves of a 2N5951 in the ohmic region. The small-signal resistance $r_{ds(on)}$ depends on the value of V_{GS}. Because $r_{ds(on)} = r_{DS(on)}$ near the origin, you can calculate $r_{ds(on)}$ by taking the ratio of drain voltage to drain current. As an example, when $V_{GS} = 0$, $I_D = 0.75$ mA and $V_{DS} = 100$ mV. Therefore,

$$r_{ds(on)} = \frac{100 \text{ mV}}{0.75 \text{ mA}} = 133 \text{ }\Omega$$

When $V_{GS} = -2$ V, $I_D = 0.4$ mA and $V_{DS} = 100$ mV. So,

$$r_{ds(on)} = \frac{100 \text{ mV}}{0.4 \text{ mA}} = 250 \text{ }\Omega$$

When $V_{GS} = -4$ V, $I_D = 0.09$ mA and $V_{DS} = 100$ mV. In this case,

$$r_{ds(on)} = \frac{100 \text{ mV}}{0.09 \text{ mA}} = 1.11 \text{ k}\Omega$$

As you see, $r_{ds(on)}$ increases when V_{GS} becomes more negative.

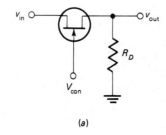

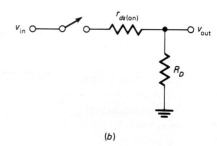

(a)

(b)

Fig. 12-22.
(a) Series switch. (b) Ac equivalent circuit.

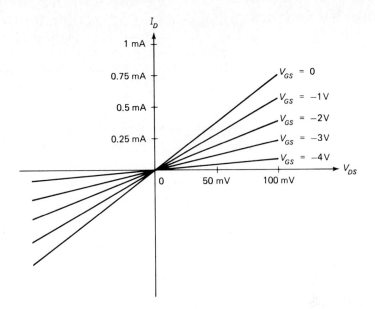

Fig. 12-23.
Drain curves are symmetrical at origin.

Since V_{GS} controls the value of $r_{ds(on)}$, we can build circuits where the ac output voltage is being controlled by V_{GS}. As a simple example, look at the circuit of Fig. 12-24a. This is similar to a shunt switch, except that V_{GS} can have any value between 0 and $V_{GS(off)}$. Figure 12-24b is the ac equivalent circuit. With the voltage-divider theorem, the output voltage is

$$v_{(out)} = \frac{r_{ds(on)}}{R_D + r_{ds(on)}} \, v_{in} \qquad \qquad \textbf{(12-23)}$$

Since V_{GS} controls the value of $r_{ds(on)}$, the output voltage can be varied continuously between its minimum and maximum value.

Later chapters will discuss op amps. At that time, you will see other applications of a JFET used as a voltage-variable resistance.

Summary

The name "field effect" is related to the depletion layers of a JFET. The more negative the gate voltage, the narrower the conducting channel

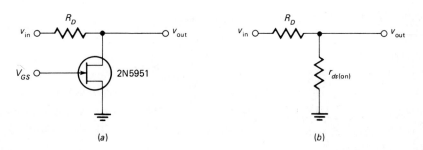

Fig. 12-24.
Using JFET as voltage-variable resistance.
(a) Circuit. (b) Ac equivalent circuit.

becomes. The input resistance of the gate ideally approaches infinity. In applications where very high input resistance is needed, the JFET is preferred to the bipolar transistor.

The pinch-off voltage is the drain voltage on the shorted-gate drain curve above which the drain current becomes almost constant. The drain current along this almost horizontal part of the curve is symbolized I_{DSS}. The gate-source voltage that reduces the drain current to approximately zero is called the gate-source cutoff voltage, symbolized $V_{GS(off)}$. The pinch-off voltage equals the negative of the gate-source cutoff voltage.

Gate bias is rarely used because the Q point is too unstable. Self-bias is a form of negative feedback that helps stabilize the Q point. Voltage-divider bias, source bias, and current-source bias are even better ways to stabilize the Q point.

Transconductance is the ratio of a change in drain current to a change in gate-source voltage. This parameter or characteristic indicates how effective gate voltage is in controlling drain current.

JFET amplifiers have three basic configurations: common-source (CS), common-drain (CD), and common-gate (CG). The CD amplifier, also called a source follower, has a voltage gain approaching unity and an input resistance approaching infinity. This circuit is used a lot near the front end of measuring instruments like oscilloscopes and electronic voltmeters.

The drain curves of a JFET are linear for small signals and extend on both sides of the origin. This area is called the ohmic region. In the ohmic region, the JFET can be used as an analog switch or a voltage-variable resistance.

Glossary

analog Continuously variable. Able to have any value in an interval. A sine wave is an example of an analog signal. The word "analog" is different from "digital," which means the signal typically has only two values: low or high. A square wave is an example of a digital signal.

analog switch One that transmits or blocks ac signals. These ac signals can be as elementary as sine waves or as complex as voice and music.

bipolar device It uses both types of charge, holes and free electrons.

gate-source cutoff voltage The gate-source voltage that reduces the drain current to zero.

ohmic region The linear part of the drain curves near the origin.

on-state resistance The resistance between the drain and source terminals of a JFET when operated in the ohmic region.

pinch-off voltage This is the approximate voltage on the shorted-gate drain curve above which drain current becomes almost constant.

shorted-gate drain current The drain current in the active region of a JFET for the condition of $V_{GS} = 0$.

source follower A widely used JFET input circuit. It has the advantage of high input resistance.

transconductance The ratio of a change in drain current to a change in gate-source voltage, equivalent to the ratio of a small ac drain current to the corresponding ac gate-source voltage.

unipolar device It uses only one type of charge, either holes or free electrons.

voltage-variable resistance When an analog voltage is applied to the gate, a JFET becomes a voltage-controlled resistance in the ohmic region.

Review Questions

1. What is the difference between a bipolar and a unipolar device?
2. Define the pinch-off voltage and the gate-source cutoff voltage.
3. Explain how self-bias stabilizes the Q point.
4. Describe how voltage-divider bias works.
5. Describe how current-source bias works.
6. Define the transconductance g_m of a JFET.
7. What does the voltage gain of a CS amplifier equal?
8. What is the major advantage of a source follower?
9. Why is the CG amplifier rarely used?
10. What does the word "analog" mean? Give an example of an analog voltage.
11. Describe the operation of a shunt switch and a series switch.
12. How does a JFET act like a voltage-variable resistance?

Problems

12-1. At room temperature a 2N4220 has a reverse gate current of 0.1 nA for a reverse gate-source voltage of 15 V. Calculate the dc resistance between the gate and the source.

12-2. A JFET has a $V_{GS(off)}$ of -5 V. What is its pinch-off voltage?

12-3. $I_{DSS} = 12$ mA and $V_{GS(off)} = -5$ V. Use Eq. (12-2) to calculate the drain current for a gate-source voltage of -2 V.

12-4. In a self-biased circuit, a drain current of 3 mA flows through a source resistor of 1.5 kΩ. What does V_{GS} equal?

12-5. We want a self-biased JFET amplifier to have $V_{GSQ} = -4$ V and $I_{DQ} = 20$ mA. What value of R_S should we use?

12-6. The JFET of Fig. 12-25a has $I_D = 10$ mA. What is the dc voltage from the drain to ground? From the source to ground? What does V_{GS} equal?

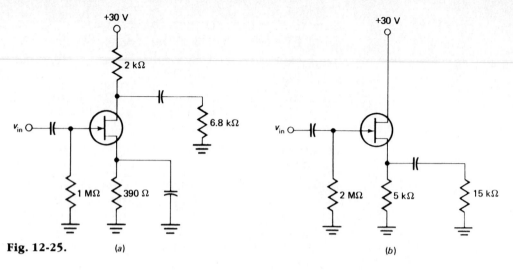

Fig. 12-25.

(a)   (b)

12-7. In Fig. 12-25b, $V_{GS} = -3.5$ V. What is the dc drain current? The dc voltage from drain to ground?

12-8. $V_{GS} = -1$ V in Fig. 12-26. Calculate the following dc voltages with respect to ground: gate voltage, source voltage, and drain voltage.

12-9. In Fig. 12-27, what is the approximate value of drain current? The dc voltage from drain to ground? If $V_{GS} = -2$ V, what is the dc collector voltage to ground?

12-10. When $V_{DS} = 10$ V, a change of 0.2 V in V_{GS} produces a change of 0.65 mA in I_D. What does g_m equal?

12-11. The g_{mo} of a JFET equals 5000 μS. If $V_{GS(off)}$ is -4 V, what does g_m equal when V_{GS} is -2 V?

12-12. If the JFET of Fig. 12-25a has a g_m of 3000 μS at the Q point, what is the voltage gain? The input resistance of the stage?

12-13. The JFET in Fig. 12-25b has a $g_m = 1500$ μS at the quiescent point. What is the voltage gain of the source follower? Its input resistance?

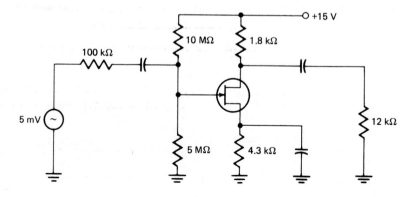

Fig. 12-26.

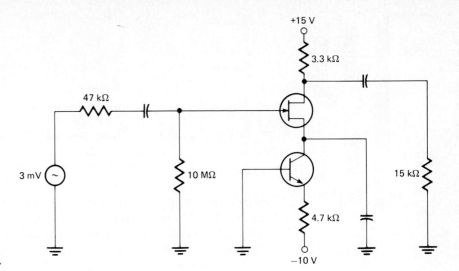

Fig. 12-27.

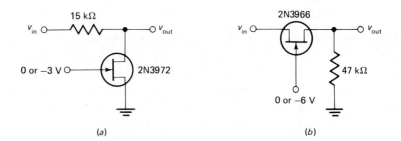

Fig. 12-28.   (a)   (b)

12-14. The JFET of Fig. 12-26 has $g_{m0} = 700 \ \mu$S, $V_{GS(off)} = -5$ V, and $V_{GS} = -2$ V. What does the voltage gain equal? The ac output voltage?

12-15. In Fig. 12-27, $I_{DSS} = 5$ mA and $g_{m0} = 6000 \ \mu$S. What does the ac output voltage equal?

12-16. A common-gate amplifier has a g_m of 4000 μS at the Q point. What is the input impedance?

12-17. A 2N5114 has an $r_{DS(on)}$ of 75 Ω. If 1 mA flows through the JFET when it's used as a closed switch, what is the voltage between the drain and the source?

12-18. The data sheet of a 2N3684 gives an $r_{DS(on)}$ of 600 Ω. What does $r_{ds(on)}$ equal if this JFET is operating in the ohmic region?

12-19. The JFET of Fig. 12-28a has an $r_{DS(on)}$ of 100 Ω and a $V_{GS(off)}$ of -3 V. If $v_{in} = 50$ mV, what does v_{out} equal when V_{GS} is 0 V? When $V_{GS} = -3$ V?

12-20. In Fig. 12-28b, the 2N3966 has an $r_{DS(on)}$ of 220 Ω and a $V_{GS(off)}$ of -6 V. What does v_{out} equal if v_{in} is 75 mV and V_{GS} is 0 V? If $V_{GS} = -6$ V, what does v_{out} equal?

MOSFETS

Maxwell: During his early years he was scorned and called "daffy" by his classmates. In spite of this, he went on to state Maxwell's equations—profound relations unifying electricity and magnetism. Maxwell's equations are essential for microwave analysis and other electromagnetic phenomena.

The metal-oxide semiconductor FET, or MOSFET, has a source, gate, and drain. Unlike a JFET, however, the gate is insulated from the channel. Because of this, gate current is extremely small whether the gate is positive or negative. The MOSFET is sometimes called an IGFET, which stands for insulated-gate FET. This chapter discusses MOSFET fundamentals, biasing, amplifiers, and switching circuits.

13-1
DEPLETION-TYPE MOSFET

Like the JFET, the MOSFET has a source, gate, and drain. The big difference is that the gate is insulated from the channel. Because of this, we can apply positive voltages as well as negative voltages to the gate. In either case, negligible gate current flows.

MOSFET Regions

To begin with, there is an n region with a source and drain as shown in Fig. 13-1a. As before, a positive voltage applied to the drain-source terminals forces free electrons to flow from the source to the drain. Unlike the JFET, the MOSFET has only a single p region, as shown in Fig. 13-1b. This region is called the *substrate*. This p region reduces the channel between the source and the drain so that only a small passage remains at the left side of Fig. 13-1b. Free electrons flowing from the source to the drain must pass through this narrow channel.

A thin layer of silicon dioxide (SiO_2) is deposited over the left side of the channel as shown in Fig. 13-1c. This material is an insulator. Finally, a metallic gate is deposited on the insulator (Fig. 13-1d). Because the gate is insulated from the channel, a MOSFET is also known as an insulated-gate FET (IGFET).

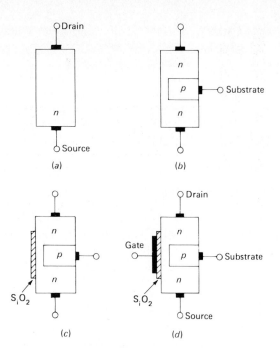

Fig. 13-1.
MOSFET structure. (a) n channel. (b) Adding the substrate. (c) Adding the silicon dioxide. (d) Adding the gate.

Depletion Mode

How does the MOSFET on Fig. 13-2a work? As usual, the V_{DD} supply forces free electrons to flow from the source to the drain. These free electrons flow through the narrow channel to the left of the p substrate. As before, the gate voltage controls the resistance of the n channel. But since the gate is insulated from the channel, we can apply either a positive or a negative voltage to the gate. Figure 13-2a shows a negative gate voltage. This voltage repels free electrons and tries to push them back to the source. This means a negative gate voltage reduces the flow between the source and the drain.

The more negative the gate voltage is, the smaller the current through the channel. Enough negative voltage on the gate cuts off the current between the source and the drain. Therefore, with negative gate voltage the action of a MOSFET is similar to that of a JFET. Because the action depends on reducing or depleting the charges in the channel, negative gate operation is known as the *depletion mode*.

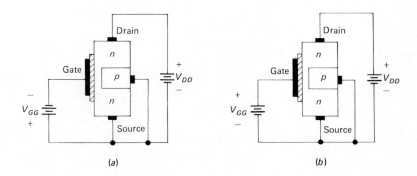

Fig. 13-2.
(a) Negative gate.
(b) Positive gate.

Enhancement Mode

In Fig. 13-2b, a positive voltage is applied to the gate. This voltage attracts free electrons and increases the current flow between the source and the drain. In other words, a positive gate voltage enhances the conductivity of the channel. The more positive the gate voltage, the greater the conduction from the source to the drain. Operating a MOSFET with a positive gate voltage is called the *enhancement mode.*

Because of the insulating layer, negligible gate current flows in either mode of operation. In fact, the input resistance of the gate is incredibly high, from 10^{10} to over 10^{14} Ω. The device in Fig. 13-2a or b is an n-channel MOSFET. The complementary device is a p-channel MOSFET.

MOSFET Curves

Figure 13-3a shows typical drain curves for an n-channel MOSFET. $V_{GS(\text{off})}$ represents the negative gate voltage that cuts off the drain current. For V_{GS} less than zero, we get depletion-mode operation. On the other hand, V_{GS} greater than zero gives enhancement-mode operation.

Figure 13-3b is the transconductance curve. I_{DSS} is the drain current with a shorted gate. Notice that I_{DSS} no longer is the maximum possible current. As you see, the transconductance curve extends to the right, so that I_D is greater than I_{DSS} in the enhancement mode. MOSFETs with a transconductance curve like Fig. 13-3b are easy to use because they require no bias voltage. As indicated in Fig. 13-3c, we can locate the Q point at the vertical intercept where $I_D = I_{DSS}$ and $V_{GS} = 0$. This means we do not have to provide any gate voltage at all, which simplifies biasing circuit.

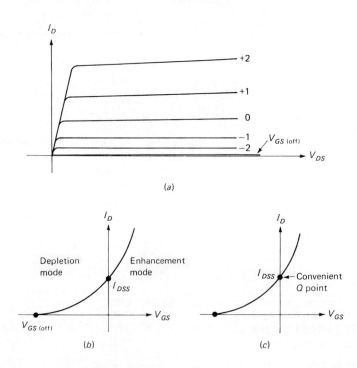

Fig. 13-3.
(a) MOSFET drain curves.
(b) Transconductance curve. (c) Zero bias.

Any MOSFET that can operate in either the depletion or the enhancement mode is called a *depletion-type* MOSFET. Since this type of MOSFET has drain current with zero gate voltage, it is also called a *normally on* MOSFET.

Schematic Symbol

Figure 13-4*a* shows the schematic symbol for a normally on MOSFET. The gate appears like a capacitor plate. Just to the right of the gate is a thin vertical line that represents the channel. The drain lead comes out the top of the channel and the source connects to the bottom. The arrow on the substrate points to the *n* material. Therefore, the device is an *n*-channel MOSFET.

Usually, the manufacturer internally connects the substrate to the source. This results in a three-terminal device whose schematic symbol is shown in Fig. 13-4*b*. By using the opposite type of doping, a manufacturer can produce a *p*-channel MOSFET, whose schematic symbol is shown in Fig. 13-4*c*.

Biasing

Because depletion-type MOSFETs can operate in the depletion mode, all the biasing methods discussed for JFETs can be used. These include gate bias, self-bias, voltage-divider bias, source bias, and current-source bias. In addition to these biasing methods, you have another option with depletion-type MOSFETs. Since a depletion-type MOSFET can operate in either the depletion or the enhancement mode, we can set this Q point at $V_{GS} = 0$, as shown in Fig. 13-5*a*. Then, an ac input signal to the gate can produce variations above and below the Q point. Being able to use zero V_{GS} is an advantage when it comes to biasing. It permits the unique biasing circuit of Fig. 13-5*b*.

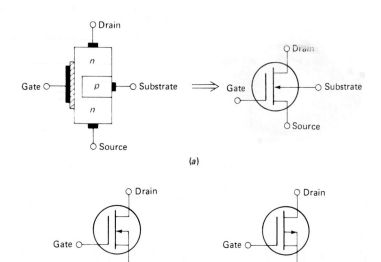

Fig. 13-4.
Depletion-type MOSFET symbols. (a) *n* channel with substrate lead. (b) *n*-channel device. (c) *p*-channel device.

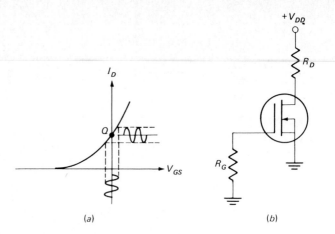

Fig. 13-5.
Zero bias.

(a)   (b)

This simple circuit has no applied gate or source voltage. Therefore, $V_{GS} = 0$ and $I_D = I_{DSS}$. The dc drain voltage is

$$V_{DS} = V_{DD} - I_{DSS} R_D \qquad (13\text{-}1)$$

The zero bias of Fig. 13-5*a* is unique with depletion-type MOSFETs. It will work with a bipolar transistor or a JFET.

Applications   After a depletion-type MOSFET is biased to a Q point, it can amplify small signals. MOSFET amplifiers are similar to JFET amplifiers, so that most of the ac analysis of the preceding chapter applies. For instance, a CS MOSFET amplifier has a voltage gain of $g_m r_D$, a source follower has a voltage gain of $g_m r_S/(1 + g_m r_S)$, and so on.

If the input resistance of a JFET is not high enough, you can use a MOSFET. Its input resistance approaches infinity because of the insulated gate. Furthermore, depletion-type MOSFETs have excellent low-noise properties, a definite advantage for any stage near the front end of a system because noise is amplified the same as a desired signal. As with a JFET, the g_m of a MOSFET decreases when V_{GS} becomes more negative. Because of this, MOSFETs can be used in *automatic-gain control circuits* (AGC).

13-2
ENHANCEMENT-
TYPE MOSFET

There is another kind of MOSFET that is very important in digital circuits. Known as an *enhancement-type* MOSFET, it can operate only in the enhancement mode.

Creating the
Inversion Layer

Figure 13-6*a* shows the different parts of an enhancement-type MOSFET. Notice that the substrate extends all the way to the silicon dioxide. Because of this, there no longer is an *n* channel between the source and the drain.

326   **CHAPTER 13**

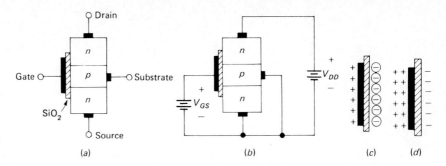

Fig. 13-6.
Enhancement-type
MOSFET. (a) Structure.
(b) Normal bias.
(c) Creation of negative
ions. (d) Creation of *n*-type
inversion layer.

How does the enhancement-type MOSFET conduct? Figure 13-6*b* shows the normal biasing polarities. When $V_{GS} = 0$, the V_{DD} supply tries to force free electrons to flow from the source and the drain, but the *p* substrate has only a few thermally produced free electrons. As a result, the current between the source and the drain is negligibly small. For this reason, the enhancement-type MOSFET is a normally off MOSFET.

To get drain current, we have to apply enough positive voltage to the gate. The gate acts like one plate of a capacitor, the silicon dioxide like a dielectric, and the *p* substrate like the other plate of a capacitor. For smaller gate voltages in Fig. 13-6*c*, the positive charges on the gate induce negative charges in the *p* substrate. These induced charges are negative ions, produced by valence electrons filling holes in the *p* substrate. With a further increase in gate voltage, the additional positive charges on the gate can put free electrons into orbit around the negative ions (see Fig. 13-6*d*). In other words, when the gate is positive enough, it can create a thin layer of free electrons stretching all the way from the source to the drain.

The created layer of free electrons is next to the silicon dioxide. This layer no longer acts like a *p*-type semiconductor. Instead, it appears like an *n*-type semiconductor because of the induced free electrons. This is why the layer of *p* material touching the silicon dioxide is called an *n-type inversion layer.*

The Threshold Voltage

The minimum gate-source voltage that creates the *n*-type inversion layer is called the *threshold voltage,* designated $V_{GS(th)}$. When the gate voltage is less than the threshold voltage, no current flows from the source to the drain. But when the gate voltage is greater than the threshold voltage, an *n*-type inversion layer connects the source to the drain, and we get current.

The threshold voltage depends on the particular type of MOSFET because $V_{GS(th)}$ can vary from less than 1 V to more than 5 V. For instance, the 3N169 has a threshold voltage of 1.5 V. When the gate voltage is less than this, the MOSFET is open. When the gate voltage is greater than 1.5 V, the MOSFET conducts.

An enhancement-type MOSFET is a natural choice for digital circuits because this type of MOSFET is normally off. When the gate

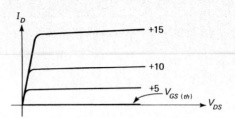

Fig. 13-7.
Drain curves for
enhancement-type MOSFET.

voltage exceeds the threshold voltage, the MOSFET turns on like a switch. Later sections discuss some digital circuits using enhancement-type MOSFETs.

Enhancement-Type Curves

Figure 13-7 shows a set of drain curves for an enhancement-type MOSFET. The lowest curve is the $V_{GS(th)}$ curve. When V_{GS} is less than $V_{GS(th)}$, the drain current is ideally zero and the MOSFET is off. When V_{GS} is greater than $V_{GS(th)}$, drain current appears. The larger V_{GS} is, the greater the drain current.

Schematic Symbols

When $V_{GS} = 0$, the enhancement-type MOSFET is off, because no conducting channel exists between the source and the drain. The schematic symbol of Fig. 13-8a has a broken channel line to indicate the normally off condition. As you know, a gate voltage greater than the threshold voltage creates an n-type inversion layer, which acts like an n channel when the device is conducting. For this reason, the advice is an n-channel enhancement-type MOSFET.

Figure 13-8b is the schematic symbol for a p-channel enhancement-type MOSFET. In this case, the threshold voltage is negative and the drain current is in the opposite direction from an n-channel device.

Maximum Gate-Source Voltage

Both depletion-type and enhancement-type MOSFETs have a thin insulating layer of silicon dioxide between the gate and the channel. This thin layer is easily destroyed by excessive gate-source voltage. For instance, a 2N3796 has a $V_{GS(max)}$ rating of ± 30 V. If the gate-source voltage becomes more positive than $+30$ V or more negative than -30 V, you can throw away the MOSFET because the thin insulating layer has been destroyed.

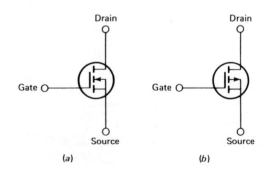

Fig. 13-8.
Enhancement-type MOSFET
symbols. (a) n channel.
(b) p channel.

Aside from directly applying an excessive V_{GS}, you can destroy the thin insulating layer in more subtle ways. Remove or insert a MOSFET into a circuit while the power is on, and transient voltages may exceed the $V_{GS(\text{max})}$ rating. Even picking up a MOSFET may deposit enough static charge to exceed the $V_{GS(\text{max})}$ rating. This is the reason MOSFETs are often shipped with a wire ring around the leads. The ring is removed after the MOSFET is connected in the circuit.

The newer MOSFETs are protected by built-in zener diodes in parallel with the gate and the source. The zener voltage is less than the $V_{GS(\text{max})}$ rating. In this way, the zener diode breaks down before any damage occurs to the thin insulating layer. The disadvantage of these built-in zener diodes is that they reduce the MOSFET's high input resistance.

13-3
BIASING ENHANCEMENT-TYPE MOSFETS

With enhancement-type MOSFETs, V_{GS} has to be greater than $V_{GS(\text{th})}$ to get current. This eliminates self-bias, current-source bias, and zero bias because all these require depletion-mode operation. That leaves gate bias, voltage-divider bias, and source bias. These will work with enhancement-type MOSFETs because they can produce the enhancement mode. Besides these three types of bias, there is one more method of biasing enhancement-type MOSFETs.

Figure 13-9a shows *drain-feedback bias,* a type of bias that you can use only with enhancement-type MOSFETs. When the MOSFET is conducting, it has a drain current of $I_{D(\text{on})}$ and a drain voltage of $V_{DS(\text{on})}$. Since the gate current is approximately zero, no voltage appears across R_D. Therefore, $V_{GS} = V_{DS(\text{on})}$. Like collector-feedback bias, the circuit of Fig. 13-9a tends to compensate for changes in FET parameters (characteristics). If $I_{D(\text{on})}$ tries to increase for some reason, $V_{DS(\text{on})}$ decreases. This reduces V_{GS}, which partially offsets the original increase in $I_{D(\text{on})}$.

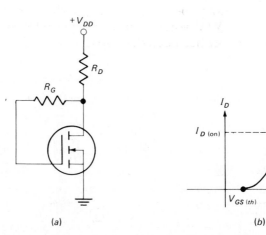

Fig. 13-9.
Drain-feedback bias.

(a)

(b)

Figure 13-9*b* shows the Q point on the transconductance curve. It has coordinates of $I_{D(on)}$ and $V_{DS(on)}$. Data sheets for enhancement-type MOSFETs usually give a value of $I_{D(on)}$ and $V_{DS(on)}$. This helps in setting up the Q point. In design, all you have to do is select a value of R_D that sets up the specified V_{DS}. In symbols,

$$R_D = \frac{V_{DD} - V_{DS(on)}}{I_{D(on)}}$$

(13-2)

For example, suppose the data sheet of an enhancement-type MOS-FET gives $V_{DS(on)} = 10$ V and $I_{D(on)} = 3$ mA. If the supply voltage is 25 V, the required drain resistance is

$$R_D = \frac{25 \text{ V} - 10 \text{ V}}{3 \text{ mA}} = 5 \text{ k}\Omega$$

DC Amplifier   A *dc amplifier* is one that can operate all the way down to zero frequency without a loss of voltage gain. One way to build a dc amplifier, or dc amp, is to leave out all coupling and bypass capacitors.

Figure 13-10 is a dc amp using MOSFETs. The imput stage is a depletion-type MOSFET with zero bias. The second and third stages use enhancement-type MOSFETs; each gate gets its V_{GS} from the drain of the preceding stage. The design of Fig. 13-10 uses MOSFETs with drain currents of 3 mA. For this reason, each drain runs at $+10$ V with respect to ground. We tap the final output voltage between the 100-kΩ resistors. Since the lower resistor is returned to -10 V, the quiescent output voltage is 0 V. When an ac voltage drives the amplifier, regardless of how low its frequency, we get an amplified output voltage.

There are other ways of designing ac amplifiers. The beauty of Fig. 13-10 is its simplicity.

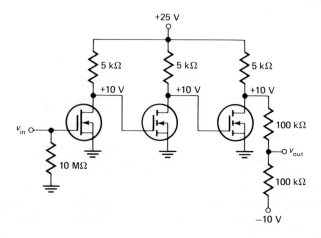

Fig. 13-10.
Dc amplifier using
MOSFETs.

13-4
ENHANCEMENT-TYPE MOSFET APPLICATIONS

Computers use integrated circuits (ICs) with thousands of transistors. These integrated circuits work remarkably well, despite transistor tolerances and changes in temperature. How is it possible? The answer is *two-state* design, using only two points on the load line of each transistor. When used in this way, the transistor acts like a switch rather than a current source. Circuits using transistor switches are called *switching* circuits, *digital* circuits, or *logic* circuits. On the other hand, circuits using transistor current sources are called *linear* circuits, analog circuits, etc.

Enhancement-Type MOSFETs Preferred

The enhancement-type MOSFET has had its greatest impact in digital circuits. One reason is its low power consumption. Another is the small amount of space it takes on a *chip* (an integrated circuit). A manufacturer can put many more enhancement-type MOSFETs on a chip than bipolar transistors. This is the reason enhancement-type MOSFETs are used in *large-scale integation* (LSI) for microprocessors, memories, and other devices requiring thousands of devices on a chip.

Passive Load

Figure 13-11a shows a MOSFET driver and a *passive* load (resistor R_D). In this switching circuit, v_{in} is either low or high, and the MOSFET acts like a switch that is either off or on. When v_{in} is low, the MOSFET is cut off and v_{out} equals the supply voltage. On the other hand, when v_{in} is high, the MOSFET conducts heavily and v_{out} drops to a low value.

Active Load

Resistors take up much more chip area than MOSFETs. For this reason, resistors are rarely used in MOS integrated circuits. Figure 13-11b shows another switching circuit with a MOSFET driver Q_2 and an *active load* Q_1. Because of the drain-feedback bias, Q_1 is always conducting. By deliberate design, this upper MOSFET has an $r_{DS(on)}$ at least 10 times greater than the $r_{DS(on)}$ of the lower MOSFET. For this reason, Q_1 acts like a resistor and Q_2 acts like a switch.

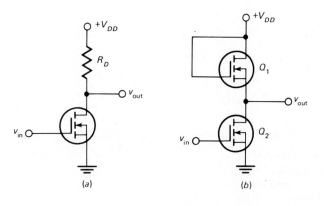

Fig. 13-11.
MOSFET driver and:
(a) passive load. (b) active load.

Using a MOS driver and a MOS load leads to much smaller integrated circuits because MOSFETs take up less room on a chip than resistors. This is why MOS technology dominates in computer applications; it allows you to get many more circuits on a chip.

The main thing to remember is the idea of active loading, using one active device as the load on another. Active loading is also used with bipolar transistors, especially with *op amps* (discussed in later chapters).

MOS Logic Circuits

One of the first semiconductor techniques used to build digital ICs was *p*-channel MOS technology. In this approach, *p*-channel enhancement-type MOSFETs act like switches and active loads. But *p*-channel MOS has a big disadvantage; its carriers are holes instead of free electrons. Holes move more slowly than free electrons, which means the switching speed of a *p*-channel device is less than that of an *n*-channel device. Because of its greater speed, *n*-channel technology dominates in memory and microprocessor applications.

CMOS Inverter

We can build *complementary* MOS (CMOS) circuits with *p*-channel and *n*-channel MOSFETs. One of the most important of all is the CMOS *inverter* shown in Fig. 13-12a. Notice that Q_1 is a *p*-channel device and Q_2 is an *n*-channel device. This circuit is analogous to the class B push-pull bipolar amplifier of Fig. 13-12b. When one device is on, the other is off, and vice versa.

For instance, when v_{in} is low in Fig. 13-12a, Q_2 is off but Q_1 is on. Therefore, the output voltage is high. On the other hand, when v_{in} is high, Q_2 is on the Q_1 is off. In this case, the output voltage is low. Since the phase of the output voltage is always opposite to that of the input voltage, the circuit is called an *inverter*.

The CMOS inverter can be modified to build complementary-type circuits. The key advantage in using CMOS design is its extremely low power consumption. Because both devices are in series, the current is determined by the leakage in the off device, which is typically in nanoamperes. This means that the total power dissipation of the cir-

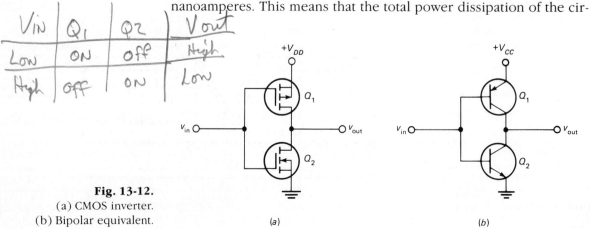

Vin	Q₁	Q₂	Vout
Low	ON	off	High
High	off	ON	Low

Fig. 13-12.
(a) CMOS inverter.
(b) Bipolar equivalent.

(a)

(b)

cuit is in nanowatts. This low power consumption is the main reason CMOS circuits are popular in pocket calculators, digital wristwatches, and satellites.

Figure 13-13a shows the structure of an enhancement-type MOSFET in an integrated circuit. The source is on the left, the gate is in the middle, and the drain is on the right. Free electrons flow horizontally from the source to the drain when V_{GS} is greater than $V_{GS(\text{th})}$. This conventional structure limits the maximum current because the free electrons must flow along the narrow inversion layer, symbolized by the dashed line. Because the channel is narrow, conventional MOS devices have small drain currents, which implies low power ratings, typically less than 1 W.

Vertical Channel

Figure 13-13b shows the structure of *vertical* MOS (VMOS). It has two sources at the top. These are usually connected. Furthermore, the substrate now acts like the drain. When V_{GS} is greater than the threshold voltage, free electrons flow vertically downward from the two

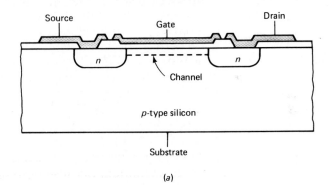

(a)

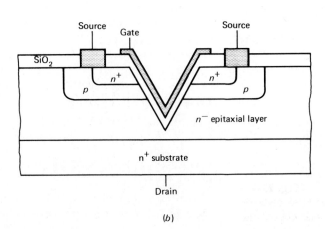

Fig. 13-13.
(a) Conventional MOSFET.
(b) VMOS.

(b)

sources to the drain. Because the conducting channel is much wider along both sides of the V groove, the current can be much larger. The overall effect is an enhancement-type MOSFET that can handle larger currents and voltages than a conventional MOSFET.

Prior to the invention of the VMOS transistor, MOSFETs could not compete with the power ratings of large bipolar transistors. But now VMOS offers a new type of MOSFET that is better than the bipolar transistor in many applications requiring high load power, including audio amplifiers, RF amplifiers, etc.

Lack of Thermal Runaway

One major advantage of VMOS transistors over bipolar transistors is the lack of *thermal runaway*. In a bipolar transistor, an increase in temperature produces a decrease in V_{BE}. With bipolar circuits, this produces more collector current and raises the temperature further. If the heat sinking is inadequate, we get thermal runaway because the V_{BE} keeps decreasing and the temperature keeps increasing until the transistor is destroyed.

Unlike a bipolar transistor, a VMOS transistor has a negative thermal coefficient. As the device temperature increases, the drain current decreases, which reduces the power dissipation. Because of this, the VMOS transistor cannot go into thermal runaway. This is a huge advantage in any power amplifier.

Parallel Connection

Bipolar transistors cannot be connected in parallel to increase the load power because their V_{BE} drops do not match closely enough. If you try to connect them in parallel, *current hogging* occurs (the one with the lower V_{BE} drop has more collector current).

Because of their negative temperature coefficients, two VMOS transistors can be connected in parallel to increase the load power. If one of the parallel VMOS transistors tries to hog the current, its negative temperature coefficient reduces the current through it, so that approximately equal currents flow through the parallel VMOS transistors.

Faster Switching Speed

When a small-signal bipolar transistor is used as a saturated switch, conservative design calls for a base current that is approximately one-tenth of the saturated collector current. Since most transistors have β_{dc} greater than 10, the excess base current guarantees saturation from one transistor to the next. The excess base carriers of a heavily saturated transistor are stored in the base region. When the transistor tries to come out of saturation, there is a small delay called the *saturation delay time* (also called storage time). For instance, the storage time t_s of a 2N3713 is 0.3 μs. This means that it takes approximately 0.3 μs for a 2N3713 to come out of saturation after the base drive is removed.

Another advantage the VMOS transistor has over the bipolar transistor is the lack of storage time. Because a VMOS transistor is a unipolar device, it does not store any charges when it is conducting. Therefore, it can come out of saturation almost immediately. Typically, a VMOS transistor can shut off amperes of current in tens of nanoseconds. This is from 10 to 100 times faster than a comparable bipolar transistor. Therefore, the VMOS transistor finds numerous applications in high-steel switching circuits, switching regulators, etc.

Driving Heavy Loads

Digital ICs are low-power devices because they can supply only small load currents. Therefore, we need to use some kind of *buffer* (a device or circuit that isolates) between a digital IC and a high-power load. The VMOS transistor is an excellent buffer because it can supply large load power to relays, motors, or lamps.

As an example, Figure 13-14 shows a digital IC driving the gate of a VMOS transistor. When the digital output is low, the VMOS is off. When the digital output is high, the VMOS transistor acts like a closed switch, and maximum current flows through the load. The VMOS transistor interfaces the low-power digital IC and the high-power load. If you tried to use the output of the digital IC to drive the load directly, the digital IC could not supply the large load power. Interfacing digital ICs to high-power loads is one of the most important applications of the VMOS transistor.

Summary

Unlike a JFET, a MOSFET has an insulated gate that allows us to use a positive or negative gate voltage. In either case, the gate current is extremely small. The depletion-type MOSFET can operate in either the depletion mode or the enhancement mode. Since this type of MOSFET has drain current with zero gate voltage, it is also known as a normally on MOSFET. The depletion-type MOSFET has a unique bias called zero bias.

The enhancement-type MOSFET, also known as a normally off MOSFET, can operate only in the enhancement mode. Furthermore, its gate voltage must be greater than its threshold voltage to produce drain

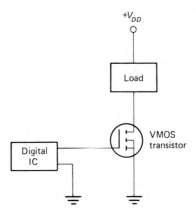

Fig. 13-14.
Interfacing a digital IC and a load.

current. Enhancement-type MOSFETs have made their greatest impact in digital electronics. With enhancement-type MOSFETs used for active loads and drivers, manufacturers can produce extremely dense integrated circuits like microprocessors and memories.

The VMOS transistor is an enhancement-type MOSFET with a vertical channel that can conduct more heavily than a conventional MOSFET. This leads to higher power ratings, negative temperature coefficients, and faster switching speeds. VMOS transistors are quite useful for interfacing low-power devices such as digital ICs to high-power loads such as relays, motors, and lamps.

Glossary

active load A bipolar transistor or FET that acts like a load resistor for another device.

dc amplifier One that can amplify signals all the way down to zero frequency.

inversion layer When the gate-source voltage of an enhancement-type MOSFET is greater than the threshold voltage, the layer of doped material next to the gate becomes conductive. This conductive layer is the inversion layer.

large-scale integration No standard definition exists, although some classify LSI as an integrated circuit whose complexity exceeds that of a circuit with 100 gates or devices.

passive load The same as an ordinary resistor.

thermal runaway A condition where increasing junction temperature reduces the base-emitter voltage, which in turn increases the collector current and the junction temperature until $T_{j(max)}$ is exceeded.

threshold voltage The gate-source voltage that just barely produces drain current in an enhancement-type MOSFET. Below the threshold voltage, drain current is approximately zero.

VMOS transistor An enhancement-type MOSFET with a V-shaped gate. This structure significantly increases the maximum current and power rating of a MOSFET.

Review Questions

1. Name two advantages of MOSFETs.
2. What modes can a depletion-type MOSFET operate in?
3. Describe how an enhancement-type MOSFET works.
4. Zero bias works only with what kind of MOSFET?
5. Why does MOS technology dominate the area of microprocessors, memories, and other LSI circuits?
6. What is the main advantage of CMOS circuits?

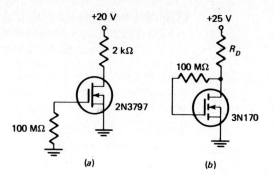

Fig. 13-15.

(a)   (b)

7. What does VMOS stand for? Is this a depletion-type or enhancement-type MOSFET?

Problems

13-1. The M113 is a depletion-type MOSFET with a gate leakage current of 100 pA when V_{GS} is 20 V. Calculate the input resistance.

13-2. A 3N169 is an enhancement-type MOSFET with a minimum threshold voltage of 0.5 V and a maximum threshold voltage of 1.5 V. If we are going to use thousands of 3N169s, what is the minimum gate voltage that ensures turn-on?

13-3. The 2N3797 has an $I_{DSS(min)}$ of 2 mA, an $I_{DSS(typ)}$ of 2.9 mA, and an $I_{DSS(max)}$ of 6 mA. If this depletion-type MOSFET is used in Fig. 13-15a, what is the typical value of V_{DS}? The lowest possible value of V_{DS}?

13-4. The 3N170 is an enhancement-type MOSFET with an $I_{D(on)}$ of 10 mA when $V_{GS} = V_{DG} = 10$ V. Select a value of R_D in Fig. 13-15b that sets up an I_D of 10 mA.

13-5. The depletion-type MOSFET of Fig. 13-16a has a $g_m = 2000$ μ S. What is the voltage gain from the gate to the drain?

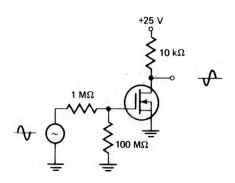

Fig. 13-16.

13-6. The input signal in Fig. 13-16a has a peak voltage of 0.1 V. If the depletion-type MOSFET has a g_m of 1500 μS, what is the peak voltage of the ac output signal?

13-7. In Fig. 13-11a, R_D equals 5 kΩ and V_{DD} is 20 V. What is the output voltage when v_{in} is zero? What is the output voltage when v_{in} is much greater than the threshold voltage?

13-8. The active load of Fig. 13-11b is equivalent to 100 kΩ. What does v_{out} equal when v_{in} is zero? When Q_2 is saturated?

Thyristors

A *thyristor* is a special kind of semiconductor switch that uses internal feedback to produce *latching* action. Unlike bipolar transistors and FETs, which operate as either linear amplifiers or switches, thyristors can operate only in the switching mode. Their main application is controlling large amounts of load power in motors, heaters, lighting systems, etc.

14-1
THE IDEAL LATCH

All thyristor devices can be explained in terms of the ideal *latch* shown in Fig. 14-1a. Notice that the upper transistor Q_1 is a *pnp* device, and the lower transistor Q_2 is *npn*.

Regeneration

Because of the unusual connection in Fig. 14-1a, we have *positive feedback*, also called *regeneration*. A change in current at any point in the loop is amplified and returned to the starting point with the same phase. For instance, if the Q_2 base current increases, the Q_2 collector current increases. This forces more base current flows through Q_1. In turn, this produces a larger Q_1 collector current, which drives the Q_2 base harder. This buildup in current will continue until both transistors are driven into saturation. In this case, the latch acts like a closed switch (Fig. 14-1b).

On the other hand, if something causes the Q_2 base current to decrease, the Q_2 collector current will decrease. This reduces the Q_1 base current. In turn, there is less Q_1 collector current, which reduces the Q_2 base current even more. This regeneration continues until

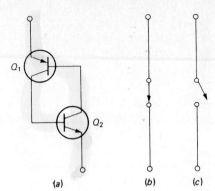

Fig. 14-1.
Complementary latch.

(a)   (b)   (c)

both transistors are driven into cutoff. At this time, the latch acts like an open switch (Fig. 14-1c).

The latch can be in either of two states, closed or open. It will remain in a state indefinitely. If closed, it stays closed until something causes the currents to decrease. If open, it stays open until something else forces the currents to increase.

Triggering One way to close a latch is by *triggering*, applying a forward-bias voltage to either base. For example, Fig. 14-2a shows a *trigger* (sharp pulse) hitting the Q_2 base. Suppose the latch is open before point A in time. Then, all the supply voltage appears across the open latch (Fig. 14-2b), and the operating point is at the lower end of the load line (Fig. 14-2d).

At point A in time, the trigger momentarily forward-biases the Q_2 base. The Q_2 collector current suddenly turns on and forces base current to flow through Q_1. In turn, the Q_1 collector current comes on and drives the Q_2 base harder. Since the Q_1 collector now supplies

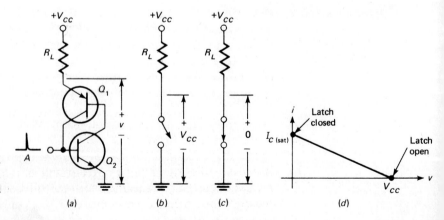

Fig. 14-2.
Latch action. (a) Circuit.
(b) Open latch. (c) Closed
latch. (d) Load line.

(a)   (b)   (c)   (d)

the Q_2 base current, the trigger pulse is no longer needed. In other words, once the regeneration starts, it will sustain itself and drive both transistors into saturation. The minimum input current needed to start the regenerative switching action is called the *trigger current*.

When saturated, both transistors look like short circuits and the latch is closed (Fig. 14-2c). Ideally, the latch has zero voltage across it when closed, and the operating point is at the upper end of the load line (Fig. 14-2d).

Breakover

Another way to close a latch is by breakover. This means using a large enough supply voltage V_{CC} to break down either collector diode. Once the breakdown begins, current comes out of one of the collectors and drives the other base. The effect is the same as if the base had received a trigger. Although breakover starts with a breakdown of one of the collector diodes, it ends with both transistors in the saturated state. This is why the term *breakover* is used instead of breakdown to describe this kind of latch closing.

Low-Current Dropout

How do we open an ideal latch? One way is to reduce the load current to zero. This forces the transistors to come out of saturation and return to the open state. For instance, in Fig. 14-2a we can open the load resistor. Alternatively, we can reduce the V_{CC} supply to zero. In either case, a closed latch will be forced to open. We call this type of opening *low-current dropout* because it depends on reducing latch current to a low value.

Reverse-Bias Trigger

Another way to open the latch is to apply a reverse-bias trigger in Fig. 14-2a. If a negative trigger is used instead of a positive one, the regeneration will drive the transistors into cutoff, and the latch is then open. As an alternative, we can drive the base of Q_1 with a positive trigger. This will reduce the Q_1 base current and start the regeneration that eventually opens the latch.

Points to Remember

Here is a summary that will help you understand how different thyristors work:

1. We can close an ideal latch by forward-bias triggering or by breakover.
2. We can open an ideal latch by reverse-bias triggering or by low-current dropout.

Some of the simpler thyristors rely on breakover to close the latch and low-current dropout to open the latch. More complex thyristors use forward-bias triggering to close the latch and reverse-bias triggering to open the latch.

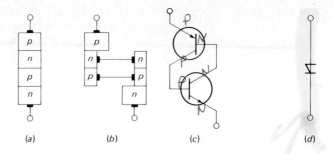

Fig. 14-3.
Four-layer diode.
(a) Structure.
(b) Equivalent structure.
(c) Equivalent circuit.
(d) Schematic symbol.

(a)   (b)   (c)   (d)

14-2
THE FOUR-LAYER
DIODE

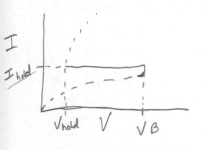

Figure 14-3a is a *four-layer diode* (also called a Shockley diode). It is classified as a diode because it has only two external leads. Because of its four doped regions, it's often called a *pnpn* diode. The easiest way to understand how it works is to visualize it separated into two halves as shown in Fig. 14-3b. The left half is a *pnp* transistor and the right half is an *npn* transistor. Therefore, the four-layer diode is equivalent to the latch shown in Fig. 14-3c.

Because there are no trigger inputs, the only way to close a four-layer diode is by breakover, and the only way to open it is by low-current dropout. With a four-layer diode it is not necessary to reduce the current all the way to zero to open the latch. The internal transistors of the four-layer diode will come out of saturation when the current is reduced to a low value called the *holding current*. Figure 14-3d shows the schematic symbol of a four-layer diode.

EXAMPLE 14-1   The 1N5158 of Fig. 14-4a has a breakover voltage of 10 V. What is the load current when the input voltage equals 15 V and the diode drop is 1 V?

SOLUTION   There is more than enough voltage to cause breakover. The load current is

$$I = \frac{15 \text{ V} - 1 \text{ V}}{100 \ \Omega} = 140 \text{ mA}$$

EXAMPLE 14-2   In Fig. 14-4a, the diode has a holding current of 4 mA. Allowing 0.5 V across the diode at the dropout point, what is the input voltage that just produces low-current dropout?

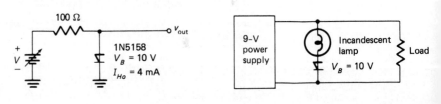

Fig. 14-4.   (a)   (b)

SOLUTION To open the four-layer diode, we have to reduce the current below the holding current of 4 mA. This means reducing the input voltage to slightly less than

$$V = 0.5 \text{ V} + (4 \text{ mA})(100 \text{ } \Omega) = 0.9 \text{ V}$$

EXAMPLE 14-3 Describe the action in Fig. 14-4b.

SOLUTION The four-layer diode has a breakover voltage of 10 V. As long as the power supply puts out 9 V, the four-layer diode is open and the lamp is dark. But if something goes wrong with the power supply and its voltage rises above 10 V, the four-layer diode latches and the lamp comes on. Even if the supply should return to 9 V, the diode remains latched as a record of the overvoltage that occurred. The only way to make the lamp go out is by turning off the supply.

The circuit is an example of an *overvoltage detector*. As long as the supply stays within normal limits, nothing happens. But if we get an overvoltage, even temporarily, the lamp comes on and stays on.

EXAMPLE 14-4 Figure 14-5a shows a sawtooth generator. Describe the circuit action.

SOLUTION If the four-layer diode were not in the circuit, the capacitor would charge exponentially and its voltage would follow the dashed curve of Fig. 14-5b. But the four-layer diode is in the circuit. Therefore, as soon as the capacitor voltage reaches 10 V, the diode breaks over and the latch closes. This discharges the capacitor, producing the *flyback* (sudden decrease) in capacitor voltage. At some point on the flyback, the current drops below the holding current and the four-layer diode opens. The next cycle then begins.

Figure 14-5a is an example of a *relaxation oscillator*, a circuit that generates an output signal whose frequency depends on the charging and discharging of a capacitor (or inductor). If we increase the RC time constant, the capacitor takes longer to charge to 10 V and the frequency of the sawtooth wave is lower. Using a rheostat as shown in Fig. 14-5c, we can get a 50:1 range in frequency.

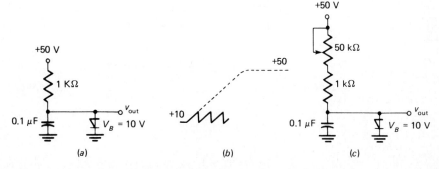

Fig. 14-5.
Sawtooth generator.
(a) Fixed frequency.
(b) Variable frequency.

14-3
THE SILICON-CONTROLLED RECTIFIER

The silicon-controlled rectifier (SCR) is far more useful than a four-layer diode because it has an extra lead connected to the base of the *npn* section as shown in Fig. 14-6*a*. We can again visualize the four doped regions separated into two transistors as shown in Fig. 14-6*b*. Therefore, the SCR is equivalent to a latch with trigger input (Fig. 14-6*c*). Schematic diagrams use the symbol of Fig. 14-6*d*. Whenever you see this, remember it's a latch with a trigger input.

Blocking Voltage

SCRs have breakover voltages from around 50 V to more than 2500 V, depending on the SCR type number. Most SCRs are designed for trigger closing and low-current opening. In other words, the SCR stays open until a trigger hits the gate (see Fig. 14-6*d*). Then, the SCR latches and remains closed, even though the trigger disappears. The only way to open an SCR is with low-current dropout

Most people think of the SCR as a device that blocks voltage until a trigger closes it. For this reason, the breakover voltage is often called the *forward blocking voltage* on data sheets. As an example, a 2N4444 is an SCR with a forward blocking voltage of 600 V. As long as the supply voltage is less than 600 V, the SCR cannot break over. The only way to close it is with a gate trigger.

High Currents

Almost all SCRs are industrial devices that can handle large currents ranging from less than 1 A to more than 2500 A, depending on the type. Because they are high-current devices, SCRs have relatively large trigger and holding currents. For example, the 2N4444 can conduct up to 8 A continuously. Its trigger current is 10 mA, and so too is its holding current. This means you have to supply the gate with at least 10 mA to control up to 8 A of anode current. (See Fig. 14-6*d* for the location of the anode and the cathode.) As another example, the C701 is an SCR that can conduct up to 1250 A with a trigger current of 150 mA and a holding current of 500 mA.

Critical Rate of Rise

Because of the capacitance inside an SCR, it is possible for a rapidly changing supply voltage to trigger the SCR. Stated another way, if the rate of rise of supply voltage is high enough, capacitive charging currents can initiate regeneration. To avoid false triggering of the

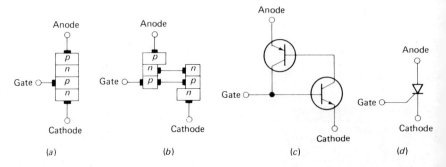

Fig. 14-6.
SCR. (a) Structure.
(b) Equivalent structure.
(c) Equivalent circuit.
(d) Schematic symbol.

SCR, the anode rate of voltage change must not exceed the *critical rate of voltage rise* listed on the data sheet.

As an example, the 2N4444 has a critical rate of voltage rise of 50 V/μs. To avoid an unwanted breakover, the anode voltage must not increase faster than 50 V/μs. As another example, the C701 has a critical rate of voltage rise of 200 V/μs. In this case, the anode voltage must not increase faster than 200 V/μs.

Switching transients on the supply lead are the main cause of exceeding the critical rate of voltage rise. One way to reduce the effects of these transients is with an *RC snubber*, shown in Fig. 14-7*a*. If a high-speed transient does appear on the supply voltage, its rate of voltage rise is reduced at the anode because of the *RC* network. The rate of increase in anode voltage depends on the load resistance, as well as the *R* and *C* values.

Larger SCRs also have a critical rate of current rise. For instance, the C701 has a critical rate of 150 A/μs. If the anode current tries to increase faster than this, hot spots can occur inside the SCR and destroy it. Including an inductor in series as shown in Fig. 14-7*b* reduces the rate of current rise, as well as helping the *RC* snubber to decrease the rate of voltage rise.

EXAMPLE 14-5

An *op amp* is a dc amplifier with an extremely large voltage gain. As shown in Fig. 14-8, this type of amplifier has a plus input and a minus input. The voltage appearing between the plus and minus terminals is called the *error voltage*. As will be discussed in Chap. 15, the output voltage of an op amp is

$$v_{out} = Av_{error}$$

where *A* is typically greater than 100,000.

Figure 14-8 includes an SCR *crowbar*, a way of protecting a load against overvoltage. Describe the circuit action.

SOLUTION

When the circuit is working properly, the load voltage is 20 V. In this case, the zener diode produces 10 V at the minus input of the op amp.

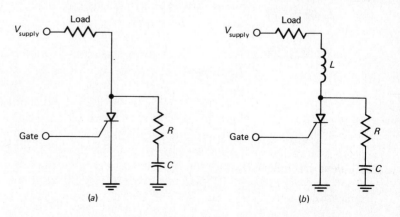

Fig. 14-7.
(a) *RC* snubber.
(b) Inductor reduces rate of current rise.

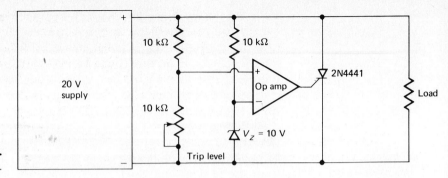

Fig. 14-8.
Crowbar circuit.

The *trip level* is adjusted to get slightly less than 10 V going into the plus input of the op amp. As a result, the error voltage is slightly negative and the op amp has a negative output. Therefore, the SCR is open and has no effect on the load voltage.

If something goes wrong with the power supply and the load voltage increases above 20 V, the plus input of the op amp becomes greater than 10 V, while the minus input stays at 10 V because of the zener diode. In this case, the error voltage is positive and the op amp delivers a positive trigger to the gate. Immediately, the SCR closes and shuts down the power supply. The action is the same as throwing a crowbar across the load terminals. Because the SCR turn-on is very fast (1 μs for a 2N4441), the load is quickly protected from the damaging effects of a large overvoltage.

Crowbarring is a drastic form of protection, but it is necessary with many digital ICs because they cannot take much overvoltage. Rather than risk overvoltage destruction of expensive ICs, we can use an SCR crowbar to shut down the power supply at the first sign of overvoltage.

14-4
VARIATIONS OF THE SCR

Other *pnpn* devices are similar to the SCR. What follows is a brief description of these SCR variations. The devices to be discussed are for low-power applications.

LASCR

Figure 14-9*a* shows the *light-activated* SCR, designated the LASCR. The arrows represent incoming light that passes through a window and hits the depletion layers. When the light is strong enough, valence electrons can be dislodged from their orbits to become free electrons. When these free electrons flow out of a collector and into a base, regeneration starts and the LASCR closes.

After a light trigger has closed the LASCR, it remains closed even though the light disappears. For maximum sensitivity to light, the gate is left open as shown in Fig. 14-9*a*. If you want an adjustable trip level, you can include the adjustment shown in Fig. 14-9*b*. The gate resistor

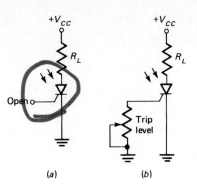

Fig. 14-9.
LASCR circuits.
(a) Maximum sensitivity.
(b) Variable trip point.

(a)   (b)

diverts some of the light-produced electrons and changes the sensitivity of the circuit to the incoming light.

GCS As mentioned earlier, low-current dropout is the normal way to open an SCR. But the gate-controlled switch (GCS) is designed for easy opening with a reverse-biased trigger. The GCS is closed by a positive trigger and opened by a negative trigger (or by low-current dropout). Figure 14-10 shows a GCS circuit. Each positive trigger closes it, and each negative trigger opens it. Because of this, we get the square-wave output shown. The GCS is useful in counting circuits, digital circuits, and other applications where a negative trigger is available for turn-off.

SCS Figure 14-11a shows the doped regions of a *silicon-controlled switch* (SCS). Now an external lead is connected to each doped region. Visualize the device separated into two halves (Fig. 14-11b). Therefore, it is equivalent to a latch with access to both bases (Fig. 14-11c). A forward-bias trigger on either base will close the SCS. Likewise, a reverse-bias trigger on either base will open the device.

Figure 14-11d shows the schematic symbol for an SCS. The lower gate is called the *cathode gate*. The upper gate is the *anode gate*. The SCS is a low-power device compared with the SCR. It handles currents in milliamperes rather than amperes.

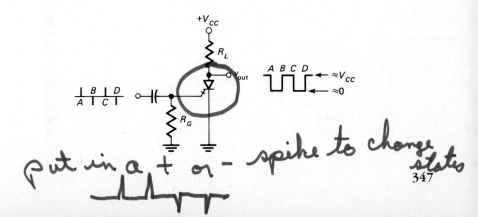

Fig. 14-10.
GCS circuit.

put in a + or - spike to change states

THYRISTORS

347

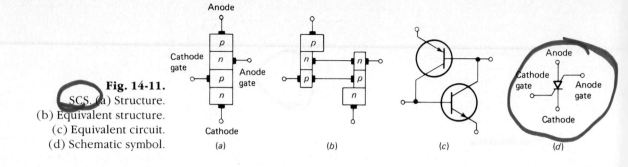

Fig. 14-11.
SCS. (a) Structure.
(b) Equivalent structure.
(c) Equivalent circuit.
(d) Schematic symbol.

(a)   (b)   (c)   (d)

14-5
BIDIRECTIONAL THYRISTORS

Up to now, all devices have been *unidirectional* because current flowed in only one direction. This section looks at *bidirectional* thyristors, where current can be in either direction.

Diac

The *diac* can have latch current in either direction. The equivalent circuit of a diac is a pair of four-layer diodes in parallel as shown in Fig. 14-12a, ideally the same as the latches in Fig. 14-12b. The diac is nonconducting until the voltage across it tries to exceed the breakover voltage in either direction.

For instance, if v has the polarity shown in Fig. 14-12a, the left latch closes when the magnitude of v tries to exceed the breakover voltage. In this case, the left latch closes as shown in Fig. 14-12c. On the other hand, if the polarity of v is opposite that of Fig. 14-12a, the right latch closes when the magnitude of v tries to exceed the breakover voltage.

Once the diac is conducting, the only way to open it is by low-current dropout. This means reducing the current below the rated holding current of the device. Figure 14-12d shows the schematic symbol of the diac.

Triac

The triac acts like two SCRs in parallel as shown in Fig. 14-13a, equivalent to the latches of Fig. 14-13b. Because of this, the triac can control current in either direction. The breakover voltage is usually high, and the normal way to turn on a triac is by applying a forward-bias trigger. If v has the polarity shown in Fig. 14-13a, we have to apply a positive trigger. This closes the left latch. When v has a polarity opposite that of Fig. 14-13a, a negative trigger is needed; it will close the right latch.

Fig. 14-12.
Diac. (a) Equivalent circuit.
(b) Latch equivalent.
(c) Left latch closed.
(d) Schematic symbol.
(e) Alternative symbol.
(f) Another symbol.

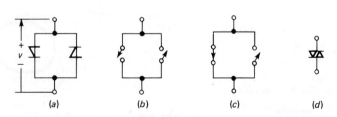

(a)   (b)   (c)   (d)

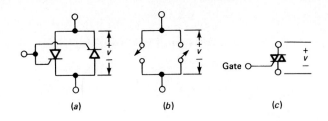

Fig. 14-13.
Triac. (a) Equivalent circuit.
(b) Latch equivalent.
(c) Schematic symbol.

(a)   (b)   (c)

Figure 14-13c shows the schematic symbol of a triac. The triac is not used as much as the SCR because it cannot handle as much current. Furthermore, there are many more commercially available SCRs than triacs.

14-6 THE UNIJUNCTION TRANSISTOR

The *unijunction transistor* (UJT) has two doped regions with three external leads (Fig. 14-14a). It has one emitter and two bases. The emitter is heavily doped, having many holes. The n region, however, is lightly doped. For this reason, the resistance between the bases is relatively high, typically 5 to 10 kΩ when the emitter is open. We call this the *interbase resistance*, symbolized R_{BB}.

Intrinsic Standoff Ratio

Figure 14-14b is the equivalent circuit of a UJT. The emitter diode drives the junction of two internal resistances, R_1 and R_2. When the emitter diode is nonconducting, R_{BB} is the sum of R_1 and R_2. When a supply voltage is between the two bases (Fig. 14-14c), the voltage across R_1 is given by

$$V_1 = \eta V \tag{14-1}$$

where

$$\eta = \frac{R_1}{R_{BB}}$$

The quantity η is called the *intrinsic standoff ratio,* which is the voltage-divider factor. The typical range of η is from 0.5 to 0.8. For

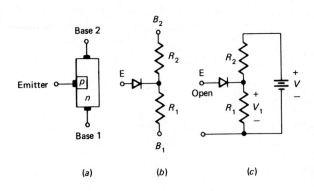

Fig. 14-14.
Unijunction transistor.
(a) Structure.
(b) Equivalent circuit.
(c) Producing standoff voltage.

(a)   (b)   (c)

instance, a 2N2646 has an η of 0.65. If this UJT is used in Fig. 14-14c with a supply voltage of 10 V,

$$V_1 = \eta V = 0.65(10\text{ V}) = 6.5\text{ V}$$

How It Works   In Fig. 14-14c, V_1 is called the *intrinsic standoff voltage* because it keeps the emitter diode reverse-biased for all emitter voltages less than V_1. If V_1 equals 6.5 V, we have to apply slightly more than 6.5 V to the emitter to turn on the emitter diode.

In Fig. 14-15a, imagine that the emitter supply voltage is turned down to zero. Then, the intrinsic standoff voltage reverse-biases the emitter diode. When we increase the emitter supply voltage, v_E increases until it is slightly greater than V_1. This turns on the emitter diode. Since the *p* region is heavily doped compared with the *n* region, holes are injected into the lower half of the UJT. The light doping of the *n* region gives these holes a long lifetime. These holes create a *p*-type inversion layer between the emitter and lower base (similar to the inversion layer in an enhancement-type MOSFET).

The flooding of the lower half of the UJT with holes drastically lowers resistance R_1 (Fig. 14-15b). Because R_1 is suddenly much lower in value, v_E suddenly drops to a low value and the emitter current increases.

Latch Equivalent   You can remember UJT action by relating it to the latch of Fig.
Circuit   14-16a. With a positive voltage from B_2 to B_1, a standoff voltage V_1 appears across R_1. This keeps the emitter diode of Q_2 reverse-biased as long as the emitter input voltage is less than the standoff voltage. When the emitter input voltage is slightly greater than the standoff voltage, however, Q_2 turns on and regeneration takes over. This drives both transistors into saturation, ideally shorting the emitter and the lower base B_1.

Figure 14-16b is the schematic symbol for a UJT. The emitter arrow reminds us of the upper emitter in a latch. When the emitter voltage exceeds the standoff voltage, the latch between the emitter and the lower base closes. Ideally, you can visualize a short between E and B_1. To a second approximation, a low voltage called the *emitter saturation voltage* $V_{E(\text{sat})}$ appears between E and B_1.

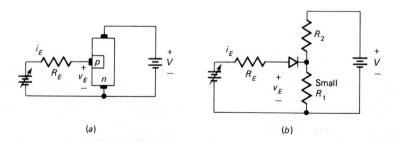

Fig. 14-15.
Turning on a UJT.

(a)   (b)

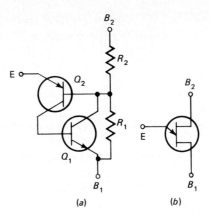

Fig. 14-16.
UJT. (a) Equivalent circuit.
(b) Schematic symbol.

(a)

(b)

The latch stays closed as long as latch current (emitter current) is greater than the holding current. Data sheets specify a *valley current* I_V equivalent to the holding current. For example, a 2N2646 has an I_V of 6 mA. To hold the latch closed, the emitter current must be greater than 6 mA.

EXAMPLE 14-6 The 2N4871 of Fig. 14-17 has an η of 0.85. What is the ideal emitter current?

SOLUTION The standoff voltage is

$$V_1 = \eta V = 0.85(10\text{ V}) = 8.5\text{ V}$$

Ideally, v_E must be slightly greater than 8.5 V to turn on the emitter diode and close the latch.

With the input switch closed, 20 V drives the 400-Ω resistor. This is more than enough voltage to overcome the standoff voltage. Therefore, the latch is closed and the emitter current ideally equals

$$I_E = \frac{20\text{ V}}{400\text{ }\Omega} = 50\text{ mA}$$

EXAMPLE 14-7 The valley current of a 2N4871 is 7 mA, and the emitter voltage is 1 V at this point. For what value of emitter-supply voltage does the UJT open in Fig. 14-17?

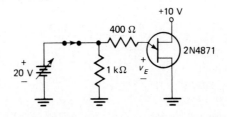

Fig. 14-17.

THYRISTORS

As we reduce the emitter-supply voltage, the emitter current decreases. At the point where it equals 7 mA, v_E is 1 V and the latch is about to open. The emitter-supply voltage is

$$V = 1 \text{ V} + (7 \text{ mA})(400 \text{ } \Omega) = 3.8 \text{ V}$$

When V is less than 3.8 V, the UJT opens. Then, it will be necessary to increase V to more than 8.5 V to close the UJT.

14-7

UJT CIRCUITS

To get a better idea of how a UJT works and how to use it, here are some practical applications. Figure 14-18a shows a UJT relaxation oscillator. The action is similar to the relaxation oscillator of Example 14-4. The capacitor charges toward V_{CC}, but as soon as its voltage exceeds the standoff voltage, the UJT latches. This discharges the capacitor until low-current dropout occurs. As soon as the UJT opens, the next cycle begins. As a result, we get a sawtooth output.

If we add a small resistor to each base circuit, we can get three useful outputs: sawtooth waves, positive triggers, and negative triggers as shown in Fig. 14-18b. The triggers appear during the flyback of the sawtooth wave because the UJT conducts heavily at this time. With the values given in Fig. 14-18b, the frequency is from approximately 50 Hz to 1 kHz, depending on the rheostat value. The width of the triggers (same as the flyback time) is in the vicinity of 20 μs.

Sharp trigger pulses out of a UJT relaxation oscillator can be used to trigger an SCR. For instance, Fig. 14-19 shows part of an automobile ignition system. When the distributor points are closed, the UJT and SCR are open. When the points open, however, the 0.1-μF capacitor charges exponentially. As soon as the capacitor voltage exceeds the standoff voltage, the UJT latch closes and produces the positive trig-

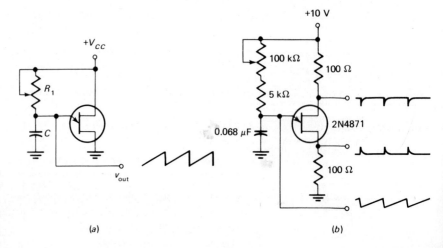

Fig. 14-18.
Relaxation oscillators.
(a) Sawtooth generator.
(b) Trigger and sawtooth
generator.

(a)   (b)

CHAPTER 14

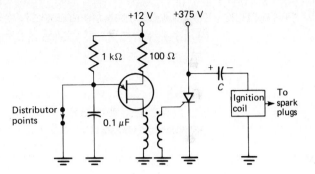

Fig. 14-19.
Car-ignition system using
UJT and SCR.

ger as previously described. This trigger is coupled to the gate of the SCR. When the SCR latches shut, the positive end of the output capacitor is suddenly grounded. As the output capacitor discharges through the ignition coil, a high-voltage pulse drives one of the spark plugs. When the points again close, the circuit resets itself in preparation for the next cycle.

Summary

A latch uses positive feedback, also called regeneration. To close a latch, we can use forward-bias triggering or breakover. We can open a latch by reverse-bias triggering or by low-current dropout.

The only way to close a four-layer diode is by breakover. To open it, you have to reduce the current below the holding current.

The SCR has a trigger input. SCRs are not intended to break over. The normal way to close an SCR is with a gate trigger. The only way to open an SCR is with low-current dropout. To avoid false triggering of an SCR, the anode rate of voltage change must not exceed the critical rate of voltage rise listed on the data sheet. Switching transients are the main cause of exceeding the critical rate of voltage rise. Larger SCRs also have a critical rate of current rise, which must not be exceeded because of possible damage to the SCR.

The diac and the triac are bidirectional thyristors because they can conduct in either direction. The triac acts like two SCRs in parallel.

The UJT has an intrinsic standoff voltage. To turn it on, the emitter voltage has to be slightly greater than the standoff voltage. To turn it off, the current has to be reduced to slightly less than the valley current.

Glossary

breakover Breakdown followed by regenerative switching that closes the latch.

critical rate of current rise The maximum rate at which current can increase without producing internal hot spots and SCR destruction.

critical rate of voltage rise The maximum rate at which voltage can increase without capacitive currents producing regeneration and premature closing of the latch.

forward blocking voltage The breakover voltage of an SCR or triac.

holding current The minimum current that keeps a latch closed.

interbase resistance The resistance between the bases of a UJT when the emitter is open.

latch A switch whose regenerative action can keep it open or closed indefinitely.

relaxation oscillator A circuit that generates a signal whose frequency depends on the time constant of a capacitive or inductive branch.

snubber An *RC* circuit across the SCR to prevent the rate of voltage rise from exceeding the critical value.

transients Temporary changes caused by switches opening or closing. A suddenly changing current, for instance, causes large voltages across inductances. These temporary induced voltages can be fed back to the supply line, where they may affect other circuits.

Review Questions

1. How many doped regions does a thyristor have? 4
2. Describe breakover. How does it differ from breakdown?
3. What are the two ways to close an ideal latch? The two ways to open it?
4. What is the only way to close a four-layer diode? The only way to open it?
5. Define the forward blocking voltage of an SCR. How many external leads does an SCR have?
6. Describe one method for reducing the rate of voltage rise at the anode of an SCR.
7. Describe the following thyristors: LASCR, GCS, and SCS.
8. What is a diac? A triac?
9. How many doped regions does a UJT have? Describe how it works.

Problems

14-1. If the 1N5160 of Fig. 14-20a is conducting, for what value of *V* will it stop conducting (ideally)? If we allow 0.7 V across the diode

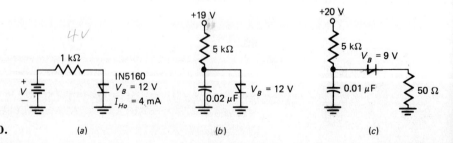

Fig. 14-20. (a) (b) (c)

at the dropout point, what is the value of V where low-current dropout occurs?

14-2. With a supply of 19 V, it takes the capacitor of Fig. 14-20b exactly one time constant to charge to 12 V, the breakover voltage of the diode. If we neglect the voltage across the diode when conducting, what is the frequency of the sawtooth output?

14-3. The current through the 50-Ω resistor of Fig. 14-20c is maximum just after the diode latches. Ideally, what is the maximum current? If we allow 1 V across the latched diode, what is the maximum current?

14-4. The 2N4216 of Fig. 14-21a has a trigger current of 0.1 mA. Ideally, what value of V turns on the SCR? If we allow 0.8 V across the gate-to-ground terminals, what is the value of V that turns on the SCR?

14-5. The four-layer diode of Fig. 14-21b has a breakover voltage of 10 V. The SCR has a trigger current of 0.1 mA. Neglecting the voltage across the gate input, what is the current through the four-layer diode just after it breaks over? The current through the 500-Ω resistor after the SCR turns on?

14-6. Figure 14-22a shows an alternative schematic symbol for a diac. The MPT32 diac breaks over when the capacitor voltage reaches 32 V. It takes one time constant for the capacitor to reach this voltage. How long after the switch is closed does the triac turn on? What is the ideal value of gate current when the diac breaks over? The load current after the triac has closed?

14-7. The frequency of the square wave in Fig. 14-22b is 10 kHz. It takes one time constant for the capacitor to reach the breakover voltage of the diac. If the MPT32 breaks over at 32 V, what is the ideal value of gate current at the instant the diac breaks over? The ideal load current?

14-8. The UJT of Fig. 14-23a has an η of 0.63. Ideally, what is the value of V that just turns on the UJT? Allowing 0.7 V across the emitter diode, what value of V just turns on the UJT?

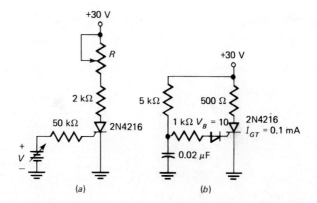

Fig. 14-21.

(a) (b)

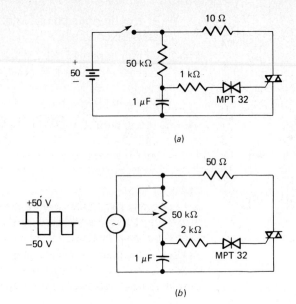

(a)

(b)

Fig. 14-22.

14-9. The valley current of the UJT in Fig. 14-23a is 2 mA. If the UJT is latched, we have to reduce V to get low-current dropout. Neglecting the drop across the emitter diode, what value of V just opens the UJT? Allowing 0.7 V across the emitter diode, what is the value of V that just opens the UJT?

14-10. The intrinsic standoff ratio of the UJT in Fig. 14-23b is 0.63. Ideally, what are the minimum and maximum output frequencies?

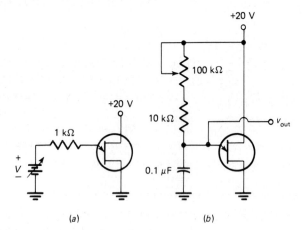

Fig. 14-23.

(a)

(b)

Op-Amp Theory

De Forest: In 1906, he invented the vacuum-tube triode, the first electronic device capable of amplifying a signal. His idea was so revolutionary that he was placed under arrest for using the mails to defraud: the charge was attempting to sell stock in a company to make "a strange device like an incandescent lamp . . . which device has proved worthless."

About a third of all linear ICs are operational amplifiers (op amps). An op amp is a high-gain dc amplifier usable from 0 to over 1 MHz. By connecting external resistors to the op amp, you can adjust the voltage gain and bandwidth to your requirements. There are over 2000 types of commercially available op amps. Almost all are linear ICs with room-temperature dissipations under a watt. Whenever you need voltage gain in a low-power application (less than 1 W), check the available op amps. In many cases, an op amp will do the job.

15-1
THE DIFFERENTIAL AMPLIFIER

Transistors, diodes, and resistors are the only practical components in an IC. Capacitors have been fabricated on a chip, but these are usually less than 50 pF. Therefore, IC designers cannot use coupling and bypass capacitors. Instead, the stages of an IC have to be direct-coupled. One of the best direct-coupled stages is the *differential amplifier* (diff amp). This amplifier is widely used as the input stage of an op amp.

Double-Ended Input and Output

Figure 15-1a shows the most general form of a diff amp. It has two inputs v_1 and v_2. Because of the direct coupling, the input voltages can have frequencies all the way down to zero. The output voltage v_{out} is the voltage between the collectors. Ideally, the circuit is symmetrical, with identical transistors and collector resistors. Therefore, when v_1 equals v_2, the output voltage is zero. When v_1 is greater than v_2, an output voltage appears with the polarity shown in Fig. 15-1a. When v_1 is less than v_2, the output voltage has the opposite polarity.

The diff amp of Fig. 15-1a has a *double-ended input* (two separate inputs). Input v_1 is called the *noninverting* input because the output voltage is in phase with v_1. On the other hand, v_2 is the *inverting* input because the output is 180° out of phase with v_2. A diff amp amplifies the difference between the two input voltages. By analyzing the ac equivalent circuit, we can derive the following formula for the output voltage:

$$v_{out} = A(v_1 - v_2) \qquad\qquad\qquad \textbf{(15-1)}$$

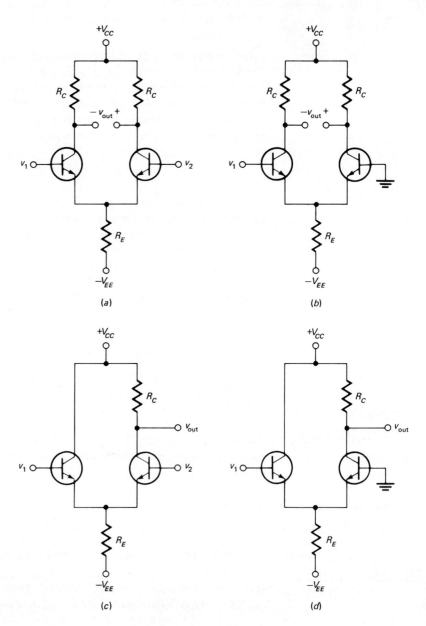

Fig. 15-1.
Differential amplifiers.
(a) Double-ended input and double-ended output.
(b) Single-ended input and double-ended output.
(c) Double-ended input and single-ended output.
(d) Single-ended input and single-ended output.

where v_{out} = voltage between collectors
$$A \cong R_C/r'_e$$
v_1 = noninverting input voltage
v_2 = inverting input voltage

Incidentally, $v_1 - v_2$ is sometimes called the *differential input*.

Single-Ended Input and Double-Ended Output

In some applications, only one of the inputs is used, with the other grounded as shown in Fig. 15-1b. This type of input is called *single-ended*. The output remains double-ended and is given by Eq. (15-1). With v_2 equal to zero, $v_{out} = Av_1$.

A double-ended output has few applications because it requires a floating load. In other words, you have to connect both ends of the load to the collectors. This is inconvenient in most applications because loads are usually single-ended, meaning that one end of the load is connected to ground.

Double-Ended Input and Single-Ended Output

Figure 15-1c shows the most practical and widely used form of diff amp. This has many applications because it can drive single-ended loads like CE amplifiers, emitter followers, and other circuits discussed in earlier chapters. The diff amp of Fig. 15-1c is the type of diff amp used for the input stage of most op amps. For this reason, the remainder of the chapter emphasizes this form of diff amp.

By analyzing the ac equivalent circuit, we can derive
$$v_{out} = A(v_1 - v_2) \qquad \text{(15-2)}$$

where v_{out} = ac voltage from collector to ground
$$A \cong R_C/2r'_e$$
v_1 = noninverting input voltage
v_2 = inverting input voltage

Notice that the voltage gain A is half the value in Eq. (15-1), a direct consequence of using only a single collector resistance R_C.

Single-Ended Input and Output

Figure 15-1d shows the final form of a diff amp. It has a single-ended input and the single-ended output. The output voltage is given by Eq. (15-2). Since v_2 is zero, v_{out} equals Av_1. A diff amp of this form is useful for direct-coupled stages where you are interested in amplifying only one input.

15-2

DC ANALYSIS OF A DIFF AMP

Figure 15-2a shows the dc equivalent circuit of a diff amp with a double-ended input and single-ended output. The bases are returned to ground through the base resistors. These may be actual resistors, or they may represent the Thevenin resistances of the circuits driving

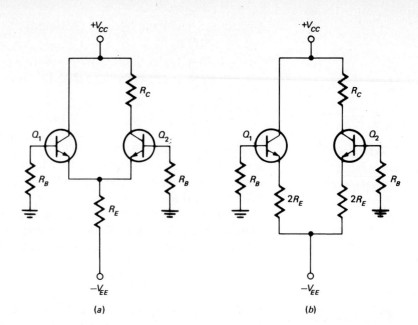

Fig. 15-2.
Differential amplifier and
dc equivalent circuit.

(a)

(b)

the diff amp. Either way, there must be a dc path from each base to ground. Otherwise, the transistors will go into cutoff.

Tail Current   A diff amp is sometimes called a *long-tail pair* because it consists of two transistors connected to a single emitter resistor (the tail). The current through this resistor is called the *tail current*. When the transistors are identical, the tail current splits equally between the Q_1 and Q_2 emitters. Because of this, we can draw the equivalent circuit of Fig. 15-2b. Notice that each emitter is biased through a resistance of $2R_E$. This equivalent circuit produces the same dc emitter currents as the original circuit.

Emitter Bias   In Fig. 15-2b, we can recognize emitter bias. Each transistor is emitter-biased by the V_{EE} supply through the $2R_E$ resistor. Therefore,

$$I_E = \frac{V_{EE} - V_{BE}}{2R_E} \tag{15-3}$$

In Fig. 15-2a, the tail current is the sum of the two emitter currents and is given by

$$I_T = 2I_E$$

or

$$I_T = \frac{V_{EE} - V_{BE}}{R_E} \tag{15-4}$$

When calculating the emitter current is either transistor, you have two choices. You can use Eq. (15-4) to calculate the tail current, then divide by 2 to get the emitter current in each transistor. Alternatively, you can use Eq. (15-3) to calculate the emitter current.

Input Offset Current

Base currents I_{B1} and I_{B2} flow to ground through the base resistors in Fig. 15-2a. The *input offset current* is defined as the difference of the base currents. In symbols,

$$I_{\text{in(off)}} = I_{B1} - I_{B2} \qquad (15\text{-}5)$$

This current indicates how closely matched the transistors are. If the transistors are identical, the input offset current is zero.

As an example, suppose $I_{B1} = 90\ \mu\text{A}$ and $I_{B2} = 70\ \mu\text{A}$. Then,

$$I_{\text{in(off)}} = 90\ \mu\text{A} - 70\ \mu\text{A} = 20\ \mu\text{A}$$

The Q_1 transistor has 20 μA more base current than the Q_2 transistor. This may cause a problem, depending on how large the base return resistors are. More will be said about this later.

Input Bias Current

The *input bias current* is the average base current, given by

$$I_{\text{in(bias)}} = \frac{I_{B1} + I_{B2}}{2} \qquad (15\text{-}6)$$

For instance, if $I_{B1} = 90\ \mu\text{A}$ and $I_{B2} = 70\ \mu\text{A}$, then the input bias current is

$$I_{\text{in(bias)}} = \frac{90\ \mu\text{A} + 70\ \mu\text{A}}{2} = 80\ \mu\text{A}$$

Input Offset Voltage

Assume that we ground both bases in Fig. 15-2a. If the transistors are identical, the quiescent dc voltage at the output is

$$V_C = V_{CC} - I_C R_C \qquad (15\text{-}7)$$

where I_C is approximately equal to the I_E of Eq. (15-3). Any deviation from this quiescent value is called the *output offset voltage*. If the transistors are not identical, the dc emitter currents are not equal and there is an output offset voltage.

The *input offset voltage* is defined as the input voltage needed to zero or null the output offset voltage. For instance, if a data sheet lists a worst-case input offset voltage of ± 5 mV, then we need to apply up to ± 5 mV to one of the inputs to reduce the output offset voltage to zero. In general, the smaller the input offset voltage, the better the diff amp because its transistors are more closely matched.

15-3
AC ANALYSIS OF A DIFF AMP

Figure 15-3a shows a diff amp with a noninverting input v_1 and an inverting input v_2. One way to derive the voltage gain is to apply the superposition theorem. This means working out the voltage gain for each input, then combining the two results to get the total gain.

Noninverting Input

Let us start by applying v_1 while grounding v_2, as shown in Fig. 15-3b. The circuit has been redrawn to emphasize the following: The input

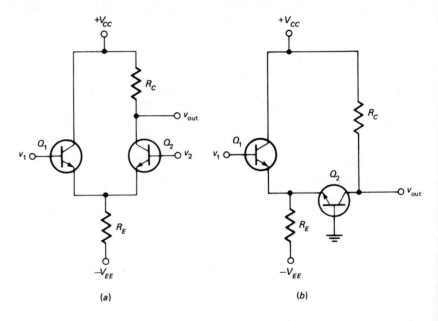

(a)   (b)

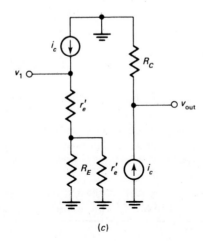   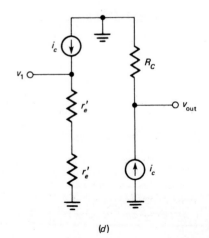

Fig. 15-3.
Ac analysis of differential amplifier. (a) Circuit. (b) CE stage drives CB stage. (c) The two emitter resistances are in series. (d) Final ac equivalent circuit.

(c)   (d)

362

CHAPTER 15

signal drives Q_1, which acts like an emitter follower. The output of the emitter follower drives Q_2, which is a common-base amplifier. Because there is no phase inversion, the final output is in phase with v_1. This is why v_1 is called the noninverting input.

Figure 15-3c shows the ac equivalent circuit. Notice that the upper r'_e is the ac emitter resistance of Q_1, and the lower r'_e is the ac input resistance of Q_2. In any practical circuit, R_E is much greater than r'_e, so that the circuit simplifies to Fig. 15-3d. This means that approximately half of the input voltage reaches the input of the CB amplifier. Therefore, the voltage gain from the noninverting input to the output is

$$\frac{v_{\text{out}}}{v_1} \cong \frac{R_C}{2r'_e} \qquad (15\text{-}8)$$

Inverting Input   Next, let us find the voltage gain for the inverting input. This means that we can ground v_1 and redraw the circuit as shown in Fig. 15-4a. Now Q_2 drives Q_1, which has an input resistance of r'_e. Figure 15-4b is the ac equivalent circuit. Again, R_E is always much greater than r'_e; so the circuit simplifies to Fig. 15-4c. In this case, the voltage gain from the inverting input to the output is

$$\frac{v_{\text{out}}}{v_2} \cong \frac{-R_C}{2r'_e} \qquad (15\text{-}9)$$

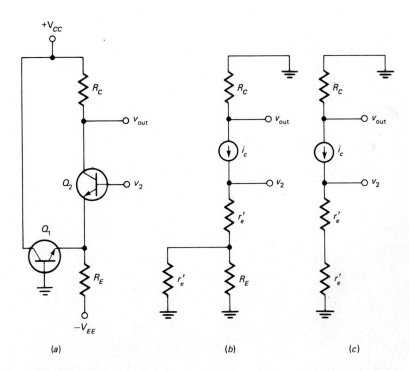

Fig. 15-4.
(a) CE stage drives CB stage. (b) Two emitter resistances are in series. (c) Final ac equivalent circuit.

where the minus sign has been included to account for the phase inversion.

Differential Gain

Compare Eqs. (15-8) and (15-9). As you can see, the magnitude of voltage gain is the same; only the phase differs. When both input signals are present, the superposition theorem says we can add the individual outputs to get the total output:

$$v_{out} \cong \frac{R_C}{2r'_e} (v_1 - v_2) \qquad \textbf{(15-10)}$$

This can be written as

$$v_{out} = A(v_1 - v_2) \qquad \textbf{(15-11)}$$

where

$$A \cong \frac{R_C}{2r'_e} \qquad \textbf{(15-12)}$$

The quantity A is called the *differential voltage gain* because it tells us how much the difference $v_1 - v_2$ is amplified.

Input Impedance

In the midband of a diff amp, the input impedance looking into either input is

$$r_{in} \cong 2\beta r'_e \qquad \textbf{(15-13)}$$

This input impedance is twice that of an ordinary CE amplifier. The factor of 2 arises because the r'_e of each transistor is in series (see Figs. 15-3*d* and 15-4*c*).

One way to get a higher input impedance with a diff amp is to use Darlington transistors. Another way is to use JFETs instead of bipolar transistors. This is the approach taken with BIFET op amps; they use JFETs for the input diff-amp stage. With BIFET op amps the input impedance at low frequencies approaches infinity.

Common-Mode Gain

A *common-mode* signal is one that drives both inputs of a diff amp equally. Most interference, static, and other kinds of undesirable pickup are common-mode. What happens is this. The connecting wires on the input bases act like small antennas. If the diff amp is operating in an environment with a lot of electromagnetic interference, each base picks up an equal amount of unwanted interference voltage. In this case,

$$v_1 = v_2$$

One of the reasons the diff amp is so popular is that it has a high voltage gain for a differential input and a low voltage gain for a common-mode input. In other words, a diff amp discriminates against common-mode signals because it refuses to amplify a common-mode input. Because of this, you don't get a lot of unwanted interference at the output.

Let us find out why the diff amp does not amplify common-mode signals. Figure 15-5a shows a common-mode signal driving a diff amp. As you can see, an equal voltage $v_{in(CM)}$ drives both inputs simultaneously. Assuming identical transistors, the equal inputs imply equal emitter currents. Therefore, we can split R_E as shown in Fig.

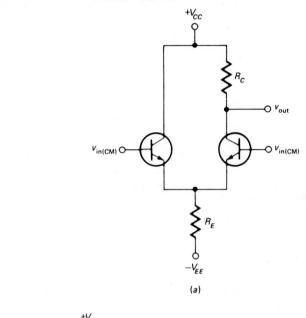

(a)

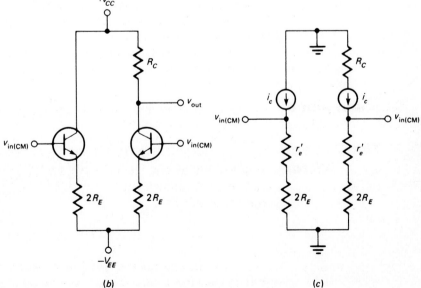

Fig. 15-5.
(a) Common-mode input means equal voltages drive the noninverting and inverting inputs.
(b) Equivalent circuit for common-mode input.
(c) Ac equivalent circuit with common-mode input.

(b)

(c)

15-5*b*. This equivalent circuit has exactly the same emitter currents as the original circuit.

Figure 15-5*c* shows the ac equivalent circuit. Can you see what this means? When a common-mode signal drives a diff amp, a large unbypassed emitter resistance appears in the ac equivalent circuit. Therefore, the voltage gain for a common-mode signal is

$$\frac{v_{out}}{v_{in(CM)}} \cong \frac{R_C}{r'_e + 2R_E}$$

Since R_E is always much greater than r'_e, we can approximate the common-mode voltage gain as $R_C/2R_E$. In symbols,

$$A_{CM} \cong \frac{R_C}{2R_E} \qquad\qquad \textbf{(15-14)}$$

For instance, if $R_C = 10$ kΩ and $R_E = 10$ kΩ, then

$$A_{CM} \cong \frac{10 \text{ k}\Omega}{20 \text{ k}\Omega} = 0.5$$

This means the diff amp attenuates a common-mode signal because the voltage gain is less than unity.

Common-Mode Rejection Ratio   Data sheets list the *common-mode rejection ratio* (CMRR), defined as the ratio of differential voltage gain to common-mode voltage gain:

$$CMRR = \frac{A}{A_{CM}} \qquad\qquad \textbf{(15-15)}$$

For example, if $A = 200$ and $A_{CM} = 0.5$, then

$$CMRR = \frac{200}{0.5} = 400$$

As a rule, the higher CMRR is, the better the diff amp.

The Current Mirror   Figure 15-6*a* shows a *current mirror,* a circuit that is extensively used in linear integrated circuits. The current directions are for electron flow. If you prefer conventional flow, visualize all currents in the opposite direction. If the compensating diode matches the emitter diode, then the diode current equals the collector current:

$$I_2 = I_C$$

At point *A*,

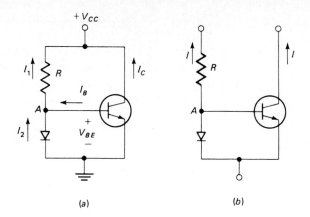

Fig. 15-6.
Current mirror.

(a)   (b)

$$I_1 = I_2 + I_B = I_C + I_B$$

Since base current is much smaller than collector current, this reduces to

$$I_1 \cong I_C$$

This is important. It says the collector current approximately equals the current flowing through R.

Figure 15-6b emphasizes the point. Current I flows through the biasing resistor. This produces an equal collector current I. Think of the circuit as a mirror; the current through R is reflected into the collector circuit, where an equal current appears. This is why the diode-transistor circuit of Fig. 15-6b is known as a current mirror.

Current-Mirror Bias

With an emitter-biased diff amp, the differential voltage gain is $R_c/2r'_e$, and the common-mode voltage gain is $R_c/2R_E$. With Eq. (15-15),

$$\text{CMRR} \cong \frac{R_C/2r'_e}{R_C/2R_E} = \frac{R_E}{r'_e}$$

From this it is clear that the higher we can make R_E, the better the CMRR.

One way to get a very high equivalent R_E is to use current-mirror bias, as shown in Fig. 15-7. This is typical for the first stage of an integrated op amp. Here you see a current mirror driving the emitters of a diff amp. With integrated circuits, the compensating diode Q_3 is actually a transistor connected as a diode (the base and collector tied together). The current through Q_3 is given by

$$I = \frac{V_{CC} + V_{EE} - V_{BE}}{R} \qquad \textbf{(15-16)}$$

This is the value of mirror current supplied by Q_4. Since Q_4 acts like a current source, it has a very high output impedance, much higher

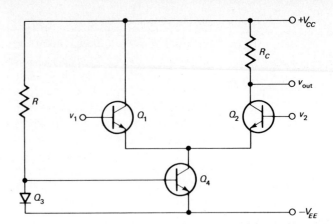

Fig. 15-7.
Current mirror sources
emitter current to diff amp.

than we can get with conventional emitter bias. This means the equivalent R_E of the diff amp is hundreds of kilohms and the CMRR is dramatically improved.

Current-Mirror Load

Earlier, we derived a differential voltage gain of $R_C/2r'_e$. The larger we can make R_C, the greater the differential voltage gain. But we have to be careful. Too large an R_C will saturate the output transistor. As a rule, the designer selects an R_C to get a quiescent voltage that is half of V_{CC}. For example, if the collector supply voltage is $+15$ V, then R_C is selected to get a V_C of $+7.5$ V. This places a limit on the size of R_C and the voltage gain.

One way around this problem is to use an active load. Figure 15-8 shows a current-mirror load. Since Q_5 is a forward-biased diode, it has a very low impedance, and the load on Q_1 still appears almost like an ac short. On the other hand, Q_6 acts like a *pnp* current source. Therefore, Q_2 sees an equivalent R_C that is hundreds of kilohms. As a result, the differential voltage gain is much higher with a current-mirror load. Active loading like this is typical of IC op amps.

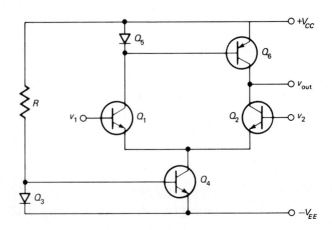

Fig. 15-8.
Diff amp with current-mirror bias drives a current-mirror load.

15-4
THE OPERATIONAL AMPLIFIER

In 1965, Fairchild Semiconductor introduced the μA709, the first widely used IC op amp. Although successful, this first-generation op amp had many disadvantages. This led to an improved op amp known as the μA741. Because it is inexpensive and easy to use, the μA741 has been an enormous success. Many other 741 designs have appeared from various manufacturers. For instance, Motorola produces the MC1741, National Semiconductor the LM741, and Texas Instruments the SN72741. All these IC op amps are equivalent because they have the same specifications on their data sheets. For convenience, most people drop the prefixes and refer to this widely used op amp simply as the 741.

The Industry Standard

The 741 has become the industry standard. As a rule, you first try to use it in your designs. In those cases where you cannot meet a design specification with a 741, you upgrade to a better op amp. Because of its great importance, we will analyze the 741 in detail. Once you understand it, you can easily branch out to other op amps.

Incidentally, the 741 has different versions, numbered 741, 741A, 741C, 741E, 741N, and so on. These differ in their voltage gain, temperature range, noise level, and other such characteristics. The 741C (the C stands for commercial grade) is the least expensive and most widely used. It has a typical input impedance of 2 MΩ, a voltage gain of 100,000, and an output impedance of 75 Ω.

Schematic Diagram of the 741

Figure 15-9 is a simplified schematic diagram of the 741. This circuit is equivalent to the 741 and many later-generation op amps. The input stage is a diff amp using *pnp* transistors (Q_1 and Q_2). To get a high

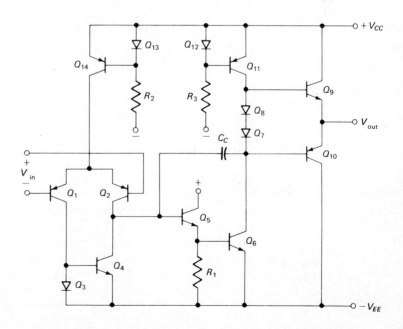

Fig. 15-9.
Simplified schematic diagram of a 741 and similar op amps.

CMRR, a current mirror (Q_{13} and Q_{14}) sources tail current to the diff amp. Also, to get as high a voltage gain as possible for the diff amp, a current-mirror load is used (Q_3 and Q_4).

The output of the diff amp (collector of Q_2) drives an emitter follower (Q_5). This stage steps up the impedance level to avoid loading down the diff amp. The signal out of Q_5 goes to Q_6, which is a class B driver. Incidentally, the plus sign on the collector of Q_5 means that it is connected to the positive V_{CC} supply. Similarly, the minus signs on the bottoms of R_2 and R_3 are connected to the negative V_{EE} supply.

The last stage is a class B push-pull emitter follower (Q_9 and Q_{10}). Because of the *split supply* (equal positive and negative voltages), the quiescent output is ideally 0 V when the input voltages are zero. Any deviation from 0 V is called the *output offset voltage*. Q_{11} and Q_{12} are a current-mirror load for the class B driver.

Notice that direct coupling is used between all stages. For this reason, there is no lower cutoff frequency. In other words, an op amp is a dc amplifier because it operates all the way down to zero frequency.

Active Loading

In Fig. 15-9, we have two examples of *active loading* (using transistors instead of resistors for loads). First, there is a current-mirror load (Q_3 and Q_4) on the diff amp. Second, there is a current-mirror load (Q_{11} and Q_{12}) on the driver stage. Because current sources have high impedances, active loads produce much higher voltage gain than is possible with resistors. These active loads produce a typical voltage gain of 200,000 for the 741.

Active loading is very popular in integrated circuits because it is easier and less expensive to fabricate transistors on a chip than it is to fabricate resistors. MOS digital integrated circuits use active loading almost exclusively. In these ICs, one MOSFET is the active load for another.

Compensating Capacitor

In Fig. 15-9, C_C is called a *compensating capacitor.* This small capacitor (typically 30 pF) has a pronounced effect on the frequency response. C_C is needed to prevent *oscillations,* unwanted signals produced within the op amp. C_C also produces *slew-rate limiting,* which is discussed later.

Floating Inputs

In Fig. 15-9, the inputs are floating. The op amp cannot possibly work unless each input has a dc return path to ground. These returns paths are usually provided by the direct-coupled circuits that drive the op amp. If the driving circuits are capacitively coupled, you have to insert separate base resistors. The key thing to remember is the dc path to ground. Each base must have such a path; otherwise, the input transistors go into cutoff.

Assuming there is a dc path to ground for each base, we still have an offset problem to worry about. Because the input transistors are not quite identical, an unwanted offset voltage will exist at the output of the typical op amp. A later section tells you how to eliminate the output offset voltage.

Input Impedance

Recall that the input impedance of a diff amp is approximately

$$r_{in} = 2\beta r'_e$$

With a small tail current in the input diff amp, an op amp can have a fairly high input impedance. For instance, the input diff amp of a 741 has a tail current of approximately 15 μA. Since each emitter gets half of this,

$$r'_e = \frac{25 \text{ mV}}{I_E} = \frac{25 \text{ mV}}{7.5 \text{ }\mu\text{A}} = 3.33 \text{ k}\Omega$$

Each input transistor has a typical β of 300; therefore,

$$r_{in} = 2\beta r'_e = 2(300)(3.3 \text{ k}\Omega) = 2 \text{ M}\Omega$$

This agrees with the data-sheet value for a 741.

BIFET Op Amps

If an extremely high input impedance is required, you can use a BIFET op amp. BIFET means that bipolar transistors and JFETs are fabricated on the same chip. A BIFET op amp uses JFETs for its input stage, followed by bipolar stages. This combination produces the high input impedance associated with JFETs and the high voltage gain of bipolar transistors. For instance, the LF355 is a popular all-purpose BIFET op amp with a typical input bias current of 0.03 nA. This is much lower than 80 nA, the typical input bias current of a 741. Therefore, if you have to upgrade a design because of input impedance, the BIFET op amp is a natural choice.

Schematic Symbol

Figure 15-10a shows the schematic symbol of an op amp with two inputs and one output. A is the unloaded voltage gain. This is the gain we get when no load resistor is connected to the output. The inverting input has a minus sign, a reminder of the phase inversion that takes place with this input. On the other hand, the noninverting input has a plus sign because no phase inversion occurs with this input.

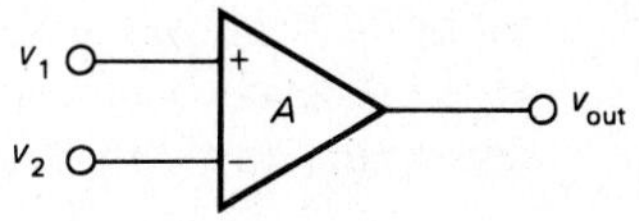

Fig. 15-10.
Schematic symbol for op amp.

15-5
OP-AMP PARAMETERS

Because an op amp is a dc amplifier, you have to consider both dc and ac characteristics when troubleshooting or designing op-amp circuits. In this section, we take a closer look at the offset problem, as well as discussing other characteristics that affect op-amp performance.

Input Offset Voltage

When the inputs of an op amp are grounded, there almost always is an output offset voltage, as shown in Fig. 15-11a, because the input transistors have different V_{BE} values. The input offset voltage is equal to this difference in V_{BE} values. For instance, a 741C has a typical input offset voltage of ± 2 mV, which means that the V_{BE} of one input transistor typically differs from the other by 2 mV. This 2 mV is amplified to produce an output offset voltage.

Theoretically, we can apply a voltage of 2 mV to one of the inputs, as shown in Fig. 15-11b. Then, the output offset voltage goes to zero for a typical 741C. (Note: the offset can have either polarity; so it might be necessary to reverse the polarity of the 2 mV.)

Input Bias Current

Suppose we get lucky and happen to use an op amp whose input transistors have equal V_{BE} values. Then the input offset voltage is zero. But a problem can still arise because of bias currents. If either input to the op amp has a resistance in its return path, an output offset voltage can exist. For instance, Fig. 15-11c shows a resistance R_B between the noninverting input and ground. Since there is a base current I_{B1} through R_B, a voltage appears at the noninverting input given by

$$v_1 = I_{B1} R_B$$

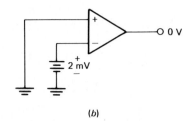

(a)   (b)

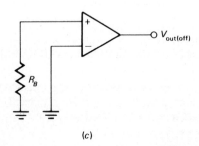

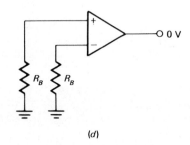

(c)   (d)

Fig. 15-11.
(a) Different in V_{BE} values causes output offset voltage when inputs are grounded.
(b) Applying 2 mV to eliminate the output offset voltage. (c) Base current through external resistors produces a dc voltage at noninverting input.
(d) Equal base resistors minimize output offset voltage.

This unwanted input voltage is amplified to produce an output offset voltage. If R_B is small enough, the resulting output offset voltage may be small enough to ignore.

If the input base currents are equal (which almost never happens), we can eliminate the output offset voltage by inserting an equal base resistor on the inverting input, as shown in Fig. 15-11d. Now the equal base currents produce equal voltage drops across the base resistors to null the output offset voltage, as shown in Fig. 15-11d.

Input Offset Current   The base currents of the input transistors are almost never equal because the β values are usually different. As mentioned earlier, the input offset current equals the difference in the base currents. Therefore, even though we use equal resistors in Fig. 15-11d, the input offset current can produce an unwanted difference voltage as follows: The input to the noninverting input is

$$v_1 = I_{R1} R_B$$

The input to the inverting input is

$$v_2 = I_{B2} R_B$$

The differential input is

$$v_1 - v_2 = I_{B1} R_B - I_{B2} R_B = (I_{B1} - I_{B2}) R_B$$

or
$$v_1 - v_2 = I_{in(off)} R_B \qquad \textbf{(15-17)}$$

Using equal base resistors as shown in Fig. 15-11d is very common. Therefore, Eq. (15-17) will be useful because you can calculate the differential input voltage produced by the input offset current and base resistance. Remember: this input voltage is amplified to produce the output offset voltage. So you have to keep R_B small enough to prevent excessive output offset voltage.

CMRR   The common-mode rejection ratio was defined earlier. For a 741C, CMRR = 30,000 at low frequencies. Given equal signals, one a difference input (desired) and the other a common-mode input (undesired), the difference signal will be 30,000 times larger at the output than the common-mode signal.

AC Compliance   The ac compliance PP is the maximum unclipped peak-to-peak output voltage that an op amp can produce. Since the quiescent output is ideally zero, the ac output voltage can swing positive or negative. How far it swings depends on the load resistance. For large load resistances, each peak can swing to within 1 or 2 V of the supply voltages. For instance, if supplies are $+15$ V and -15 V, the ac compliance is approximately 27 V.

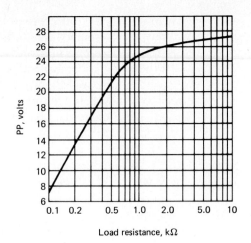

Fig. 15-12.
Ac compliance decreases
when load resistance
decreases.

As the load resistance decreases, the slope of the ac load line changes and the ac compliance of a 741C decreases. Figure 15-12 shows the typical variation of ac compliance with load resistance. Notice that PP is approximately 27 V for an R_L of 10 kΩ, 25 V for 1 kΩ, and 7 V for 100 Ω. As you see, the ac compliance approaches zero for small load resistances.

Short-Circuit Output Current

In some applications, an op amp may drive a load resistance that is approximately zero. Because an IC op amp is a low-power device, its output current is limited. For instance, the 741C can supply a maximum short-circuit output current of only 25 mA. If you are using small load resistors (less than 75 Ω), don't expect to get a large output voltage because the voltage cannot be greater than 25 mA times the load resistance.

Frequency Response

Negative feedback means sacrificing some voltage gain in exchange for a stable voltage gain, less distortion, and other improvements in amplifier performance. When an op amp uses negative feedback, the operation is called *closed-loop*. If the op amp is running wide open without negative feedback, the operation is known as *open-loop*.

Figure 15-13 shows the small-signal frequency response of a 741C. At low frequencies, the open-loop voltage gain is 100,000. The 741C has an open-loop cutoff frequency f_{OL} of 10 Hz. As indicated, the voltage gain is 70,700 at this frequency. Beyond cutoff, the voltage gain decreases by a factor of 10 for each decade increase in frequency. This frequency response is caused by the compensating capacitor C_C, discussed earlier.

The unit-gain frequency f_{unity} is the frequency where the voltage gain equal 1. In Fig. 15-13, f_{unity} is 1 MHz. Data sheets specify the value of f_{unity} because it represents the upper limit on the useful gain of an op amp. For instance, the data sheet of a 318 lists an f_{unity} of 15 MHz.

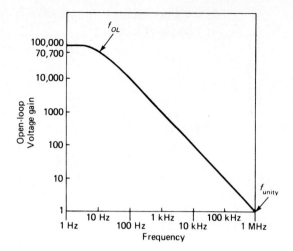

Fig. 15-13.
Typical open-loop response
of a 741.

This means a 318 can amplify signals at much higher frequencies than can the 741C. Naturally, you have to pay considerably more for a 318.

As shown in Fig. 15-13, the 741C has a low-frequency voltage gain of 100,000 and a cutoff frequency of 10 Hz for open-loop operation. As a rule, you never run an op amp without negative feedback because it is too unstable. Later, you will learn how to connect a few external resistors to an op amp to get closed-loop operation. This results in less voltage gain but greater bandwidth. In fact, you will learn how to trade off voltage gain for increased bandwidth.

Slew Rate   Of all the specifications affecting the ac operation of an op amp, *slew rate* is one of the most important because it limits the ac compliance at higher frequencies. To understand slew rate, we have to discuss some basic circuit theory. The charging current in a capacitor is given by

$$i = C\frac{dv}{dt}$$

where i = current into capacitor
C = capacitance
dv/dt = rate of capacitor voltage change

We can rearrange this to get

$$\frac{dv}{dt} = \frac{i}{C} \qquad\qquad (15\text{-}18)$$

This says that the rate of voltage change equals the charging current divided by the capacitance.

As an example, suppose $i = 60$ μA and $C = 30$ pF. Then, the rate of voltage change is

$$\frac{dv}{dt} = \frac{60\ \mu A}{30\ pF} = 2\ V/\mu s$$

This says the voltage across the capacitor changes at a rate of 2 V/μ_s. The voltage cannot change faster than this unless we can increase the charging current or decrease the capacitance.

The *slew rate,* symbolized S_R, of an op amp is defined as the maximum rate of output voltage change. When a large positive step of input voltage drives the op amp of Fig. 15-9, Q_1 saturates and Q_2 cuts off. Therefore, all of the tail current I_T passes through Q_1 and Q_3. Because of the current mirror, the current through Q_4 equals I_T. Since Q_2 is cut off, all Q_4 current passes on to the next stage. Approximately all of it goes to C_C because the Q_5 base current is negligible.

As C_C charges, the output voltage rises. Assuming unity voltage gain for the output stage, the rate of output voltage change is equal to the rate of voltage change across C_C. With Eq. (15-18), the maximum rate of output voltage change is

$$S_R = \frac{dv}{dt} = \frac{i}{C}$$

or
$$S_R = \frac{I_T}{C_C} \qquad \qquad \textbf{(15-19)}$$

This says the output voltage of an op amp can change no faster than the ratio of tail current to compensating capacitance.

Here's an example. In Fig. 15-9, $I_T = 15$ μA and $C_C = 30$ pF. Therefore, the slew rate of a 741C is

$$S_R = \frac{15\mu A}{30\ pF} = 0.5\ V/\mu s$$

This is the ultimate speed of a 741C; its output voltage can change no faster than 0.5 V/μs. Figure 15-14 illustrates the idea. If we overdrive a 741C with a large step input (Fig. 15-14a), the output slews as shown in Fig. 15-14b. It takes 20 μs for the output voltage to change from 0 to 10 V (nominal swing with 12-V supplies)

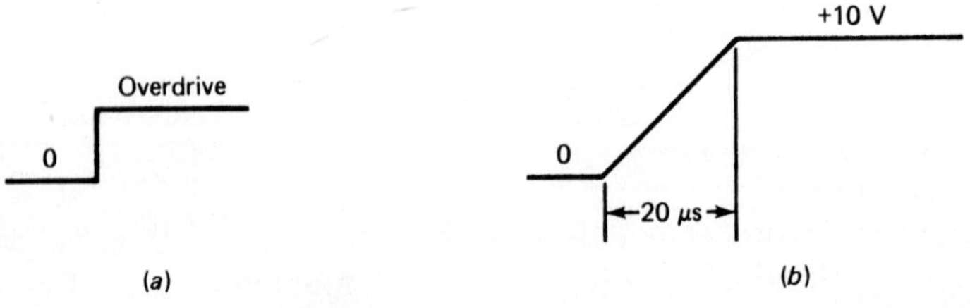

Fig. 15-14.
A step input produces slew-rate limiting of output.

(a)

+10 V

0

←20 μs→

(b)

We can also get slew-rate limiting with a sinusoidal signal. Figure 15-15*a* shows the maximum sinusoidal output from a 741C when the peak voltage is 10 V. As long as the initial slope of the sine wave is less than S_R, there is no slew-rate limiting. But when the initial slope is greater than S_R, we get the slew-rate distortion shown in Fig. 15-15*b*. The output begins to look triangular (solid curve). The higher the frequency, the smaller the swing and the more triangular the waveform. (The dashed curve is what we would get if there were no slew-rate distortion.)

Power Bandwidth

Slew-rate distortion of a sine wave starts at the point where the initial slope of the sine wave equals the slew rate of the op amp. With calculus, we can derive this useful equation:

$$f_{max} = \frac{S_R}{2\pi V_P} \qquad (15\text{-}20)$$

where f_{max} = highest undistorted frequency
$\quad\quad S_R$ = slew rate of op amp
$\quad\quad V_P$ = peak of output sine wave

As an example, if a 741C has $V_P = 10$ V and $S_R = 0.5$ V/μs, the maximum undistorted frequency for large-signal operation is

$$f_{max} = \frac{0.5 \text{ V/}\mu\text{s}}{2\pi(10 \text{ V})} = 7.96 \text{ kHz}$$

$$\frac{\dfrac{.5v}{\mu s}}{2\pi\,(10)v}$$

Above this frequency, you will begin to notice slew-rate distortion on an oscilloscope.

Frequency f_{max} is called the *power bandwidth* (also the large-signal bandwidth) of an op amp. We have just found that the 10-V power bandwidth of a 741C is approximately 8 kHz. This means that the undistorted bandwidth for large-signal operation is 8 kHz. If we try to amplify higher frequencies of the same peak value, the output voltage drops off, as shown in Fig. 15-15*b*.

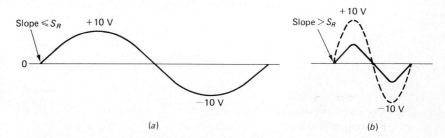

Slope ≤ S_R +10 V

0

−10 V

(a)

Slope > S_R +10 V

−10 V

(b)

Trade-off One way to increase the power bandwidth is to accept less output voltage. Figure 15-16 is a graph of Eq. (15-20) for three different slew rates. By trading off amplitude for frequency, we can improve the power bandwidth. For instance, if peak amplitudes of 1 V are acceptable in an application, the power bandwidth of a 741C increases to 80 kHz (the bottom curve). If a peak amplitude of 0.1 V is all right, the power bandwidth increases to 800 kHz.

On the other hand, if you really want a peak amplitude of 10 V, you need to use a better op amp than a 741C. In Fig. 15-16, notice that the 10-V power bandwidth increases to 80 kHz for an S_R of 5 V/μs and to 800 kHz for an S_R of 500 V/μs.

Summary

The differential amplifier is one of the best direct-coupled stages. The most widely used form of diff amp has a double-ended input and single-ended output. This kind of diff amp is usually the input stage of an op amp. It has noninverting and inverting inputs.

The input offset current of a diff amp is the difference of the two base currents, while the input bias current is the average of the two base currents. The input offset voltage of a diff amp is the input voltage needed to zero or null the output offset voltage.

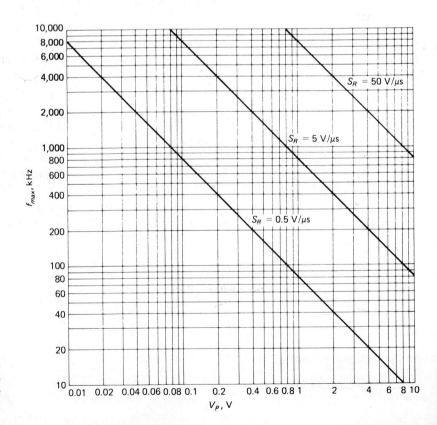

Fig. 15-16.
Power bandwidth decreases when peak voltage increases.

One reason the diff amp is important is that it discriminates against common-mode signals such as interference, static, and other undesirable pickup. The common-mode rejection ratio CMRR is the ratio of differential voltage gain to common-mode voltage gain. The higher the CMRR, the better the diff amp because it means desired signals are being amplified more than unwanted signals.

An op amp is a high-gain dc amplifier usable from 0 to over 1 MHz. By connecting external resistors to the op amp, you can adjust the voltage gain and bandwidth to your requirements. The 741 has become the industry standard. Of the different versions, the 741C (commercial grade) is used most often. It has typical parameters of $A = 100,000$, $r_{in} = 2\ M\Omega$, $r_{out} = 75\ \Omega$, $I_{in(bias)} = 80\ nA$, $I_{in(off)} = 20\ nA$, $V_{in(off)} = 2\ mV$, $S_R = 0.5\ V/\mu s$, and $f_{unity} = 1\ MHz$. If any of these specifications is inadequate for your applications, you can upgrade to a better op amp.

Slew rate limits the ac compliance at higher frequencies. As the frequency increases, we reach a point where slew-rate distortion begins because the output of the op amp can change no faster than the slew rate. For a typical 741C, the output voltage can change no faster than $0.5\ V/\mu s$ because this is the typical value of slew rate. Given a fixed slew rate, the power bandwidth (highest undistorted frequency) decreases when the peak voltage increases. One way to increase the power bandwidth is to accept less output voltage.

Glossary

BIFET op amp This is an IC op amp using JFETs for the input stage and bipolar transistors for the remaining stages.

closed-loop gain The voltage gain of a feedback amplifier when there is a closed loop or signal path all the way around the circuit.

open-loop gain the voltage gain of a feedback amplifier when the feedback path is opened. It is the same as the differential voltage gain of the op amp.

power bandwidth For a specified output peak voltage, it is the highest frequency an op amp can handle without slew-rate distortion.

slew rate This is the maximum rate of output voltage change for an op amp. With a 741 and similar op amps, this limiting rate is determined by the internal compensating capacitor, which has to be charged before its voltage can change.

unity-gain frequency The frequency where the open-loop gain equals unity. This frequency equals the gain-bandwidth product of the op amp.

Review Questions

1. What kind of input stage does an op amp normally have?
2. Define the input offset current and the input bias current.

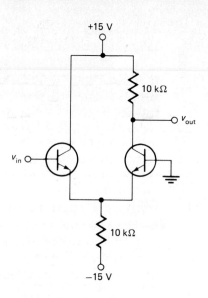

+15 V

10 kΩ

v_{out}

v_{in}

10 kΩ

−15 V

Fig. 15-17.

3. What is a common-mode signal? The CMRR?
4. Describe a current mirror.
5. What are the typical values of these 741C parameters: A, r_{in}, r_{out}, $I_{in(off)}$, $I_{in(bias)}$, $V_{in(off)}$, S_R, and f_{unity}?
6. What is a BIFET op amp?
7. Define the slew rate and power bandwidth of an op amp.

Problems

15-1. If v_{in} is zero in Fig. 15-17, what does the tail current equal?   1.43 ma

15-2. In Fig. 15-17, v_{in} = 0. What does v_{out} equal?

15-3. Suppose I_{B1} = 120 nA and I_{B2} = 95 nA in Fig. 15-17. What do $I_{in(off)}$ and $I_{in(bias)}$ equal?

15-4. If v_{in} has a peak-to-peak value of 2 mV in Fig. 15-17, what is the peak-to-peak value of v_{out}?

15-5. In Fig. 15-17, each transistor has a β of 200. What does r_{in} equal?

15-6. A 741C has $I_{in(off)}$ = 20 nA and $I_{in(bias)}$ = 80nA. Find the values of I_{B1} and I_{B2} by a simultaneous solution of Eqs. (15-5) and (15-6).

15-7. An op amp has a slew rate of 10 V/μs. Calculate the power bandwidth for a peak-to-peak output voltage of 4 V.

398 KHz.

Op-Amp Negative Feedback

Black: In 1928, he tried to patent a negative-feedback amplifier. Unfortunately, the patent office classified his idea as "another one of those perpetual-motion follies." As it has turned out, negative feedback has become one of the most valuable ideas ever discovered in electronics.

When an amplifier uses feedback, the output is sampled and a fraction of it is returned to the input. This feedback signal can produce remarkable changes in the circuit performance. *Negative feedback* means that the returning signal has a phase that opposes the input signal. The advantages of negative feedback are stable gain, less distortion, and more bandwidth.

16-1 NONINVERTING VOLTAGE FEEDBACK

The most basic type of negative feedback is *noninverting voltage feedback*. The word "noninverting" refers to the input voltage driving the noninverting input of an amplifier. The words "voltage feedback" are used because a fraction of the output voltage is fed back to the inverting input. An amplifier with noninverting voltage feedback tends to act like a perfect voltage amplifier, one with infinite input impedance, zero output impedance, and constant voltage gain.

Error Voltage

Figure 16-1 shows an amplifier with a double-ended input. This amplifier is usually an op amp, but it may be a discrete amplifier with one or more stages. Notice that the output voltage is sampled by a voltage divider. Therefore, the feedback voltage v_2 is proportional to the output voltage.

In a feedback amplifier, the difference between the noninverting and inverting input voltages is called the *error voltage*. In symbols,

$$v_{error} = v_1 - v_2$$

This error voltage is amplified to get an output voltage of

$$v_{out} = Av_{error}$$

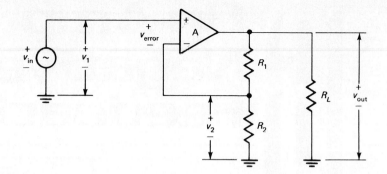

Fig. 16-1.
Noninverting voltage
feedback.

Typically, A is very large. To avoid saturation of the output transistors, therefore, v_{error} must be very small. For instance, if the differential voltage gain is 100,000, an error voltage of only 0.1 mV produces an output voltage of 10 V.

Stable Voltage Gain

In Fig. 16-1, the overall voltage gain is approximately constant, even though the differential voltage gain may change. Why? Suppose A increases for some reason such as temperature change or op-amp replacement. Then the output voltage will try to increase. This means that more voltage is fed back to the inverting input, causing the error voltage to decrease. This almost completely offsets the attempted increase in output voltage.

A similar argument applies to a decrease in differential voltage gain. If A decreases, the output voltage tries to decrease. In turn, the feedback voltage decreases, causing v_{error} to increase. This almost completely offsets the attempted decrease in A.

Remember the key idea. When the input voltage is constant, an attempted change in output voltage is fed back to the inverting input, producing an error voltage that automatically compensates for the attempted change.

Mathematical Analysis

Most op amps have an extremely large voltage gain A, a very high input impedance r_{in}, and a very low output impedance r_{out}. For instance, the 741C has typical values of $A = 100,000$, $r_{in} = 2$ MΩ, and $r_{out} = 75\ \Omega$.

In Fig. 16-1, the voltage divider returns a sample of the output voltage to the inverting input. Ignoring the high r_{in} of the op amp, the feedback voltage is

$$v_2 = \frac{R_2}{R_1 + R_2}\, v_{out}$$

This is usually written as

$$v_2 = Bv_{out}$$

where B is the fraction of output voltage fed back to the input. In symbols,

$$B = \frac{R_2}{R_1 + R_2} \qquad (16\text{-}1)$$

The error voltage is

$$v_{\text{error}} = v_1 - v_2 = v_{\text{in}} - Bv_{\text{out}}$$

This is amplified to get an output voltage of

$$v_{\text{out}} = Av_{\text{error}} = A(v_{\text{in}} - Bv_{\text{out}})$$

By rearranging, we get

$$\frac{v_{\text{out}}}{v_{\text{in}}} = \frac{A}{1 + AB} \qquad (16\text{-}2)$$

Approximate Voltage Gain The product AB is called the *loop gain*. For noninverting voltage feedback to be effective, the designer deliberately makes AB much greater than unity, so that Eq. (16-2) reduces as follows:

$$\frac{v_{\text{out}}}{v_{\text{in}}} = \frac{A}{1 + AB} \cong \frac{A}{AB}$$

or

$$\frac{v_{\text{out}}}{v_{\text{in}}} \cong \frac{1}{B} \qquad (16\text{-}3)$$

Why is this result important? Because it says that the overall voltage gain of the circuit equals the reciprocal of B, the feedback fraction. In other words, the overall voltage gain no longer depends on the differential voltage gain of the op amp but rather depends on the feedback fraction of the voltage divider. Since we can use precision resistors with tolerances of 1 percent for the voltage divider, B is a precise and stable value that is independent of amplifier characteristics. Therefore, the voltage gain of a feedback amplifier becomes a rock-solid value approximately equal to $1/B$.

Simplified Viewpoint Here is a simple way to remember Eq. (16-3). Because the error voltage is very small, the inverting input voltage is approximately equal to the noninverting input voltage:

$$v_1 = v_2$$

This can be written as

$$v_{in} = Bv_{out}$$

or

$$\frac{v_{out}}{v_{in}} = \frac{1}{B}$$

This is the same result as before, derived with a lot less work.

The key idea to remember about op-amp circuits is the very small error voltage. This allows us to say that v_2 is approximately equal to v_1. If v_1 increases, v_2 will increase by the same amount, so that the two voltages remain almost equal. This follow-the-leader action is called *bootstrapping*. In other words, the inverting input is bootstrapped to the noninverting input, which means the inverting input voltage follows the noninverting input voltage and remains approximately equal to it.

Closed-Loop Voltage Gain

As discussed in Chap. 11, the open-loop voltage gain is voltage gain of a negative-feedback amplifier when the feedback path is opened. In Fig. 16-1, the open-loop voltage gain is equal to A, the differential voltage gain of the op amp. The closed-loop voltage gain A_{CL} is equal to voltage gain with the feedback path closed. Therefore, Eq. (16-2) can be written as

$$A_{CL} = \frac{A}{1 + AB} \qquad \textbf{(16-4)}$$

where A_{CL} = closed-loop voltage gain
A = open-loop voltage gain
B = feedback fraction

In most negative-feedback amplifiers, the loop gain AB is much greater than unity and Eq. (16-4) simplifies to

$$A_{CL} = \frac{1}{B} \qquad \textbf{(16-5)}$$

Since $B = R_2/(R_1 + R_2)$, an alternative form is

$$A_{CL} = \frac{R_1 + R_2}{R_2}$$

which is often written as

$$A_{CL} = \frac{R_1}{R_2} + 1 \qquad \textbf{(16-6)}$$

EXAMPLE 16-1 If the 741C of Fig. 16-2 has an open-loop gain of 100,000, what is the closed-loop gain? If $v_{in} = 1$ mV, what do the output and error voltages equal?

SOLUTION The voltage divider has a feedback fraction of

$$B = \frac{R_2}{R_1 + R_2} = \frac{2 \text{ k}\Omega}{100 \text{ k}\Omega} = 0.02$$

The closed-loop gain is approximately

$$A_{CL} = \frac{1}{B} = \frac{1}{0.02} = 50$$

A more accurate gain is given by

$$A_{CL} = \frac{A}{1 + AB} = \frac{100,000}{1 + 100,000(0.02)} = 49.975$$

Look at how close this is to 50. The point is that $1/B$ is an accurate approximation for the closed-loop voltage gain of an amplifier that uses noninverting voltage feedback.

If $v_{in} = 1$ mV, the output voltage is

$$v_{out} = A_{CL}v_{in} = 50(1 \text{ mV}) = 50 \text{ mV}$$

The error voltage is

$$v_{error} = \frac{v_{out}}{A} = \frac{50 \text{ mV}}{100,000} = 0.5 \text{ } \mu\text{V}$$

Notice how small the error voltage is. This is typical of op-amp feedback amplifiers because the open-loop voltage gain is quite high.

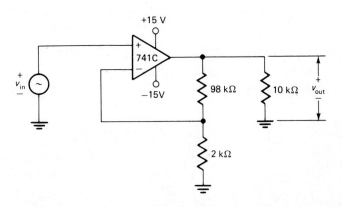

Fig. 16-2.

EXAMPLE 16-2   Suppose the 741C of the preceding example is replaced by another 741C that has a voltage gain of only 20,000 (worst-case value on data sheet). Recalculate the value of A_{CL}, v_{out}, and v_{error}.

SOLUTION   The closed-loop voltage gain is

$$A_{CL} = \frac{A}{1 + AB} = \frac{20,000}{1 + 20,000(0.02)} = 49.875$$

As you see, the closed-loop voltage gain is still extremely close to 50, despite the huge drop in open-loop voltage gain. Without negative feedback, the overall voltage gain would drop from 100,000 to 20,000, a decrease of 80 percent. Using negative feedback, we have less overall voltage gain, but in return we get a fabulously stable closed-loop voltage gain. In this example, the closed-loop voltage gain decreases from 49.975 to 49.875, a decrease of only 0.2 percent. Therefore, the closed-loop voltage gain is nearly independent of the op-amp voltage gain.

Since A_{CL} is approximately 50, v_{out} is approximately 50 mV when v_{in} is 1 mV. But the error voltage changes to

$$V_{error} = \frac{v_{out}}{A} = \frac{50 \text{ mV}}{20,000} = 2.5 \ \mu V$$

Compared with the preceding example, the error voltage has increased from 0.5 to 2.5 μV.

Do you understand what has happened? When the open-loop gain drops by a factor of 5, the error voltage increases by a factor of 5. Therefore, the output voltage remains at approximately 50 mV. This echoes the earlier explanation of negative feedback. Attempted changes in output voltage are fed back to the input, producing an error voltage that automatically compensates for the output change.

16-2

OTHER BENEFITS OF NONINVERTING VOLTAGE FEEDBACK

Stable voltage gain is not the only benefit of noninverting voltage feedback. It also improves input impedance, output impedance, non-linear distortion, and output offset voltage. Without feedback, the input voltage would appear across the r_{in} of the op amp. With feedback, only the error voltage appears across r_{in}, which implies less input current, equivalent to a higher input impedance.

By a mathematical derivation, we can show that

$$r_{in(CL)} = (1 + AB)r_{in} \qquad (16\text{-}7)$$

where $r_{in(CL)}$ = closed-loop input impedance
r_{in} = open-loop input impedance
AB = loop gain

In most negative-feedback amplifiers, AB is much larger than unity, which means that $r_{in(CL)}$ is much larger than r_{in}.

The use of noninverting voltage feedback with op amps leads to input impedances that approach infinity. This means that an op amp using noninverting voltage feedback approximates an ideal voltage amplifier.

Output Impedance

When the load resistance decreases, more current flows through the output impedance of the op amp. Because of the voltage drop across the output impedance, the output voltage tries to decrease. Since less voltage is then fed back to the input, the error voltage increases. This almost completely offsets the attempted decrease in output voltage. In effect, negative feedback makes the output impedance of the overall circuit appear smaller than the output impedance of the op amp.

It can be shown that

$$r_{out(CL)} = \frac{r_{out}}{1 + AB} \qquad (16\text{-}8)$$

where $r_{out(CL)}$ = closed-loop output impedance
r_{out} = open-loop output impedance
AB = loop gain

When the loop gain is much greater than unity, $r_{out(CL)}$ is much smaller than r_{out}. In fact, noninverting voltage feedback with op amps results in output impedances that approach zero, the ideal value for a voltage amplifier.

Nonlinear Distortion

The final stage has some nonlinear distortion when the signal swings over most of the ac load line. Large swings in current cause the r'_e of a transistor to change during the cycle, equivalent to saying that open-loop gain is changing. This changing voltage gain is the cause of the nonlinear distortion.

Noninverting voltage feedback reduces nonlinear distortion because the feedback stabilizes the closed-loop voltage gain, making it almost independent of the changes in open-loop voltage gain. As long as the loop gain is much greater than unity, the output voltage equals $1/B$ times the input voltage. This implies that the output will be a more faithful reproduction of the input. And this is exactly what happens when we use noninverting voltage feedback.

We can show that

$$v_{dist(CL)} = \frac{v_{dist}}{1 + AB} \qquad (16\text{-}9)$$

where $v_{dist(CL)}$ = closed-loop distortion voltage
v_{dist} = open-loop distortion voltage
AB = loop gain

When the loop gain is much greater than unity, the closed-loop distortion is much smaller than the open-loop distortion. Again, non-inverting voltage feedback produces a real improvement in the quality of the amplifier.

Reduced Output Offset Voltage

As discussed earlier, an output offset voltage may exist even though the input voltage is zero. There are three causes for this unwanted output offset voltage: input offset voltage, input bias current, and input offset current. Noninverting voltage feedback will reduce the output offset voltage as follows: Some of the output offset voltage is fed back to the inverting input, reducing the error voltage. In turn, this means less output offset voltage.

With a mathematical derivation, we can show that

$$V_{oo(CL)} = \frac{V_{oo}}{1 + AB} \qquad (16\text{-}10)$$

where $V_{oo(CL)}$ = closed-loop output offset voltage
V_{oo} = open-loop output offset voltage
AB = loop gain

When the loop gain is much greater than unity, the closed-loop output offset voltage is much smaller than the open-loop output offset voltage.

Desensitivity

The closed-loop voltage gain with noninverting voltage feedback is

$$A_{CL} = \frac{A}{1 + AB} \qquad (16\text{-}11)$$

The denominator $1 + AB$ is called the *desensitivity* because it indicates how much the voltage gain is reduced by the negative feedback. For instance, if $A = 100,000$ and $B = 0.02$, then

$$1 + AB = 1 + 100,000(0.02) = 2001$$

The desensitivity is 2001, meaning that the voltage gain is reduced by a factor of 2001.

We can rearrange Eq. (16-11) to get

$$1 + AB = \frac{A}{A_{CL}} \qquad (16\text{-}12)$$

This says the desensitivity equals the ratio of open-loop voltage gain to closed-loop voltage gain. For example, if $A = 100,000$ and $A_{CL} = 250$, the desensitivity is

$$1 + AB = \frac{100,000}{250} = 400$$

Equation (16-12) is convenient when you know the values of A and A_{CL}, but not B.

Table 16-1 summarizes the effects of noninverting voltage feedback. As you can see, the desensitivity appears in all of the formulas. This is why it is important to remember how to calculate desensitivity. You can use either $1 + AB$ or A/A_{CL}.

EXAMPLE 16-3 Figure 16-3 shows a 741C with pin numbers. If $A = 100,000$, $r_{in} = 2\ \text{M}\Omega$, and $r_{out} = 75\ \Omega$, what are the closed-loop input and output impedances?

SOLUTION The feedback fraction is

$$B = \frac{R_2}{R_1 + R_2} = \frac{100\ \Omega}{100,100\ \Omega} = 0.000999$$

and the desensitivity is

$$1 + AB = 1 + 100,000(0.000999) = 101$$

The closed-loop input impedance is

$$r_{in(CL)} = (1 + AB)r_{in} = 101(2\ \text{M}\Omega) = 202\ \text{M}\Omega$$

and the closed-loop output impedance is

$$r_{out(CL)} = \frac{r_{out}}{1 + AB} = \frac{75\ \Omega}{101} = 0.743\ \Omega$$

EXAMPLE 16-4 Figure 16-4 shows a circuit called a *voltage follower*. What is its closed-loop voltage gain? Its closed-loop input and output impedances? The output offset voltage? Use typical 741C parameters: $A = 100,000$, $r_{in} = 2\ \text{M}\Omega$, $r_{out} = 75\ \Omega$, $V_{in(off)} = 2\ \text{mV}$, $I_{in(bias)} = 80\ \text{nA}$, and $I_{in(off)} = 20\ \text{nA}$.

TABLE 16-1. Noninverting Voltage Feedback

Quantity	Symbol	Effect	Formula
Voltage gain	A_{CL}	Stabilizes	$A/(1 + AB)$
Input impedance	$r_{in(CL)}$	Increases	$(1 + AB)r_{in}$
Output impedance	$r_{out(CL)}$	Decreases	$r_{out}/(1 + AB)$
Distortion	$v_{dist(CL)}$	Decreases	$v_{dist}/(1 + AB)$
Output offset	$V_{oo(CL)}$	Decreases	$V_{oo}/(1 + AB)$

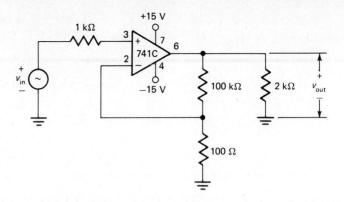

+15 V

1 kΩ

741C

−15 V

100 kΩ   2 kΩ v_{out}

v_{in}

100 Ω

Fig. 16-3.

SOLUTION All of the output voltage is fed back to the input because R_1 is zero and R_2 is infinite. Therefore, $B = 1$. This is massive negative feedback, the most you can have. In this case, the closed-loop voltage gain equals 1 to a very close approximation.

The closed-loop input impedance of a voltage follower is

$$r_{in(CL)} = (1 + AB)r_{in} = (1 + A)r_{in} = Ar_{in}$$
$$= 100,000(2 \text{ M}\Omega) = 202 \text{ M}\Omega$$

The closed-loop output impedance is

$$r_{out(CL)} = \frac{r_{out}}{1 + AB} = \frac{r_{out}}{1 + A} = \frac{r_{out}}{A}$$
$$= \frac{75}{100,000} = 0.00075 \ \Omega$$

As you see, $r_{in(CL)}$ approaches infinity and $r_{out(CL)}$ approaches zero. A voltage follower is very useful because of its high input impedance, low output impedance, and unity voltage gain.

Since the closed-loop voltage gain is unity, the desensitivity is

$$1 + AB = \frac{A}{A_{CL}} = A = 100,000$$

The input bias current is 80 nA and the input offset current is 20 nA. By a simultaneous solution of Eqs. (15-5) and (15-6), the two base currents

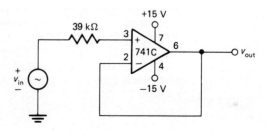

+15 V

39 kΩ

741C

−15 V

v_{in}

v_{out}

Fig. 16-4.
Voltage follower.

are 90 nA and 70 nA. In the worst case, the 90 nA flows through the 39 kΩ and the maximum input offset voltage is

$$v_1 - v_2 = V_{in(off)} + I_{B_1}R_B$$
$$= 2 \text{ mV} + (90 \text{ nA})(39 \text{ k}\Omega) = 5.51 \text{ mV}$$

So, the output offset voltage is

$$V_{oo(CL)} = \frac{V_{oo}}{1 + AB} = \frac{100,000(5.51 \text{ mV})}{100,000} = 5.51 \text{ mV}$$

In other words, the voltage follower is almost immune to offset problems. Because it has only unity voltage gain, the output offset voltage can be no more than total input offset voltage.

16-3 INVERTING VOLTAGE FEEDBACK

Figure 16-5 shows *inverting voltage feedback.* "Inverting" means the input voltage drives the inverting input of the op amp through a series resistance R_S. "Voltage feedback" again refers to a fraction of the output voltage being fed back to the input. With this type of negative feedback, the output voltage is 180° out of phase with the input voltage.

Virtual Ground

Here are two ideas that simplify the analysis of inverting voltage feedback. First, notice that the noninverting input is grounded. Therefore, the inverting input is bootstrapped to ground, which means its voltage is approximately zero. Second, realize that an op amp has a very high input resistance, equivalent to saying it draws negligible input current from the source.

These two key ideas are summarized by the concept of *virtual ground,* which is any point in a circuit that has zero voltage and at the same time has no current flowing into it. The inverting input of Fig. 16-5 approaches a virtual ground because it has almost zero voltage and almost no input current. An *ordinary ground* has zero voltage, but it can have a lot of current flowing into it (known as

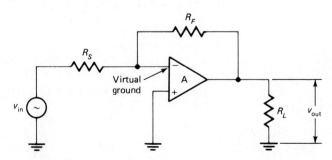

Fig. 16-5.
Inverting voltage feedback.

sinking current). A virtual ground has zero voltage but can sink no current. The word "virtual" is used to distinguish this new kind of ground from the ordinary type of ground.

Closed-Loop Voltage Gain

In the following discussion, assume the inverting input is a virtual ground, one that has zero voltage and can sink no current. Look at Fig. 16-5 and notice the following. Because the virtual ground draws no current, all of the input current through R_s has to pass through R_F. The voltage across R_F therefore equals

$$v_F = i_{in}R_F$$

Since the virtual ground has zero voltage to ground, the output voltage must equal the voltage across R_F:

$$v_{out} = i_{in}R_F$$

Furthermore, because the inverting input is at virtual ground, we may write

$$v_{in} = i_{in}R_S$$

Taking the ratio of output voltage to input voltage gives the closed-loop voltage gain:

$$A_{CL} = \frac{v_{out}}{v_{in}} = \frac{i_{in}R_F}{i_{in}R_S}$$

or

$$A_{CL} = \frac{R_F}{R_S} \qquad \qquad (16\text{-}13)$$

Although this approximation assumes the inverting input is a virtual ground, it is highly accurate for IC op amps because the error voltage is quite small and the input resistance is very high.

Impedances

Because of the virtual ground, the right end of R_s appears grounded, and the source sees a closed-loop input impedance of

$$r_{in(CL)} = R_S \qquad \qquad (16\text{-}14)$$

This means a designer can control the input impedance which equals the series resistance. One of the reasons the inverting voltage amplifier is popular is that the designer can set up a precise value of input impedance as well as voltage gain.

In Fig. 16-5, the feedback fraction is

$$B = \frac{R_S}{R_F + R_S} \qquad \qquad (16\text{-}15)$$

It can be shown that the closed-loop output impedance is the same as with noninverting voltage feedback:

$$r_{out(CL)} = \frac{r_{out}}{1 + AB} \qquad \text{(16-16)}$$

Since the loop gain AB is usually much greater than unity, $r_{out(CL)}$ approaches zero.

Inverting voltage feedback produces other benefits, similar to noninverting voltage feedback. Table 16-2 summarizes inverting voltage feedback. This type of negative feedback stabilizes the voltage gain and input impedance. Also, the desensitivity $1 + AB$ reduces output impedance, distortion, and output offset voltage.

EXAMPLE 16-5   Calculate the closed-loop voltage gain and the desensitivity in Fig. 16-6. Use an open-loop voltage gain of 100,000 for the 741C.

SOLUTION   The voltage gain is

$$A_{CL} = \frac{R_F}{R_S} = \frac{22 \text{ k}\Omega}{1 \text{ k}\Omega} = 22$$

The desensitivity is

$$1 + AB = \frac{A}{A_{CL}} = \frac{100,000}{22} = 4545$$

Therefore, the output impedance, nonlinear distortion, and output offset voltage is reduced by a factor of 4545.

16-4
BANDWIDTH

An op amp has no lower cutoff frequency because its stages are direct-coupled. But it does have an *open-loop upper cutoff frequency*, designated f_{OL}. At this frequency the open-loop voltage gain equals 0.707 of its maximum value. In a typical negative-feedback amplifier,

TABLE 16-2. Noninverting Voltage Feedback

Quantity	Symbol	Effect	Formula
Voltage gain	A_{CL}	Stabilizes	R_F/R_S
Input impedance	$r_{in(CL)}$	Stabilizes	R_S
Output impedance	$r_{out(CL)}$	Decreases	$r_{out}/(1 + AB)$
Distortion	$V_{dist(CL)}$	Decreases	$v_{dist}/(1 + AB)$
Output offset	$V_{oo(CL)}$	Decreases	$V_{oo}/(1 + AB)$

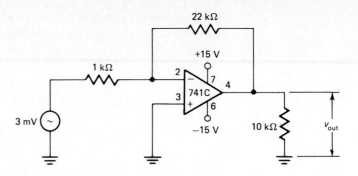

Fig. 16-6.

the loop gain at f_{OL} is still very large. Therefore, the closed-loop voltage gain is still approximately equal to $1/B$. This means that the frequency can increase well above f_{OL} before the output voltage decreases to 0.707 of its maximum value, which is equivalent to saying that the closed-loop cutoff frequency is greater than the open-loop cutoff frequency.

Mathematical Analysis

The mathematical derivation is too complicated to go into here, but it can be shown that

$$f_{CL} = (1 + AB)f_{OL} \qquad (16\text{-}17)$$

where f_{CL} = closed-loop upper cutoff frequency
f_{OL} = open-loop upper cutoff frequency
AB = loop gain

When the loop gain AB is much greater than unity, f_{CL} is much greater than f_{OL}.

With an op amp, there is no lower cutoff frequency because the stages are direct-coupled. Therefore, the bandwidth equals f_{CL}. Equation (16-17) tells us that negative feedback increases the bandwidth by a factor of $1 + AB$.

Open-Loop Gain-Bandwidth Product

The desensitivity is

$$1 + AB = \frac{A}{A_{CL}}$$

Therefore, Eq. (16-17) may be written as

$$f_{CL} = \frac{A}{A_{CL}} f_{OL}$$

or

$$A_{CL}f_{CL} = Af_{OL} \qquad (16\text{-}18)$$

The right-hand side of this equation is called the *open-loop gain-bandwidth product* because it is the product of open-loop gain and bandwidth. For a 741C, the typical voltage gain is 100,000 and the cutoff frequency is 10 Hz. Therefore,

$$Af_{OL} = 100,000(10 \text{ Hz}) = 1 \text{ MHz}$$

This says that a 741C has an open-loop gain-bandwidth product of 1 MHz.

Closed-Loop Gain-Bandwidth Product

The left-hand side of Eq. (16-18) is called the *closed-loop gain-bandwidth product* because it is the product of closed-loop gain and bandwidth. Because of Eq. (16-18), the gain-bandwidth product is the same for open-loop or closed-loop conditions. Given a typical 741C, the product of A_{CL} and f_{CL} always equals 1 MHz, regardless of the values of the external resistors.

Equation (16-18) implies that the gain-bandwidth product is a constant. Therefore, even though A_{CL} and f_{CL} change when we change external resistors, the product of these two quantities remains constant and equal to Af_{OL}.

Figure 16-7 summarizes the idea for a 741C. At low frequencies, the top graph indicates an open-loop voltage gain of 100,000. The gain decreases to 70,700 at 10 Hz, which is the open-loop cutoff frequency. The open-loop gain then decreases until it reaches an f_{unity} of 1 MHz.

The closed-loop voltage gain is different. If we select resistors to get a closed-loop voltage gain of 1000 at low frequencies (middle graph), then the voltage gain of Fig. 16-7 is down to 707 at 1 kHz, the closed-loop cutoff frequency. Beyond this frequency, the closed-loop response superimposes the open-loop response until both cross the horizontal axzis at f_{unity}.

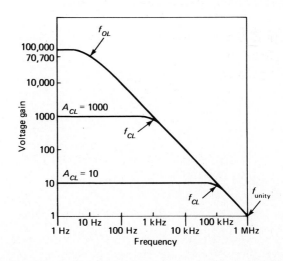

Fig. 16-7.
Bandwidth increases when closed-loop voltage gain decreases.

If you reduce the closed-loop voltage gain, you can get more bandwidth. With a voltage gain of 10 at low frequencies (bottom graph), the gain decreases to 7.07 at 100 kHz, the closed-loop cutoff frequency. By looking at the three graphs, you can see that bandwidth increases when the low-frequency voltage gain decreases.

Unity-Gain Frequency

If $A_{CL} = 1$, Eq. (16-18) simplifies to

$$f_{unity} = Af_{OL} \qquad \text{(16-19)}$$

This says the unity-gain frequency equals the gain-bandwidth product. Data sheets usually list the value of f_{unity} because it equals the gain-bandwidth product. The higher the f_{unity}, the larger the gain-bandwidth product of the op amp. For instance, a 741C has an f_{unity} of 1 MHz. The LM318 has an f_{unity} of 15 MHz. Although it costs more, the LM318 may be a better choice if you need the higher gain-bandwidth product. With an LM318 you get 15 times more voltage gain than a 741C with the same bandwidth.

The gain-bandwidth product gives us a fast way to compare IC op amps. The greater the gain-bandwidth product, the higher we can go in frequency and still have usable gain. With Eqs. (16-18) and (16-19), we can get this useful formula for the closed-loop cutoff frequency:

$$f_{CL} = \frac{f_{unity}}{A_{CL}} \qquad \text{(16-20)}$$

EXAMPLE 16-6 Calculate f_{CL} for each of the following values of A_{CL}: 1000, 100, 10, and 1. Use an f_{unity} of 1 MHz.

SOLUTION An A_{CL} of 1000 gives

$$f_{CL} = \frac{f_{unity}}{A_{CL}} = \frac{1 \text{ MHz}}{1000} = 1 \text{ kHz}$$

When A_{CL} is 100,

$$f_{CL} = \frac{f_{unity}}{A_{CL}} = \frac{1 \text{ MHz}}{100} = 10 \text{ kHz}$$

When A_{CL} is 10,

$$f_{CL} = \frac{f_{unity}}{A_{CL}} = \frac{1 \text{ MHz}}{10} = 100 \text{ kHz}$$

When A_{CL} is 1,

$$f_{CL} = \frac{f_{unity}}{A_{CL}} = \frac{1 \text{ MHz}}{1} = 1 \text{ MHz}$$

The point is this. We can trade off closed-loop gain for bandwidth. By changing the external resistors, we can tailor a negative-feedback voltage amplifier for a particular application. The values in this example give these choices of gain and bandwidth:

$$A_{CL} = 1000, \; f_{CL} = 1 \text{ kHz}$$
$$A_{CL} = 100, \; f_{CL} = 10 \text{ kHz}$$
$$A_{CL} = 10, \; f_{CL} = 100 \text{ kHz}$$
$$A_{CL} = 1, \; f_{CL} = 1 \text{ MHz}$$

Summary

Noninverting voltage feedback is the most common type of negative feedback. Noninverting means the input voltage drives the noninverting input of an op amp. Voltage feedback means a fraction of the output voltage is fed back to the inverting input. With noninverting voltage feedback, the closed-loop voltage gain equals $1/B$, where B is the fraction of output voltage fed back to the input. The desensitivity equals $1 + AB$. Although voltage gain is reduced by $1 + AB$, input impedance, output impedance, nonlinear distortion, output offset voltage, and bandwidth are improved by a factor of $1 + AB$.

Inverting voltage feedback is another popular type of negative feedback. Inverting means the input voltage drives the inverting input through a series resistor. Voltage feedback again refers to a fraction of output voltage being fed back to the input. Usually, the noninverting input is grounded, while the inverting input is at virtual ground because its voltage and current are approximately zero.

The gain-bandwidth product is a constant. Because of this, bandwidth is inversely proportional to voltage gain. This means you can trade off voltage gain for bandwidth. If you cut the voltage gain in half, you double the bandwidth. The 741C has a typical f_{unity} of 1 MHz, while the 318 has a typical f_{unity} of 15 MHz.

Glossary

closed-loop gain The voltage gain of a feedback amplifier when there's a closed loop or signal path all the way around the circuit.

open-loop gain The voltage gain of a feedback amplifier when the feedback path is opened. It is the same as the differential voltage gain of the op amp.

error voltage This is the voltage between the noninverting and inverting inputs of an op amp. Because of the high gain, the error voltage approaches zero.

gain-bandwidth product This is a way to compare the frequency response of different op amps. The higher it is, the better the frequency response for a given voltage gain.

negative feedback Whenever the feedback signal has a phase that opposes the input signal, you have negative feedback.

op amp A high-gain dc amplifier usable from 0 to over 1 MHz. It has a high input impedance and a low output impedance. It is not meant to be used open-loop.

virtual ground Any point in a circuit where the voltage to ground is zero, and yet no current can flow into the point. The inverting input of an op amp approaches a virtual ground when the noninverting input is grounded.

voltage follower A noninverting op amp with negative feedback and a closed-loop voltage gain of unity.

Review Questions

1. What do the words "voltage feedback" mean? Name two types of voltage feedback.
2. What effect does noninverting voltage feedback have on the input impedance? The output impedance?
3. What is virtual ground? How does it differ from an ordinary ground?
4. How is the gain-bandwidth product of an op amp related to its f_{unity}?
5. When the gain-bandwidth product is constant, how is the bandwidth related to the voltage gain?

Problems

16-1. If $v_{in} = 2$ mV in Fig. 16-8, what does v_{out} equal? For a typical 741C, what does v_{error} equal?

16-2. Repeat Prob. 16-1 for a 741C whose $A = 20,000$.

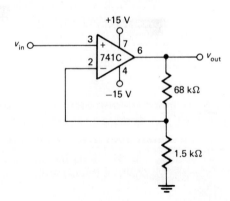

Fig. 16-8.

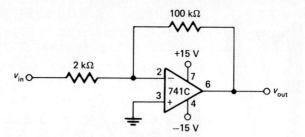

Fig. 16-9.

16-3. For a typical 741C, what does the desensitivity equal in Fig. 16-8? The closed-loop input and output impedances?

16-4. What is the closed-loop voltage gain in Fig. 16-9? The value of $r_{in(CL)}$?

16-5. If v_{in} = 2 mV in Fig. 16-9, what does v_{out} equal?

16-6. What is the bandwidth in Fig. 16-8? (Assume a typical 741C.)

16-7. Repeat Prob. 16-6 for an f_{unity} of 800 kHz.

16-8. What is the bandwidth in Fig. 16-9 for a typical 741C?

16-9. An op amp has an f_{unity} of 1 MHz. If used in a noninverting voltage feedback amplifier, what is the bandwidth for each of these values of A_{CL}: 100, 50, 20, 10, 5, 2, and 1?

Op-Amp Circuits

Oersted: During a classroom lecture in 1819, he accidentally laid a compass near a conductor in which charges were flowing. Rather than pointing to the earth's north pole, the compass needle pointed to the conductor. Oersted immediately grasped the importance of this phenomenon because it meant that electricity and magnetism were related.

IC op amps are inexpensive, reliable, and easy to use. For this reason, they can be used not only for voltage amplifiers but also for current boosters, current sources, active filters, comparators, waveshaping, and many other applications. This chapter discusses a variety of circuits to give you an idea of what can be done with the amazing and versatile IC op amp.

17-1 VOLTAGE AMPLIFIERS

The preceding chapter introduced noninverting and inverting voltage feedback. Now you will see a number of different amplifier circuits using each type of negative feedback.

AC Amplifier

In some applications you don't need a response that extends down to zero frequency because only ac signals drive the input. In this case, you can insert *coupling capacitors* on the input and output sides as shown in Fig. 17-1. While you're at it, you can minimize the output offset voltage by inserting a *bypass capacitor* in the feedback loop, as shown. In the middle range of frequencies, the bypass capacitor appears shorted, and the closed-loop voltage gain is $R_1/R_2 + 1$.

But at zero frequency the bypass capacitor appears open and the feedback fraction increases to unity. Therefore, the desensitivity for dc signals is $1 + A$, the maximum value it can have. This reduces the output offset voltage to a minimum, producing maximum ac compliance.

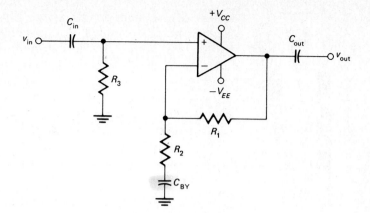

Fig. 17-1.
Ac amplifier.

The bypass capacitor produces a cutoff frequency of

$$f_{BY} = \frac{1}{2\pi R_2 C_{BY}}$$

(17-1)

At this frequency the closed-loop voltage gain equals 0.707 of its maximum value. Ten times above this frequency, A_{CL} is within half a percent of its maximum value.

Audio Amplifier Most op amps use dual or split supplies, such as $+15$ V and -15 V. But sometimes you will see an op amp running off of a single supply voltage, as shown in Fig. 17-2. Notice that the V_{EE} input is grounded. To get maximum ac compliance, we need to bias the noninverting input at half the positive supply voltage. In Fig. 17-2, the collector of the bipolar stage typically has a quiescent voltage of approximately half of V_{CC}. Therefore, it can supply the noninverting input of the op amp with the required dc voltage for single-supply operation.

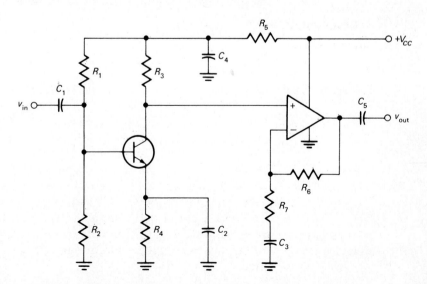

Fig. 17-2.
Audio amplifier.

Most of the components in the bipolar stage should be familiar from earlier discussions. For instance, R_1 and R_2 provide voltage-divider bias, with C_2 bypassing the emitter to ground for maximum voltage gain. The only new components are R_5 and C_4. This is a *decoupling network*. It has as low cutoff frequency as possible to prevent unwanted positive feedback between the op amp and the bipolar stage through the supply line.

An audio amplifier covers the frequencies from 20 Hz to 20 kHz. If a 741C is used for the op amp, a closed-loop voltage gain of 50 produces a typical upper cutoff frequency of 20 kHz. With a supply of 15 V, the bipolar stage will have a voltage gain in the vicinity of 200. Therefore, the audio amplifier of Fig. 17-2 has an overall voltage gain of approximately 10,000.

JFET-Switched Voltage Gain

Some applications require a change in closed-loop voltage gain. Figure 17-3 shows a JFET-controlled amplifier. The control voltage of the JFET switch comes from another circuit that produces a two-level output, either 0 V or a voltage that is equal to $V_{GS(off)}$. When the control voltage equals $V_{GS(off)}$, the JFET switch is open and the closed-loop voltage gain is $R_1/R_2 + 1$. When the control voltage is zero, the JFET switch is closed and the closed-loop voltage gain is

$$A_{CL} = \frac{R_1}{R_2 \parallel R_3} + 1 \qquad (17\text{-}2)$$

A typical JFET for an application like this is the 2N4860, which has a maximum $r_{ds(on)}$ of 40 Ω. In most designs, R_3 is made much larger than $r_{ds(on)}$ to prevent $r_{ds(on)}$ from affecting the closed-loop voltage gain. Sometimes, you will see several JFET switches and resistors connected in parallel with R_2 to provide a selection of closed-loop voltage gains. (TTL control is typically used. TTL is a type of digital integrated circuit with a two-level output: high and low.)

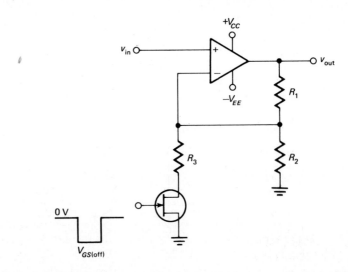

Fig. 17-3.
JFET controls closed-loop voltage gain.

Audio AGC   AGC stands for *automatic gain control.* Figure 17-4 shows an audio AGC circuit. Q_1 is a JFET used as a voltage-variable resistance. For small-signal operation with drain voltages near zero, the JFET operates in the ohmic region and has a resistance of $r_{ds(on)}$ to ac signals. The $r_{ds(on)}$ of a JFET can be controlled by gate voltage. The more negative V_{GS} is, the larger $r_{ds(on)}$ becomes. With a JFET like a 2N4681, $r_{ds(on)}$ can vary from 100 Ω to more than 10 MΩ. If R_3 is around 100 kΩ, the R_3-Q_1 combination acts like a voltage divider whose output varies between $0.001v_{in}$ and R_3-Q_{1in}. Therefore, the noninverting input voltage is between $0.001v_{in}$ and v_{in}, a 1000:1 range. The amplified output voltage is $R_1/R_2 + 1$ times this.

In Fig. 17-4, the output voltage is coupled to the base of Q_2. For peak-to-peak outputs less than 1.4 V, Q_2 is cut off because there is no bias on it. In this case, capacitor C_2 is uncharged and the gate of Q_1 is at $-V_{EE}$, enough to cut off the JFET. This means that almost all of the input voltage reaches the noninverting input.

When the output has a peak-to-peak voltage greater than 1.4 V, Q_2 conducts during part of the negative half cycle. This charges capacitor C_2 and raises the gate voltage above the quiescent level of $-V_{EE}$. When this happens, $r_{ds(on)}$ decreases. As it does, the output of the R_3-Q_1 voltage divider decreases, and so less input voltage reaches the noninverting input. Stated another way, the overall voltage gain of the circuit decreases when the peak-to-peak output voltage gets above 1.4 V.

The whole purpose of the AGC circuit is to change the voltage gain when the input signal changes, so that the output stays approximately constant. One reason for doing this is to prevent the sudden increases

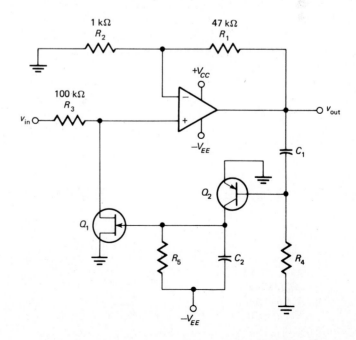

Fig. 17-4.
Audio AGC.

in signal level from overdriving a loudspeaker. If you're listening to a radio, for instance, you don't want an unexpected increase in signal to bombard your hearing. In summary, even though the input voltage of Fig. 17-4 varies over a 1000:1 range, the peak-to-peak output voltage is restricted to slightly more than 1.4 V.

Switchable Inverter

Figure 17-5 shows an op amp that can function as either an inverter or a noninverter with a voltage gain of unity. With the switch in the lower position, the noninverting input is grounded. Since the feedback and source resistors are equal, we have an inverting voltage amplifier with a closed-loop voltage gain of unity.

When the switch is moved to the upper position, the input voltage drives the noninverting input. Since the inverting input is bootstrapped to the noninverting input, there is approximately zero current through the R on the left. For this reason, we can visualize this resistor as open. Therefore, the circuit is equivalent to a noninverting voltage-feedback amplifier with a feedback fraction of unity. This means that the closed-loop voltage gain is approximately unity.

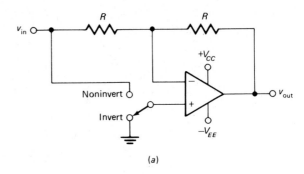

(a)

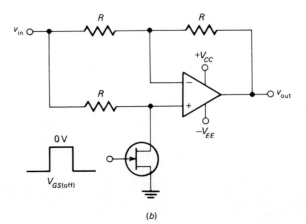

(b)

FIG. 17-5.
Switchable noninverter/
inverter.

JFET-Controlled Switchable Inverter

Figure 17-5b is a modification of Fig. 17-5a. This time, a JFET is used as a switch, a voltage-variable resistance with a very low or a very high resistance. Recall that the drain curves of a JFET extend on both sides of the origin. For this reason, no dc supply voltage is needed. The ac signal voltage on the drain is sufficient.

When the gate of the JFET is at 0 V, the JFET is closed and the circuit is an inverter with a voltage gain of unity. On the other hand, when the gate voltage is at $V_{GS(off)}$, the JFET switch is open and the circuit is a noninverter with a voltage gain of unity. For proper operation, R should be at least 100 times greater than $r_{ds(on)}$ when the JFET switch is closed.

Adjustable Bandwidth

Sometimes we would like to change the bandwidth of an inverting voltage amplifier without changing the voltage gain. Sound impossible? Not at all. Figure 17-6a shows an adjustable resistor R connected between the inverting input and ground. Figure 17-6b shows an equivalent circuit after applying Thevenin's theorem. The effective source resistance driving the inverting input is R_S in parallel with R. For this reason, the feedback fraction is

$$B = \frac{R_S \parallel R}{R_S \parallel R + R_F}$$

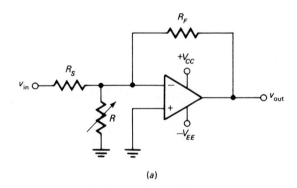

(a)

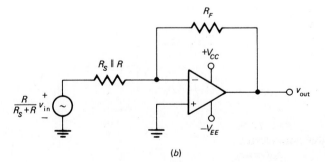

(b)

Fig. 17-6.
Fixed voltage gain with variable bandwidth.

The closed-loop bandwidth is

$$f_{CL} = Bf_{unity}$$

Since R is adjustable, we can change B and control f_{CL}.
The magnitude of output voltage is

$$v_{out} = \frac{R_F}{R_S \| R} \frac{R}{R_S + R} v_{in}$$

which reduces to

$$v_{out} = \frac{R_F}{R_S} v_{in}$$

This means that the output voltage is constant even though the bandwidth changes.

The Summing Amplifier   Another advantage of an inverting amplifier is its ability to handle more than one input at a time. Figure 17-7a shows a *summing amplifier*. Because of the virtual ground,

$$i_1 = \frac{v_1}{R_1}$$

$$i_2 = \frac{v_2}{R_2}$$

and

$$v_{out} = (i_1 + i_2)R_3 = i_1 R_3 + i_2 R_3$$

or

$$v_{out} = \frac{R_3}{R_1} v_1 + \frac{R_3}{R_2} v_2 \qquad\qquad (17\text{-}3)$$

This means we can have a different voltage gain for each input. The output is the sum of the amplified inputs. The same idea applies to

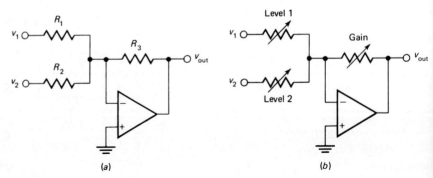

Fig. 17-7.
Summing amplifier.

any number of inputs because we can add another input resistor for each new input signal.

Figure 17-7*b* shows a convenient way to additively mix two signals. The adjustable input resistors allow us to set the level of each input, and the gain control lets us control the output. A circuit like this is useful for combining audio signals.

17-2
CURRENT BOOSTERS FOR VOLTAGE AMPLIFIERS

The maximum output current of a typical op amp is limited. For instance, the 741C has a maximum output current of 25 mA. If the load requires more than this, you can add a *current booster* to the output.

Unidirectional Load Current

If a unidirectional load current is all right, then add an emitter follower to the output of an op amp, as shown in Fig. 17-8. With the transistor inside the feedback loop, the negative feedback automatically adjusts V_{BE} to the required value. Since the circuit is a noninverting voltage-feedback amplifier, the closed-loop voltage gain

$$A_{CL} = \frac{R_1}{R_2} + 1 \qquad (17\text{-}4)$$

and the output impedance is

$$r_{out(CL)} = \frac{r_{out}}{1 + AB} \qquad (17\text{-}5)$$

where r_{out} is the open-loop output impedance looking back into the emitter. Because of the emitter follower, the op amp has only to

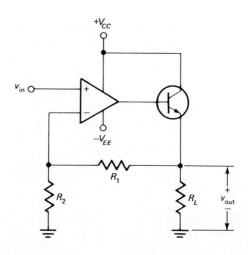

Fig. 17-8.
Emitter follower is a
current booster for op amp.

supply the base current rather than the load current. As a result, the current booster of Fig. 17-8 allows us to use smaller load resistances.

Bidirectional Load Current The main disadvantage of the Fig. 17-8 is its unidirectional load current. One way to get bidirectional load current is with a class B push-pull emitter follower, as shown in Fig. 17-9. In this case, the closed-loop voltage gain is

$$A_{CL} = \frac{R_F}{R_S} \qquad\qquad (17\text{-}6)$$

The output impedance is still $r_{out}(1 + AB)$ because of the voltage feedback.

Again, the negative feedback automatically adjusts the V_{BE} values to whatever is needed. Furthermore, there's no need to provide bias to eliminate crossover distortion because the negative feedback reduces it by a factor of $1 + AB$. When the input voltage goes positive, the lower transistor is conducting and the load voltage is negative. On the other hand, when the input voltage goes negative, the upper transistor conducts and the output voltage is positive.

17-3
VOLTAGE-CONTROLLED CURRENT SOURCES

Figure 17-10a shows how to stabilize the current through a load. Here is how it works. Since the inverting input is bootstrapped to the noninverting input, the voltage across R equals v_{in} and the current through R is

$$i = \frac{v_{in}}{R}$$

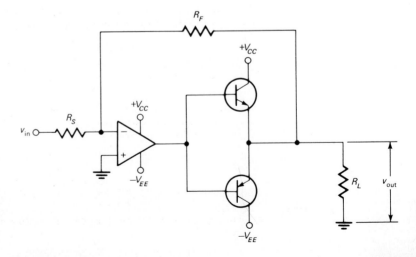

Fig. 17-9.
Class B push-pull emitter follower is a bidirectional current booster.

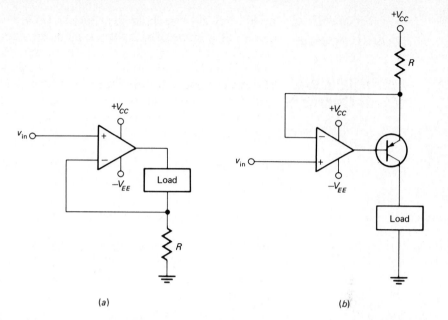

Fig. 17-10.
Current sources.

(a)

(b)

Since the inverting input has a very high input resistance, the fore-going current must flow through the load, which means

$$i_{out} = \frac{v_{in}}{R} \qquad (17\text{-}7)$$

The value of load resistance does not appear in this equation. There-fore, the output current is independent of the value of load resistance. This is equivalent to saying that the load is being driven by a current source.

Grounded Load If a floating load is all right, then a circuit like Fig. 17-10a works quite well. If the load has to be grounded on one end (the usual case), we can modify the basic circuit as shown in Fig. 17-10b. Because of the bootstrap effect, the inverting input voltage equals v_{in}. Therefore, the voltage across R is $V_{CC} - v_{in}$, and the emitter current is

$$I_E = \frac{V_{CC} - v_{in}}{R}$$

Since the collector current approximately equals the emitter current, the current flowing through the load resistor is

$$i_{out} = \frac{V_{CC} - v_{in}}{R} \qquad (17\text{-}8)$$

Again, the output current is independent of the value of load resis-tance.

Grounded Voltage-to-Current Converter

In Eq. (17-8), the load current decreases when the input voltage increases. Figure 17-11 shows a circuit where the load current increases when the input voltage increases. Because of the bootstrap effect, the inverting input of the first op amp has a voltage of v_{in}. The current flowing through the first transistor is

$$i = \frac{v_{in}}{R}$$

This current produces a collector voltage of

$$V_C = V_{CC} - iR = V_{CC} - \frac{v_{in}}{R}R = V_{CC} - v_{in}$$

Since this voltage drives the noninverting input of the second op amp, the inverting input voltage is

$$v_2 = V_{CC} - v_{in}$$

to a close approximation. This implies that the voltage across the final R is

$$V_{CC} - v_2 = V_{CC} - (V_{CC} - v_{in}) = v_{in}$$

and the output current is

$$i_{out} = \frac{v_{in}}{R} \qquad\qquad \textbf{\textit{(17-9)}}$$

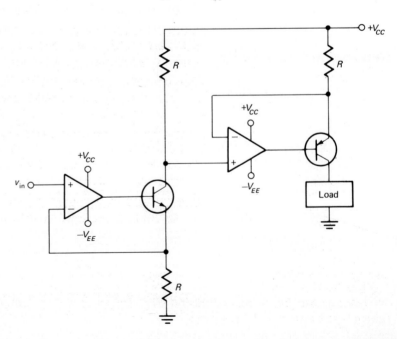

Fig. 17-11.
A voltage-to-current
converter.

The circuit of Fig. 17-11 is sometimes called a *voltage-to-current converter* because the input voltage controls the output current. A grounded load is used a lot more than a floating load; so try to remember this circuit, because it is useful and practical whenever you want to convert an input voltage to an output current.

17-4 ACTIVE FILTERS

Figure 17-12a shows how virtual ground can be used to build *active filters*. The closed-loop voltage gain is given by

$$A_{CL} = \frac{Z_2}{Z_1} \qquad (17\text{-}10)$$

where Z_1 and Z_2 are complex impedances. These impedances determine the frequency response of the circuit. For instance, Fig. 17-12b

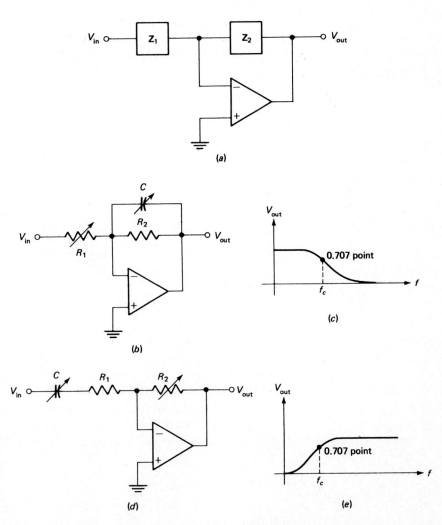

Fig. 17-12.
Active filters. (a) General case. (b) Low-pass filter. (c) Low-pass response. (d) High-pass filter. (e) High-pass response.

is an *active low-pass filter*. At low frequencies the capacitor appears open, and the circuit acts like an inverting amplifier with a voltage gain of R_2/R_1. As the frequency increases, the capacitive reactance decreases, causing the voltage gain to drop off.

Figure 17-12c illustrates the frequency response of a low-pass filter. When the frequency reaches the cutoff frequency, the output decreases to 0.707 times the low-frequency value. It can be shown that the cutoff frequency is given by

$$f_c = \frac{1}{2\pi R_2 C} \qquad \textbf{(17-11)}$$

The adjustable C of Fig. 17-12b controls the cutoff frequency, and the adjustable R_1 varies with the voltage gain.

The *active high-pass filter* of Fig. 17-12d is different. At low frequencies the capacitor appears open, and the voltage gain approaches zero. At high frequencies the capacitor appears shorted, and the circuit becomes an inverting amplifier with a voltage gain of R_2/R_1. Figure 17-12e is the frequency response. By further analysis, we can show that the cutoff frequency is given by

$$f_c = \frac{1}{2\pi R_1 C} \qquad \textbf{(17-12)}$$

17-5
ACTIVE DIODE CIRCUITS

Op amps can enhance the performance of diode circuits. For one thing, an op amp with negative feedback reduces the effect of the diode offset voltage, allowing us to rectify, peak-detect, clip, and clamp low-level signals (those with peak values less than the diode offset voltage). What follows is a discussion of these active diode circuits.

Half-Wave Rectifier

Figure 17-13a shows an *active half-wave rectifier,* one that includes an op amp. When the input voltage goes positive, the output voltage goes positive and turns on the diode. The circuit then acts like a voltage follower, and the positive half cycle appears across the load resistor. On the other hand, when the input goes negative, the op-amp output goes negative and turns off the diode. Since the diode is open, no voltage appears across the load resistor. This is why the output is almost a perfect half-wave signal.

The high gain of the op amp almost eliminates the effect of diode offset voltage. For instance, if the offset voltage is 0.7 V and the open-loop voltage gain is 100,000, the input voltage that turns on the diode is

$$v_{in} = \frac{0.7\ V}{100,000} = 7\ \mu V$$

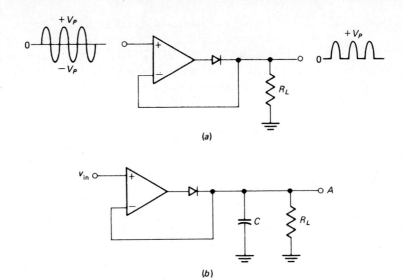

Fig. 17-13.
(a) Active half-wave
rectifier. (b) Active peak
detector.

Therefore, an input voltage as small as 7 μV will turn on the diode. The effect is equivalent to reducing the offset voltage by a factor of A. In symbols,

$$\phi' = \frac{\phi}{A} \qquad (17\text{-}13)$$

where ϕ' is the offset potential seen by the input voltage. This means we can rectify low-level signals.

Active Peak Detector

To peak-detect small signals, we can use an *active peak detector* like Fig. 17-13*b*. Because of the op-amp gain, the input offset potential ϕ is in the microvolt region. This means the circuit can peak-detect signals whose peak voltages are much less than a volt. Furthermore, when the diode is on, the heavy negative feedback produces a Thevenin output impedance approaching zero. For this reason, the charging time constant shrinks to a very small value, eliminating source effects.

Active Positive Clipper

Figure 17-14*a* illustrates an *active positive clipper.* With the wiper all the way to the left, v_{ref} is zero and the noninverting input is grounded. When v_{in} goes positive, the error voltage drives the op-amp output negative and turns on the diode. This means that v_{out} is at virtual ground for any positive value of v_{in}. When v_{in} goes negative, the op-amp output is positive, which turns off the diode and opens the loop. As this happens, the virtual ground is lost, and the final output v_{out} is free to follow the negative half cycle of input voltage. This is why the negative half cycle appears at the output. By adjusting v_{ref}, we can change the clipping level.

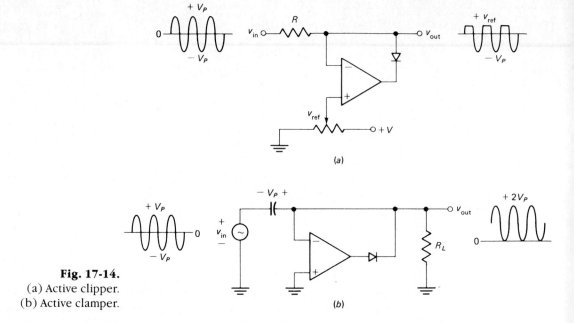

Fig. 17-14.
(a) Active clipper.
(b) Active clamper.

Active Positive Clamper

Figure 17-14b is an *active positive clamper.* The first negative half cycle produces a positive op-amp output that turns on the diode. This allows the capacitor to charge to the peak value of the input voltage. Just beyond the negative peak, the diode turns off, the loop opens, and the virtual ground is lost. Since V_p is being added to a sinusoidal input voltage, the final output waveform is shifted positively through V_p volts. In other words, we get the positive clamped output shown. It swings from 0 to $2V_p$. Because the input offset potential is in the microvolt region, the circuit can positively clamp low-level signals.

17-6 COMPARATORS

Often, we want to compare one voltage with another to see which is larger. All we need is a yes/no answer. A *comparator* is a circuit with two input voltages (noninverting and inverting) and one output voltage. When the noninverting input voltage is larger than the inverting input voltage, the comparator produces a high output voltage. When the noninverting input voltage is less than the inverting input voltage, the output is low. The high output symbolizes the "yes" answer, and the low output stands for the "no" answer.

Basic Circuit

The simplest way to build a comparator is to use an op amp without feedback resistors as shown in Fig. 17-15a. Since the inverting input is grounded, the slightest input voltage is enough to saturate the op amp. For instance, if the supplies are $+15$ V and -15 V, the ac compliance is from approximately $+13$ V to -13 V. With a 741C, the

CHAPTER 17

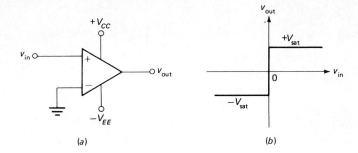

Fig. 17-15.
(a) Comparator.
(b) Transfer characteristic.

(a)

(b)

open-loop voltage gain is typically 100,000. Therefore, the minimum input voltage that produces saturation is

$$v_{in} = \frac{13 \text{ V}}{100,000} = 0.13 \text{ mV}$$

This is so small that the transfer characteristic of Fig. 17-15*b* has what appears to be a vertical transition at $v_{in} = 0$. This is not actually vertical. With a 741C it takes at least +0.13 mV of input voltage to produce positive saturation and −0.13 mV to get negative saturation.

Because the input voltages needed to produce saturation are so small, the transition of Fig. 17-15*b* appears to be vertical. As an approximation, we will treat it as vertical. This means that a positive input voltage produces positive saturation, while a negative input voltage produces negative saturation.

The *trip point* (also called the threshold or reference) of a comparator is the value of input voltage where the output switches states (low to high, or vice versa). In Fig. 17-15*a*, the trip point is zero. When v_{in} is greater than the trip point, the output is high. When v_{in} is less than the input, the output is low. A circuit like this is called a *zero-crossing detector*.

Moving the Trip Point

In Fig. 17-16*a*, a *reference* voltage is applied to the inverting input:

$$v_{ref} = \frac{R_2}{R_1 + R_2} V_{CC} \qquad\qquad (17\text{-}14)$$

When v_{in} is less than v_{ref}, the error voltage is negative and the output is low. When v_{in} is greater than v_{ref}, the error voltage is positive and the output is high.

Incidentally, a bypass capacitor is typically used on the inverting input, as shown in Fig. 17-16*a*. This reduces the power supply ripple and noise appearing at the inverting input. To be effective, the cutoff frequency of this bypass circuit should be much lower than the ripple frequency.

Figure 17-16*b* shows the transfer characteristic. The trip point is now equal to v_{ref}. When v_{in} is slightly more than v_{ref}, the output of the

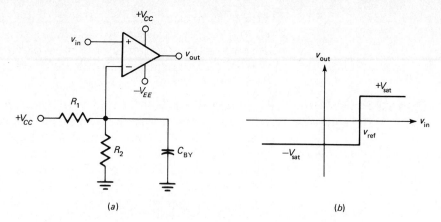

Fig. 17-16.
(a) Limit detector.
(b) Transfer characteristic.

(a)

(b)

comparator goes into positive saturation. When v_{in} is less than v_{ref}, the output goes into negative saturation. A comparator like this is sometimes called a *limit detector* because a positive output indicates that the input voltage exceeds a specific limit. With different values of R_1 and R_2, we can set the positive trip point anywhere between zero and V_{CC}. If a negative trip point is preferred, then you can connect a negative supply voltage to the voltage divider.

Single-Supply Comparator As you know, a typical op amp like the 741C can be run from a single positive supply by grounding the $-V_{EE}$ point, as shown in Fig. 17-17a. Now the output voltage has only one polarity, either a low or a high positive voltage. For instance, with V_{CC} equal to +15 V, the ac compliance is from approximately 1 or 2 V (low state) to around 13 or 14 V (high state).

The reference voltage applied to the inverting input is positive and equal to

$$v_{ref} = \frac{R_2}{R_1 + R_2} V_{CC}$$

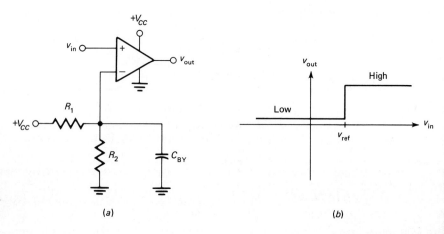

Fig. 17-17.
Single-supply comparator.

(a)

(b)

When v_{in} is greater than v_{ref}, the output is high, as shown in Fig. 17-17b. When v_{in} is less than v_{ref}, the output is low. In either case, the output has a positive polarity. For most digital applications, this is the kind of comparator output that is preferred.

Speed Problems An op amp like a 741C can be used as comparator, but it has speed limitations. As you know, the slew rate limits the rate of output voltage change. With a 741C, the output can change no faster than 0.5 V/μs. Because of this, a 741C takes more than 50 μs to switch between a low output of -13 V and a high output of $+13$ V. One approach to speeding up the switching action is to use an op amp with a faster slew rate, such as the 318. This op amp has a slew rate of 70 V/μs; so it can switch from -13 V to $+13$ V in approximately 0.3 μs.

17-7
THE SCHMITT TRIGGER

Since the input to a comparator contains noise, the output may be erratic when v_{in} is near the trip point. For instance, with a zero-crossing detector, the output is high when v_{in} is positive and low when v_{in} is negative. If the input contains a noise voltage with a peak of 1 mV or more, then the comparator will detect the zero crossings produced by the noise. We can avoid this unwanted noise triggering by using a *Schmitt trigger*, a comparator with positive feedback.

Basic Circuit Figure 17-18a shows one way to build a Schmitt trigger. Because of the voltage divider, we have positive feedback. When the output voltage is positively saturated, a positive voltage is fed back to the noninverting input. This positive input holds the output in the high state. On the other hand, when the output voltage is negatively saturated, a negative voltage is fed back to the noninverting input, holding the output in the low state. In either case, the positive feedback reinforces the existing output state.

The feedback fraction is

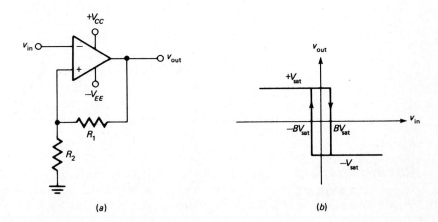

Fig. 17-18.
Schmitt trigger.

(a)   (b)

$$B = \frac{R_2}{R_1 + R_2} \qquad \textbf{\textit{(17-15)}}$$

When the output is positively saturated, the reference voltage applied to the noninverting input is

$$v_{\text{ref}} = +BV_{\text{sat}} \qquad \textbf{\textit{(17-16)}}$$

When the output is negatively saturated, the reference voltage is

$$v_{\text{ref}} = -BV_{\text{sat}} \qquad \textbf{\textit{(17-17)}}$$

The output will remain in a given state until the input exceeds the reference voltage for that state. For instance, if the output is positively saturated, the reference voltage is $+BV_{\text{sat}}$. The input voltage v_{in} must be increased to slightly more than $+BV_{\text{sat}}$. Then the error voltage reverses and the output voltage changes to the low state, as shown in Fig. 17-18b. Once the output is in the negative state, it will remain there indefinitely until the input voltage becomes more negative than $-BV_{\text{sat}}$. Then the output switches from negative to positive (Fig. 17-18b).

Hysteresis Positive feedback has an unusual effect on the circuit. It forces the reference voltage to have the same polarity as the output voltage. The reference voltage is positive when the output is high and negative when the output is low. This is why we get an *upper trip point* (UTP) and a *lower trip point* (LTP). The trip points are given by

$$\text{UTP} = BV_{\text{sat}}$$
$$\text{LTP} = -BV_{\text{sat}}$$

In a Schmitt trigger, the difference between the two trip points is called *hysteresis*. Because of the positive feedback, the transfer characteristic has the hysteresis shown in Fig. 17-18b. If there were no positive feedback, B would equal zero and the hysteresis would disappear, because the trip points would both equal zero. But there is positive feedback, and this spreads those trip points as shown.

Some hysteresis is desirable because it prevents noise from causing false triggering. Imagine a Schmitt trigger with no hysteresis. Then any noise at the input would cause the Schmitt trigger to randomly jump between states. Next, visualize a Schmitt trigger with hysteresis. If the peak-to-peak noise is less than the hysteresis, there is no way the noise can produce false triggering. A circuit with enough hysteresis is immune to noise triggering.

Moving the Trip Points Figure 17-19a shows how to move the trip points. An additional resistor R_3 is connected between the noninverting input and $+V_{CC}$. This sets up the center of the hysteresis loop:

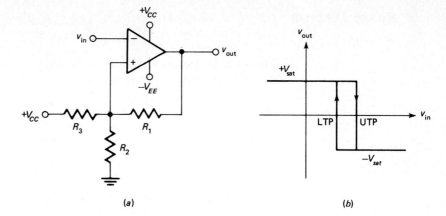

Fig. 17-19.
Schmitt trigger with
movable trip points.

(a)

(b)

$$v_{cen} = \frac{R_2}{R_2 + R_3} V_{CC} \qquad (17\text{-}18)$$

The positive feedback spreads the trip points on each side of this center voltage. By applying Thevenin's theorem, we can show that the feedback fraction is

$$B = \frac{R_2 \parallel R_3}{R_1 + R_2 \parallel R_3} \qquad (17\text{-}19)$$

When the output is positively saturated, the noninverting reference voltage is

$$\text{UTP} = v_{cen} + BV_{sat} \qquad (17\text{-}20)$$

When the output is negatively saturated,

$$\text{LTP} = v_{cen} - BV_{sat} \qquad (17\text{-}21)$$

Figure 17-19*b* is the transfer characteristic. You can calculate the trip points with Eqs. (17-18) through (17-21).

17-8
THE INTEGRATOR

An *integrator* is a circuit that performs the mathematical operation called integration. A common application of an integrator is to use a constant input voltage to produce a *ramp* of output voltage. (A ramp is a linearly increasing or decreasing voltage.) For instance, if you drive a 741C with a voltage step, the output slews at a rate of 0.5 V/μs. This means the output voltage changes 0.5 V during each microsecond. This is an example of a *ramp,* a voltage that changes linearly with time. With an op amp, we can build an integrator, a circuit that produces a ramp output from a rectangular input.

Basic Circuit Figure 17-20*a* is an op-amp integrator. The typical input is a rectangular pulse like Fig. 17-20*b*. The capitalized V_{in} represents a constant voltage during pulse time *T*. Visualize V_{in} applied to the left end of *R*. Because of the virtual ground, the input current is a constant and equals

$$I_{in} = \frac{V_{in}}{R}$$

Approximately all of this current goes to the capacitor. The basic capacitor law says that

$$C = \frac{Q}{V}$$

or
$$V = \frac{Q}{C} \qquad\qquad (17\text{-}22)$$

Since a constant current is flowing into the capacitor, the charge *Q* increases linearly. This means the capacitor voltage increases linearly with the polarity shown in Fig. 17-20*a*. Because of the phase inversion of the op amp, the output voltage is a negative ramp, as shown in Fig. 17-20*c*. At the end of the pulse period, the input voltage returns to zero, and the charging current stops. Because the capacitor holds its charge, the output voltage remains constant at a negative level.

To get a formula for the output voltage, divide both sides of Eq. (17-22) by *T*:

$$\frac{V}{T} = \frac{Q/T}{C}$$

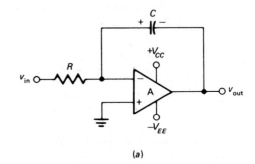

(a)

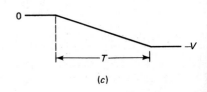

V_{in}

0

(b)

0

$-V$

(c)

Fig. 17-20.
Integrator.

Since the charging current is constant, we can write

$$\frac{V}{T} = \frac{I}{C}$$

or
$$V = \frac{IT}{C}$$
(17-23)

This is the voltage across the capacitor with the polarity shown in Fig. 17-20a. Because of the phase inversion, $v_{out} = -V$.

For example, if $I = 4$ mA, $T = 2$ ms, and $C = 1$ μF, then the capacitor voltage at the end of the charging period is

$$V = \frac{(4 \text{ mA})(2 \text{ ms})}{1 \ \mu F} = 8 \text{ V}$$

Because of the phase reversal, the output voltage is -8 V after 2 ms.

Practical Integrator The circuit of Fig. 17-20a needs a slight modification to make it practical. Because the capacitor acts like an open to dc voltage, the closed-loop voltage equals the open-loop voltage gain at zero frequency. This will produce too much output offset voltage. Without negative feedback at zero frequency, the circuit will treat input offsets the same as a valid input voltage. In other words, the input offsets will eventually charge the capacitor and force the output into positive or negative saturation.

One way to reduce the effect of input offsets is to insert a resistor in parallel with the capacitor, as shown in Fig. 17-21a. This resistor should be at least 10 times larger than the input resistor. If the added resistance equals $10R$, the closed-loop voltage gain is -10 and the output offset voltage is greatly reduced. The integrator works approx-

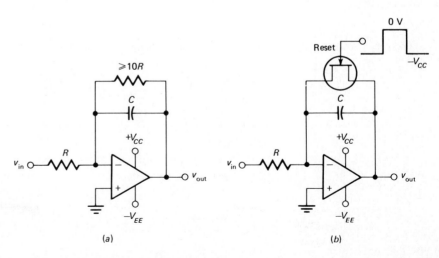

Fig. 17-21.
Practical integrators.

(a) (b)

imately as previously described because the bulk of the input current still goes to the capacitor.

JFET Reset   Another way to suppress the effect of input offsets is to use a JFET reset switch, as shown in Fig. 17-21b. This allows us to discharge the capacitor just before the pulse is applied to the input. When the gate voltage is $-V_{CC}$, the JFET switch is open and the circuit works as previously described. When the gate voltage is changed to 0 V, the JFET switch closes and discharges the capacitor. When the gate voltage gain goes negative, the JFET opens, and the capacitor can be recharged by the next input pulse.

17-9

WAVEFORM CONVERSION   With op amps we can convert sine waves to rectangular waves, rectangular waves to triangular waves, and so on. This section is about some basic circuits that convert an input waveform to an output waveform of a different shape.

Sine to Rectangular   Figure 17-22a shows a Schmitt trigger, and Fig. 17-22b is the transfer characteristic. When the input signal is *periodic* (repeating cycles), the Schmitt trigger produces a rectangular-wave output. This assumes that the input voltage is large enough to pass through both trip points of Fig. 17-22c. When the input voltage exceeds the UTP on the upward swing of the positive half cycle, the output voltage switches to $-V_{sat}$. Half a cycle later, the input voltage becomes more negative than the LTP, and the output switches back to $+V_{sat}$.

A Schmitt trigger always produces a rectangular output, regardless of the shape of the input signal. As an example, Fig. 17-22d shows a Schmitt trigger with trip points of approximately UTP = 0.1 V and LTP = -0.1 V. If the input voltage is repetitive and has a peak-to-peak value greater than 0.2 V, then the output is rectangular with a peak-to-peak value of approximately 20 V (2 times V_{sat}).

Rectangular to Triangular   In Fig. 17-23a, a rectangular wave is the input to an integrator. Since the input has a dc or average value of zero, the dc or average value of the output is also zero (assuming negligible output offset). As shown in Fig. 17-23b, the ramp is decreasing during the positive half cycle of input voltage and increasing during the negative half cycle. Therefore, the output is a periodic triangular wave with the same frequency as the input.

By further mathematical analysis, we can derive this formula for the peak-to-peak output:

$$v_{out(pp)} = \frac{v_{in(pp)}}{4fRC} \qquad (17\text{-}24)$$

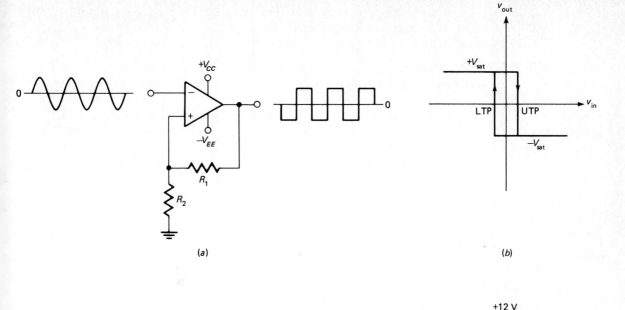

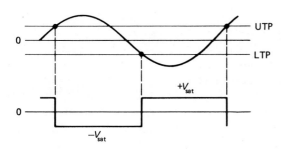

Fig. 17-22.
Periodic input to Schmitt
trigger produces
rectangular output.

(c)

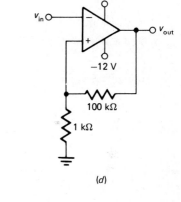

(d)

For instance, suppose a square wave drives the integrator of Fig. 17-23c. If the frequency is 1 kHz and the peak-and-peak input voltge is 10 V, then the output is a triangular wave with a peak-to-peak value of

$$v_{out(pp)} = \frac{10 \text{ V}}{4(1 \text{ kHz})(1 \text{ k}\Omega)(10 \text{ } \mu\text{F})} = 0.25 \text{ V}$$

Triangular to
Pulse

In some applications we would like to produce a pulse with a variable width. Figure 17-24a shows one way this can be done, with a limit detector. This is a comparator with an adjustable reference voltage. This allows us to move the trip point from zero to some positive level. As long as the input triangular voltage exceeds the reference

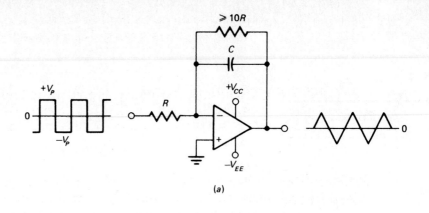

(a)

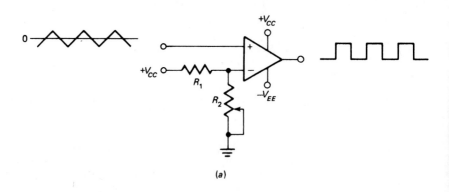

Fig. 17-23.
Rectangular input to
integrator produces
triangular output.

(b)

(c)

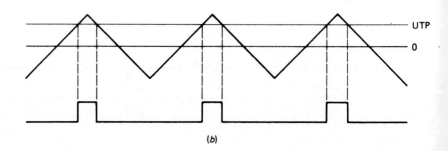

(a)

Fig. 17-24.
Generating pulse output
with variable width.

(b)

voltage, the output is high, as shown in Fig. 17-24b. Since v_{ref} is adjustable, we can vary the width of the output pulse.

17-10 WAVEFORM GENERATION

With positive feedback it is possible to build *oscillators,* circuits that generate an output signal with no external input signal. This section discusses some op-amp circuits that can generate nonsinusoidal signals.

Relaxation Oscillator

In Fig. 17-25a, there is no input signal. Nevertheless, the circuit generates an output rectangular wave. How does it work? Assume that the output is in positive saturation. The capacitor will charge exponentially toward $+V_{sat}$. It never reaches $+V_{sat}$ because its voltage hits the UTP (Fig. 17-25b). When this happens, the output switches to

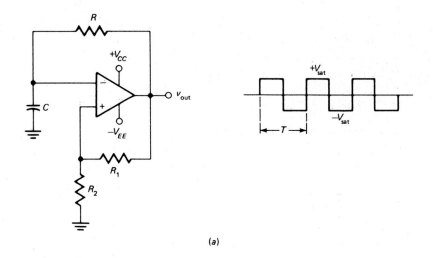

(a)

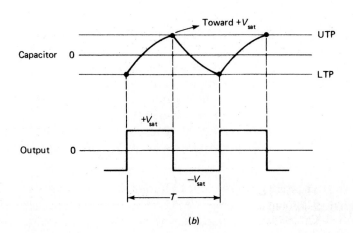

Fig. 17-25.
Relaxation oscillator.

(b)

$-V_{sat}$. Now, a negative voltage is fed back, and the capacitor reverses its charging direction. The capacitor voltage decreases as shown. When the capacitor voltage hits the LTP, the output switches back to $+V_{sat}$. Because of the continuous charging and discharging of the capacitor, the output is a rectangular wave.

With a mathematical analysis, we can derive this formula for the period of the output rectangular wave:

$$T = 2RC \ln \frac{1 + B}{1 - B} \qquad (17\text{-}25)$$

where $B = R_2/(R_1 + R_2)$. Also note that ln is the logarithm to the base e. To calculate the frequency of the output, take the reciprocal of the period:

$$f = \frac{1}{T} \qquad (17\text{-}26)$$

As an example, suppose $R = 1 \text{ k}\Omega$, $C = 0.1 \ \mu\text{F}$, $R_1 = 2 \text{ k}\Omega$, and $R_2 = 18 \text{ k}\Omega$. Then the feedback fraction is

$$B = \frac{18 \text{ k}\Omega}{20 \text{ k}\Omega} = 0.9$$

The RC time constant is

$$RC = (1 \text{ k}\Omega)(0.1 \ \mu\text{F}) = 100 \ \mu\text{s}$$

The period of the output signal is

$$T = 2(100 \ \mu\text{s}) \ln \frac{1.9}{0.1} = 589 \ \mu\text{s}$$

and the frequency is

$$f = \frac{1}{589 \ \mu\text{s}} = 1.7 \text{ kHz}$$

Figure 17-25a is an example of a *relaxation oscillator,* a circuit that generates an output signal whose frequency depends on the charging and discharging of a capacitor (or inductor). If we increase the RC time constant, it takes longer for the capacitor voltage to reach the trip points. Therefore, the frequency is lower. By making R adjustable, we can typically get a 50:1 tuning range.

Generating Triangular Waves

One way to generate triangular waves is to use the output of a relaxation oscillator to drive an integrator as shown in Fig. 17-26. The rectangular waves out of the relaxation oscillator swing between $+V_{sat}$ and $-V_{sat}$. You can calculate its frequency with Eqs. (17-25) and

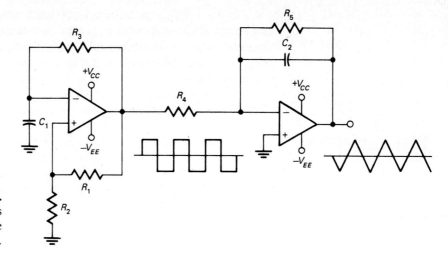

Fig. 17-26.
Relaxation oscillator drives integrator to produce triangular output.

(17-26). The triangular wave has the same frequency; you can calculate its peak-to-peak value with Eq. (17-24).

Summary

If you don't need a response that extends down to zero frequency, you can use coupling and bypass capacitors with an op amp. If you prefer, you can drive an op amp with a single power supply. One method for varying the voltage gain is to use JFETs to switch in different external resistances. AGC decreases the voltage gain as the input signal increases; this tends to hold the output constant.

A switchable inverter can act as either an inverter or a noninverter. The summing amplifier allows us to add several inputs.

If you need more output current than an op amp can provide, add a class A emitter follower to the output; this produces a unidirectional load current. If you need bidirectional load current, you can use a class B push-pull emitter follower. With op amps, we can build voltage-controlled current sources. Op amps also allow us to build active filters.

Op amps can enhance the performance of diode circuits because they reduce the equivalent offset voltage of the diode. With op amps, we can build active half-wave rectifiers, peak detectors, clippers, and clampers that can process low-level voltages.

A comparator is a circuit that produces either a low or a high output. The trip point is the value of input voltage where the output changes states. One way to speed up a comparator is use an op amp with a higher slew rate such as the 318. A Schmitt trigger is a comparator with positive feedback. This results in hysteresis, important when noise may cause false triggering.

When a rectangular pulse drives an integrator, the output is a ramp. This linearly changing voltage equals the negative of the voltage across the capacitor. Often, a resistor is shunted across the capacitor to reduce the closed-loop voltage gain to approximately 10. This prevents excessive output offset voltage.

A Schmitt trigger converts any periodic input signal to a rectangular output signal. An integrator converts may rectangular input voltage to a triangular output voltage. A relaxation oscillator produces a rectangular output voltage; this type of circuit requires no input signal.

Glossary

active diode circuit By including an op amp in a diode circuit, we can greatly reduce the effect of diode offset or knee voltage. Such a circuit can process low-level voltages in the millivolt region.

active filter Before the advent of the IC op amp, filters were passive, built with resistances, inductors, and capacitors. Inductors are undesirable because of their bulk and expense. Since the IC op amp became commercially available, simpler filters using resistances, capacitances, and op amps have eliminated the need for inductors.

automatic gain control A circuit whose voltage gain decreases when the input voltage increases, so that the final output voltage is approximately constant.

comparator A circuit that produces either a low or a high output voltage, depending on whether the noninverting input is smaller or greater than the inverting input.

current booster An emitter follower added to the output of an op amp to increase the maximum load current.

decoupling network An *RC* circuit in the power-supply line between two stages. This circuit prevents undesirable positive feedback.

hysteresis The difference between the two trip points of a Schmitt trigger.

integrator A circuit that produces a ramp output when driven by a rectangular input.

limit detector A circuit that detects when the noninverting input is greater than a predetermined limit.

periodic wave A signal whose cycles are repetitive.

relaxation oscillator A circuit that produces a rectangular output whose frequency depends on the time it takes to charge or discharge a capacitor.

Schmitt trigger A comparator with positive feedback. This results in hysteresis or a difference between the trip points.

trip point The value of input voltage where the output of a comparator switches states.

voltage-controlled current source Also called a voltage-to-current converter, this circuit acts like a current source that is controlled by an input voltage.

zero-crossing detector A comparator with a trip point of zero.

Review Questions

1. What is AGC?
2. What can you connect to the output of an op amp if it is unable to produce enough load current?
3. What is a voltage-to-current converter?
4. What components do you need to build an active filter?
5. What advantage do active diode circuits have over ordinary diode circuits?
6. What is a comparator?
7. How many trip points does a Schmitt trigger have? Why is hysteresis sometimes desirable?
8. Define a ramp of voltage.
9. Given any periodic input signal, how can you produce a rectangular output signal with the same frequency?
10. Describe one way to produce a triangular output voltage.
11. Describe how a relaxation oscillator works.

Problems

$$\frac{R_F}{R}$$

17-1. In Fig. 17-6a, $R_F = 47$ kΩ, and $R = 2.2$ kΩ. What is the closed-loop voltage gain? If $R_S = 1$ kΩ, what does f_{CL} equal for a typical 741C?

17-2. In Fig. 17-8, $R_1 = 100$ kΩ and $R_2 = 1$ kΩ. What is the closed-loop voltage gain? If $R_L = 10$ Ω, what is the load current when v_{in} is 10 mV?

17-3. All resistors equal 1 kΩ in Fig. 17-11. The supply voltages are $+15$ V and -15 V. If $v_{in} = 4$ V and $R_L = 100$ Ω, what does the voltage across the load resistor equal?

17-4. In Fig. 17-12b, $R_1 = 2.2$ kΩ, $R_2 = 3.3$ kΩ, and $C = 1000$ pF. What is the closed-loop voltage gain? The cutoff frequency?

17-5. The input voltage of Fig. 17-13b has a peak-to-peak value of 10 V. What is the output voltage?

17-6. If the input voltage of Fig. 17-14b had a peak-to-peak value of 10 V, what is the peak-to-peak output voltage?

17-7. In Fig. 17-16, $R = 10$ kΩ, $R_2 = 2.2$ kΩ, and $V_{CC} = 15$ V. What does the trip point equal?

17-8. The input voltage of Fig. 17-20a is a rectangular pulse like Fig. 17-20b. If $V_{in} = 5$ V, $R = 1$ kΩ, $T = 25$ μs, and $C = 10$ μF, what is the capacitor voltage at the end of the input pulse?

17-9. A rectangular wave with a frequency of 500 kHz and a peak-to-peak voltage of 8 V drives the integrator of Fig. 17-23a. If $R = 3.3$ kΩ and $C = 20$ μF, what is the peak-to-peak output voltage?

17-10. The relaxation oscillator of Fig. 17-25a has $R = 3.3$ kΩ, $C = 0.22$ μF, $R_1 = 2.2$ kΩ, and $R_2 = 15$ kΩ. What is frequency of the output signal?

Answers to Odd-Numbered Problems

CHAPTER 1 SEMICONDUCTOR PHYSICS 1-1. The diagram looks like Fig. 1-2b, except that the core is labeled Au core instead of Cu core. 1-3. This is like Fig. 1-6b except that one of the covalent bonds is left out. 1-5. Draw Fig. 1-8b with 12 free electrons in the conduction band and 3 holes in the valence band. 1-7. The majority carriers are holes, so they flow to the right. The minority carriers are free electrons, which flow to the left. 1-9. conductor-1 valence electron; semiconductor-4 valence electrons; hole-acceptor atom; ambient-surrounding; covalent bond-shared electron; forbidden gap-unstable orbits; recombination-free electron falls into hole; silicon-widely used semiconductor; larger orbit-higher energy level; conduction band-energy levels of free electrons; atomic path for holes-valence band; extrinsic-doped

CHAPTER 2. RECTIFIER DIODES 2-1. 0.3 A   2-3. 0.293 A, 0.318 A   2-5. 50 mA   2-7. Positive peak is 0 V, negative peak is -14 V   2-9. 0.93 mA, 0.651 mW   2-11. 12.5 μA   2-13. 80 nA   2-15. The 1N4001   2-17. The most likely trouble is b, diode open. Theoretically, trouble d, load resistor shorted is also correct if you ignore the Thevenin resistance of the source.

CHAPTER 3. SPECIAL DIODES 3-1. 9.76 mA   3-3. 10.2 mA   3-5. Ideally, the output is a half-wave rectified sine wave with a peak value of 10 V. To a second approximation, the output has a peak of 9.75 V.   3-7. 2.52 MHz, 5.03 MHz   3-9. 10 mA, 30 mA   3-11. 11.7 mA, 7.56 mA, 4.14 mA   3-13. The only trouble that produces the symptoms is c, series resistor is shorted. As a consequence of this short, the zener diode must be open because it must have been destroyed by excessive current when the short first appeared.

CHAPTER 4. DIODE APPLICATIONS 4-1. 146 V, 179 V   4-3. 6.23 V, 19.6 V, 0.0916 A   4-5. 6.23 V, 19.6 V, 0.514 A   4-7. 18 V, 28.3 V, 1.2 A   4-9. 1.48 V, 17.8 V, 17.8 V, 0.089 A   4-11. 0.967 V, 25 V, 25.5 V, 0.058 A   4-13. 14.1 mA, 16.6 mA, 5.1 mA   4-15. 1527 V   4-17. b

CHAPTER 5. BIPOLAR TRANSISTORS 5-1. 0.987, 49   5-3. 0.99, 0.997   5-5. 3.04 mA, 9.12 mA   5-7. 1.42 mA, 16.6 V, 6.85 mA, 13.2 V   5-9. 5.79 mA, 13.4 V   5-11. Figure 5-34a: $I_{C(\text{sat})}$ = 0.824 mA, $V_{CE(\text{cutoff})}$ = 15 V; Fig. 5-34b: $I_{C(\text{sat})}$ = 12.5 mA, $V_{CE(\text{cutoff})}$ = 25 V; Fig. 5-34c: $I_{C(\text{sat})}$ = 6.56 mA, $V_{CE(\text{cutoff})}$ = 40 V 5-13. 0.151 mA, 4.98 V 5-15. 13.1 mW, 48.2 mW   5-17. a. Decrease b. Increase c. Same d. Same

CHAPTER 6. COMMON-EMITTER APPROXIMATIONS 6-1. 0.49 mA   6-3. The total current has a dc component of 0.49 mA and an ac component of 0.06 μA rms   6-5. 2.1 mA, 11.9 Ω 6-7. 0.168 mA   6-9. 0.084 mA   6-11. 120, 240 mV   6-13. 159 Ω, 159 Ω, 1.59 Ω 6-15. 66.3, 116   6-17. 4.83 kΩ 6-19. a. Same. b. Decrease. c. Same. d. Decrease.

CHAPTER 7. COMMON-COLLECTOR APPROXIMATIONS 7-1. 0.966   7-3. 0.993, 210 kΩ, 3.28 kΩ 7-5. 124, 198   7-7. 550 mV   7-9. 6.8 V, 10 mA, 9.2 mA   7-11. Either transistor may be shorted or open, either coupling capacitor may be open, bypass capacitor shorted, 2.2 kΩ shorted, 10 kΩ open, etc.

CHAPTER 8. COMMON-BASE APPROXIMATIONS 8-1. 0.477 mA, 2.85 V   8-3. 38.6 kΩ 8-5. 152, 152 mV   8-7. 28.4 mV   8-9. No source signal, open 50 Ω, open input coupling capacitor, shorted or open 5.6 kΩ, defective transistor, shorted or open 2.4 kΩ, open output coupling capacitor, shorted 1.5 kΩ, no supply voltages, etc.

CHAPTER 9. CLASS A POWER AMPLIFIERS 9-1. $i_{c(sat)}$ = 10.6 mA, $v_{ce(cutoff)}$ = 11 V   9-3. $i_{c(sat)}$ = 15.1 mA and $v_{ce(cutoff)}$ = 9.71 V; $i_{c(sat)}$ = 9.44 mA and $v_{ce(cutoff)}$ = 11.6 V; becomes steeper   9-5. 9.74 V   9-7. 19.5 V, 19.4 V, 18.8 V, 14.6 V, 4.53 V   9-9. Using the nearest standard values (5 percent), R_1 = 82 kΩ, R_2 = 91 kΩ, R_E = 5.1 kΩ 9-11. 24.8 mW, 2.17 mW   9-13. 4.43%, 4.06%, 4.04%   9-15. 99 mW   9-17. 21.5 W

CHAPTER 10. OTHER POWER AMPLIFIERS 10-1. $i_{c(sat)}$ = 0.625 A and $v_{ce(cutoff)}$ = 10 V   10-3. 23.8 mA   10-5. 70.2%   10-7. 9.84 W, 71.4%   10-9. 20 A, 13.6 V   10-11. 30 A, 1.6 mA, 2.99 V

CHAPTER 11. CASCADING STAGES, FREQUENCY RESPONSE, AND H PARAMETERS 11-1. 56.6   11-3. 10 kΩ, 250, 25   11-5. 12 kHz, 1414   11-7. About 250 MHz; slightly better answer is 208 MHz   11-9. 39.8406, 482.626, 14,536.7, 229,412

CHAPTER 12. JFETS 12-1. 150(10^9) Ω 12-3. 1.92 mA   12-5. 200 Ω 12-7. 0.7 mA, 30   12-9. 1.98 mA, 8.47 V, 2 V   12-11. 2500 μS   12-13. 0.849, 2 MΩ 12-15. 30.6 mV   12-17. 75 mV   12-19. 0.331 mV, 50 mV

CHAPTER 13. MOSFETS 13-1. 2(10^{11}) Ω 13-3. 14.2 V, 8 V   13-5. 20   13-7. 20 V, 0 V

CHAPTER 14. THYRISTORS 14-1. 4 V, 4.7 V   14-3. 0.18 A, 0.16 A   14-5. 10 mA, 60 mA   14-7. 16 mA, 1 A   14-9. 2 V, 2.7 V

CHAPTER 15. OP-AMP THEORY 15-1. 1.43 mA   15-3. 25 nA, 108 nA   15-5. 14 kΩ 15-7. 796 kHz

CHAPTER. 16. OP-AMP NEGATIVE FEEDBACK 16-1. 92.6 mV, 0.926 μV   16-3. 2160, 4320 M Ω, 0.0347 Ω 16-5. 100 mV   16-7. 17.3 kHz   16-9. 10 kHz, 20 kHz, 50 kHz, 100 kHz, 200 kHz, 500 kHz, 1 MHz

CHAPTER 17. OP-AMP CIRCUITS 17-1. 21.4, 18.3 kHz   17-3. 400 mV   17-5. 5 V   17-7. 2.7 V   17-9. 60.6 μV

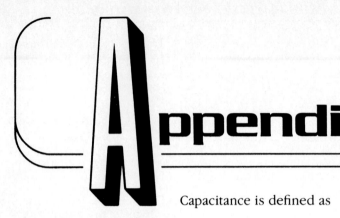

Appendix

A-1.
Derivation of Eq. (4–20)

Capacitance is defined as

$$C = \frac{Q}{V}$$

which can be written as

$$V = \frac{Q}{C}$$

Suppose the capacitor discharge begins at $t = T_1$. Then the initial voltage can be written as

$$V_1 = \frac{Q_1}{C}$$

If the capacitor discharge ends at $t = T_2$, the final voltage is

$$V_2 = \frac{Q_2}{C}$$

The peak-to-peak ripple equals the difference of the foregoing voltages:

$$V_1 - V_2 = \frac{Q_1 - Q_2}{C}$$

Divided both sides by the discharge time:

$$\frac{V_1 - V_2}{T_1 - T_2} = \frac{Q_1 - Q_2}{C(T_1 - T_2)}$$

When the time constant is much longer than the period of the ripple, the discharge time approximately equals T, the period of the ripple. Therefore,

$$\frac{V_1 - V_2}{T} = \frac{Q_1 - Q_2}{CT}$$

Because the load voltage is almost constant, the load current I_{dc} is approximately constant and the preceding equation simplifies to

$$\frac{V_1 - V_2}{T} = \frac{I_{dc}}{C}$$

To get the final formula for peak-to-peak ripple, let V_{rip} represent $V_1 - V_2$ and notice that the ripple frequency f is the reciprocal of T. With the foregoing in mind, we can write

$$V_{rip} = \frac{I_{dc}}{fC}$$

This gives reasonable results for a long discharging time constant, also known as the *lightly loaded case*.

A-2.
Heavily Loaded Case for Capacitor-input Filter

For the heavily loaded case, the analysis is very complicated because the operation becomes nonlinear. Figure A-1 shows the relation of dc load voltage to the discharging time constant of a bridge rectifier. In these graphs, R_{TH} is the Thevenin resistance facing the filter capacitor. This resistance includes the bulk resistance of the diodes and the winding resistance. We have been using an R_{TH} of zero, the top curve. The lightly loaded case is along the top curve for discharging time constants greater than approximately 50 ms. This gives us dc load voltages of at least $0.95V_{2(peak)}$.

A-3.
Beta Sensitivity

When β_{dc} changes, I_C changes. In fact, the percent change in I_C is directly proportional to the percent change in β_{dc}. In symbols,

$$\frac{\Delta I_C}{I_C} = K\frac{\Delta \beta_{dc}}{\beta_{dc}}$$

where K is the constant of proportionality, a value between 0 and 1. K is called the *beta sensitivity*. It determines how sensitive the collector current is to variations in β_{dc}. When K is near zero, the collector current is almost independent of β_{dc}. On the other hand, when K is

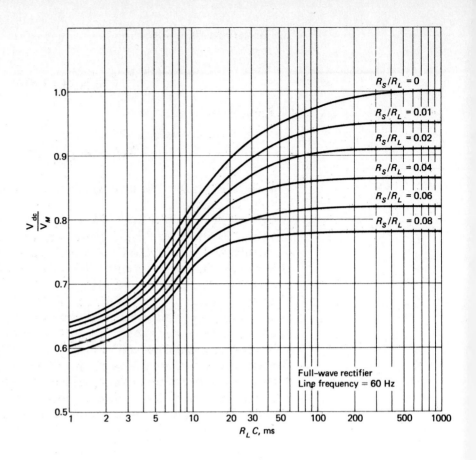

near unity, the collector current is higly sensitive to changes in β_{dc}. The best biasing circuits have a K near zero.

With calculus it is possible to derive the following formulas for the value of K:

1. Base bias:

$$K = 1$$

2. Emitter-feedback bias

$$K = \frac{1}{1 + \beta_{dc}R_F/R_B}$$

3. Collector-feedback bias:

$$K = \frac{1}{1 + \beta_{dc}R_C/R_B}$$

4. Voltage-divider bias:

$$K = \frac{1}{1 + \beta_{dc}R_E/(R_1 R_2)}$$

5. Emitter bias:

$$K = \frac{1}{1 + \beta_{dc}R_E/R_B}$$

With these formulas you can calculate the value of K for any of the biasing circuits.

Figure A-2 shows the value of K for the different biasing circuits. As you see, K is 1 for base bias, 0.75 for emitter feedback, 0.5 for collector feedback, 0.01 for voltage-divider bias, and 0.01 for emitter bias. This means that a 10 percent change in dc produces a change in collector current of 10 percent for base bias, 7.5 percent for emitter feedback, 5 percent for collector feedback, 0.1 percent for voltage-divider bias, and 0.1 percent for emitter bias. These values were calculated for the following conditions: $R_C = R_E$ for emitter feedback; $R_B = \beta_{dc}R_C$ for collector feedback; $R_B = R_E$ and $\beta_{dc} = 100$ for voltage-divider bias; $R_1 \| R_2 = R_E$ and $\beta_{dc} = 100$ for emitter bias. Figure A-2 illustrates the overwhelming superiority of voltage-divider bias and emitter bias. You will see these circuits used more than any others because the Q point is almost independent of the value of β_{dc}.

A-4.
Derivation of Eq.
(6–1)

The starting point is the rectangular *pn* junction equation derived by Shockley:

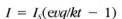

$$I = I_s(e^{vq/kt} - 1)$$

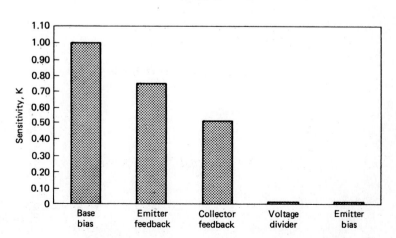

At room temperature, q/kT is approximately 40, and the equation reduces to

$$I = I_s(e^{40V} - 1) \qquad \text{(A-1)}$$

Differentiate to get

$$\frac{dI}{dV} = 40 I_s e^{40V}$$

With Eq. (A-1), the foregoing can be rewritten as

$$\frac{dI}{dV} = 40(I + I_s)$$

Take the reciprocal to get

$$\frac{dV}{dI} = \frac{1}{40(I + I_s)} = \frac{25 \text{ mV}}{I + I_s}$$

I_s is the saturation current in the reverse direction and is usually much smaller than I. Therefore, the equation reduces to

$$\frac{dV}{dI} = \frac{25 \text{mV}}{I}$$

Since this derivative is the ratio of voltage to current, we define it as the ac resistance of a diode and write

$$r_{ac} = \frac{25 \text{ mV}}{I}$$

Index